Atmospheric Icing of Power Networks

Atmospheric Icing of Power Networks

Masoud Farzaneh
Editor

Université du Québec à Chicoutimi, Canada

Springer

CIGELE

UQAC
UNIVERSITÉ DU QUÉBEC
À CHICOUTIMI

Dr. Masoud Farzaneh
Université du Québec à Chicoutimi
555 Boulevard de l'Université
Chicoutimi G7H 2B1
Canada
farzaneh@uqac.ca

ISBN: 978-1-4020-8530-7 e-ISBN: 978-1-4020-8531-4

Library of Congress Control Number: 2008927555

Cover Illustration: Fig. 8.4 from this book.

Printed on acid-free paper

9 8 7 6 5 4 3 2 1

springer.com

Foreword

Atmospheric ice takes a wide range of forms, usually quite beautiful and harmless. But it may, on occasion, pose severe risks to the security of many types of man-made structures, including power networks and transportation systems. As ice or sticky snow accumulates on network equipment and structures, it adds weight which, if combined with wind, can upset the precarious balance of these systems, sometimes leading to partial or total collapse. Other factors can also come into play; for example, ice or wet snow formation along insulators can eventually bridge the shed spacing, which can cause flashovers and, consequently, power outages. Serious damage and even loss of life can result from severe ice storms, as has been noted in the recent past, and efforts to mitigate their effects are on-going. This brings us to the purpose of this book.

First of all, let us mention that, despite the existence of many technical reports and papers in specialized journals and conference proceedings, none are assembled as a comprehensive study of the atmospheric icing phenomenon, and the results are not sufficiently distilled to support power line design. With its clear and tight focus, this book aims to fill that gap in the field of atmospheric icing. Furthermore, standards-based, deterministic approaches to overhead line design are currently used in the field, while international standards are striving to incorporate probabilistic design methods. Design experts need to understand where the probability distributions come from and know how to apply them.

Consequently, a team of internationally acclaimed experts in various aspects of atmospheric icing was invited to produce a compendium of their respective expertise. This compilation gives a detailed account of the fundamentals of atmospheric icing and it moves through a survey of the state of the art in design, modelling, prevention, and more, all in a richly illustrated format. In essence, we wanted to arrange the book in a logical sequence, from the meteorological aspect, moving on through various subjects, and finally leading to design. Accordingly, Chapter 1, *Modern Meteorology and Atmospheric Icing*, looks at how meteorology can help engineers and designers to better plan power-line routes or situate wind-turbine parks, through better understanding of weather patterns in a given region. In the next chapter, *Statistical Analysis of Icing Event Data for Transmission Line Design Purposes*, the authors describe how data from ice storms is gathered by monitoring systems and is used to establish design parameters for lines crossing regions where severe icing events occur. The third chapter, *Numerical Modelling of Icing on Power Network Equipment*, discusses how numerical icing models have become such essential tools in the field, as they use observations and measurements to produce simulations of extreme events that may be beyond our empirical experience. This is followed by *Wet Snow Accretion on Overhead Lines*, which deals with the physics of snow, particularly wet snow accretion on power line conductors, both in the wind tunnel and under natural conditions, in terms of overload hazards. Chapter 5, *Effects of Ice and Snow on the Dynamics of Transmission Line Conductors*, deals with the reliability and lifespan of iced conductors under such stresses as galloping, or wind-induced oscillations and aeolian vibrations, the mechanisms involved, and prevention methods. This is followed by a review of mitigation methods in *Anti-icing and De-icing Techniques for Overhead Lines,* which describes the various methods used by utilities, or under development, to combat ice accretion, by either removing already accreted ice or preventing it from sticking to surfaces. Then, *Effects of Ice and Snow on the Electrical Performance of Power Network Insulators* is a detailed look at the electrical performance of line and station insulators covered with ice or snow; it takes us through the modelling, testing, design and mitigation stages. Finally, Chapter 8, *Design of Transmission Lines for Atmospheric Icing*, is the ABC of structural design for adverse winter conditions – a thorough description of transmission line design, taking into account snow and ice overloads and other extreme weather effects. All in all, the book is a comprehensive and exhaustive examination of atmospheric icing, its causes, effects, and how to best mitigate the various hazards it poses.

The work is intended as a useful tool for utilities, first and foremost, looking to implement or adjust company-wide design policies with regard to severe wind and ice loads on overhead lines, and utility maintenance engineers and operators, who try to balance the costs and benefits of mitigation options when addressing specific icing problems. As well, professionals involved with the IEEE Power Engineering Society (PES), CIGRE and IEC, in their efforts to develop international icing standards, will find the book useful in their detailed studies of specific areas of research and consulting. The volume is also intended to be used as a fundamental text for students and researchers in the area of high voltage power transmission in university and college programs, who will find in it many worked examples for evaluating network reliability under various load conditions.

In the end, we hope that this book will, first of all, fill the need for up-to-date knowledge about the progress of research in the field of atmospheric icing of power network equipment and other sensitive man-made structures in recent years. Secondly, we hope we have achieved the purpose we had in mind, by compiling, in a single volume, much essential information that would otherwise remain dispersed throughout various technical journals and workshop proceedings.

As Editor, I would like to sincerely thank everyone who contributed to the publishing of this endeavour, and particularly the authors, who put in countless hours to provide us with the core of their research and developments. These utility and academic experts jointly participate in a biennial conference series called the International Workshop on Atmospheric Icing of Structures (IWAIS), where they are motivated to discuss ways to reduce the devastation from atmospheric icing at a practical cost. The rich content of these workshops, two of which I have had the honour to Chair, in Chicoutimi in 1996 and in Montreal in 2005, is at the root of the idea for this book. Indeed, on the occasion of the 11th IWAIS in Montreal, I invited keynote speakers to head the individual sessions of the conference and I subsequently asked them to expand their presentations for inclusion in this book. Once again, I thank them and I hope that the fruit of their efforts will find its place everywhere that atmospheric icing issues need to be managed.

Masoud Farzaneh
Editor

Contents

About the Editor

Professor Masoud Farzaneh is an internationally renowned expert in the field of power engineering, including atmospheric icing of power network equipment, insulation and corona-induced vibration (CIV). He is currently Director of the International Centre on Icing and Power Network Engineering (CENGIVRE), as well as Chairholder of the NSERC/Hydro-Quebec Industrial Chair on Atmospheric Icing of Power Network Equipment (CIGELE) and of the Canada Research Chair on Engineering of Power Network Atmospheric Icing (INGIVRE) at the University of Quebec in Chicoutimi, Canada. His fruitful and long-term collaboration with Hydro-Quebec, which led to the creation of the most complete icing research laboratory worldwide, was officially recognized when he received the prestigious NSERC Leo-Deriks award in 2005. In 2008, he received the prestigious Charles Biddle Award highlighting his exceptional contribution to the scientific development of Quebec. He has authored as many as 600 scientific publications, including 360 refereed papers, as well as several books and chapters in the areas of high voltage, insulation, CIV and atmospheric icing.

He is Associate Editor of IEEE Transactions on Dielectrics and Electrical Insulation, Chair of IEEE DEIS Outdoor Insulation Committee, as well as Convenor of CIGRE WG B2.29 on HV and UHV overhead line anti-icing and de-icing systems. He is also Chairman or member of several IEEE and CIGRÉ task forces dealing with atmospheric icing of HV equipment.

Dr Farzaneh is Chartered Engineer of the Engineering Council (U.K.), Charter Member of International Society of Offshore and Polar Engineers (ISOPE), as well as member of Conseil international des grands réseaux électriques (CIGRÉ). He is Fellow of IEEE, Fellow of the Institution of Electrical Engineers (IEE), Fellow of the Engineering Institute of Canada (EIC), member of the New York Academy of Sciences and the American Association for the Advancement of Sciences.

Contributors

Pierre Admirat
Meteorology Consultant, 96 Chemin des Sept Laux, 38330 Saint Ismier, France, tarimda@wanadoo.fr

William A. Chisholm
499 Millwood Road, Toronto, Ontario, Canada M4S 1K6, w.a.chisholm@ieee.org

Masoud Farzaneh
University of Québec at Chicoutimi, 555 Boulevard de l'Université, Chicoutimi, Canada G7H 2B1, farzaneh@uqac.ca

Svein M. Fikke
Meteorology Consultant, Lindeveien 1, 1470 Lørenskog, Norway, fikke@metconsult.no

Anand Goel
AG Engineering Innovations, 76 Pathlane Road, Richmond Hill, Ontario, Canada L4B 4C7, anandgoel@rogers.com

David G. Havard
Havard Engineering, 3142 Lindenlea Drive, Mississauga, Ontario, Canada L5C 2C2, dhavard@rogers.com

Jón Egill Kristjánsson
Department of Geosciences, University of Oslo, P.O. Box 1022, Blindern, 0315 Oslo, Norway, j.e.kristjansson@geo.uio.no

André Laneville
Université de Sherbrooke, Département de génie mécanique, 2500, boul. de l'Université, Sherbrooke (Québec), Canada J1K 2R1, andre.laneville@usherbrooke.ca

André Leblond
Hydro-Québec TransÉnergie, 800, boul. De Maisonneuve Est, 21st Floor, Montreal, Quebec, Canada H2L 4M8, leblond.andre.2@hydro.qc.ca

Edward P. Lozowski
Department of Earth and Atmospheric Sciences, University of Alberta, Edmonton, Canada T6G 2E3, edward.lozowski@ualberta.ca

Lasse Makkonen
Technical Research Centre of Finland, 02044 VTT, Finland, lasse.makkonen@vtt.fi

Bjørn Egil Kringlebotn Nygaard
Norwegian Meteorological Institute, P.O. Box 43, Blindern 0313 Oslo, Norway, bjorn.e.nygaard@met.no

Konstantin Savadjiev
University of Québec at Chicoutimi, 555 Boulevard de l'Université, Chicoutimi, Canada G7H 2B1, ksavadji@uqac.ca

Pierre Van Dyke
Hydro-Québec Research Institute – IREQ, 1800 boul. Lionel-Boulet, Varennes (Québec), Canada J3X 1S1, van_dyke.pierre@ireq.ca

Christophe Volat
University of Québec at Chicoutimi, 555 Boulevard de l'Université, Chicoutimi, Canada G7H 2B1, Christophe_Volat@uqac.ca

Chapter 1
Modern Meteorology and Atmospheric Icing

Svein M. Fikke, Jón Egill Kristjánsson and Bjørn Egil Kringlebotn Nygaard

1.1 Introduction

Atmospheric icing affects a wide variety of man-made structures in many countries. It is generally well known to occur in northern countries like Japan (Admirat and Sakamoto 1988), Canada (Farzaneh and Savadjiev 2001), United Kingdom (Wareing and Chetwood 2000), Iceland (Thorsteins and Elíasson 1998), Finland (Lehtonen et al. 1986), Hungary (Krómer 1993), Norway (Fikke and Johansen 1987), Czech Republic (Popolanský 2000), Romania (Goia 2000) and Russia (Golikova et al. 1989), as well as many other countries in both hemispheres.

Man-made structures at the top of mountains are often exposed to rime icing. In other areas, wet snow or freezing rain likewise affect infrastructures at lower altitudes. Therefore, power lines, wind turbines, telecommunication towers or high masts, ski lifts and other buildings are designed to withstand the loads and other adverse effects due to icing, as well as ice loads affecting their mechanical strength or operational reliability in many ways. Most countries have their own standards to take care of ice loads on their structures. At the international level, efforts are made to establish and improve standards and methodologies for handling the impacts of icing on various structures in the most economical and rational manner by both the International Electrotechnical Commission (IEC 1997; IEC 2003), the International Standardisation Organisation (ISO 2000) and the International Council on Large Electric Systems (Cigré 2001).

Some examples of icing are illustrated in Figs. 1.1, 1.2 and 1.3. Figure 1.1 shows the largest ice loading ever recorded on an overhead power line. This accretion was observed in Norway in April 1961, and the greatest elliptic cross-section diameter was measured at 1.4 m and the smallest at 0.95 m. A one-metre length of the accretion was collected and weighed 305 kg.

Figure 1.2 shows a wet snow incidence in Iceland. The cross-section accretion is in this case quite uniform in physical appearance, without a pronounced pattern showing the elliptic build-up.

S.M. Fikke
Meteorology Consultant – Overhead lines, Lindeveien 1, 1470 Lørenskog, Norway
e-mail: fikke@metconsult.no

M. Farzaneh (ed.), *Atmospheric Icing of Power Networks*,

Fig. 1.1 Rime icing on a 22 kV electric power line in Norway April 1961, 1 400 m above sea level. The ice load was measured to 305 kg/m (Photo: O. Wist, reproduced by permission of S. M. Fikke)

Figure 1.3 is from a Swiss test station on the mountain Gütsch, near Andermatt, in the Alps. Together with an operating wind turbine, there is a test site where a variety of meteorological instruments as well as icing detectors and devices for measuring ice loads are installed for the purpose of performance and feasibility testing. The project is a part of the European Cooperation in the field of Scientific and Technical Research (COST) Action 727: "Atmospheric Icing on Structures Measurements and Data Collection on Icing", operating through the years 2004–2009. The project also generates data sets to be used for calibrating atmospheric models for icing forecasts, see (Fikke 2005a, 2007a,b).

During the last century, when societies expanded their economic developments and new infrastructures had to be established in hitherto unknown places, experience

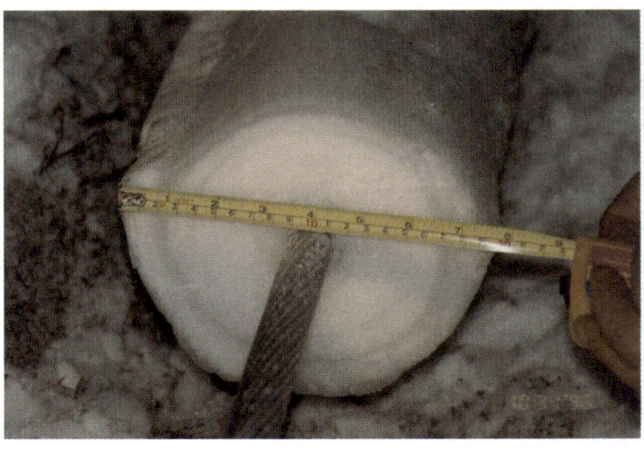

Fig. 1.2 Wet snow accretion on a collapsed power line in Iceland (Reproduced by permission of Á. Elíasson, Landsnet, Iceland)

Fig. 1.3 Rime icing on a wind turbine blade at the Gütsch test station, Switzerland (Reproduced by permission of Meteotest, Switzerland)

in many countries showed that it was necessary to cope with other types of weather impacts than considered before. As well, many attempts were made worldwide to establish understanding of various icing conditions in remote areas, especially in the mountains. Probably the first attempt to establish a 3D atmospheric model for icing in remote areas was suggested by Ervik and Fikke (1982).

Over the past 30–50 years, such knowledge was built up from field observations and measurements, laboratory studies and model development. Laboratory studies and the variety of modelling tools available today will be discussed in other chapters of this book. However, despite this better understanding, actual weather conditions at a remote location are always a critical question for a potential overhead power line, TV tower, wind turbine or ski lift. Also, for a given meteorological weather station, it is not always as easy as desired be to determine the wind speed and wind direction when all anemometers and wind vanes are stuck in thick ice layers.

A comprehensive survey of atmospheric icing was published by Poots (2000). The most recent update of international knowledge and research activities on atmospheric icing was published by Cigré (2006).

Parallel to the other dramatic developments in natural sciences and technology over the past decades, related to computers, instruments, remote monitoring, etc., there have also been vast developments within the science of meteorology, which can be more or less directly implemented, to the benefit of all who enjoy much improved and reliable weather forecasts today, compared with the situation 5–10 years ago. Accordingly, weather forecasts can now be extended for longer periods of time; even forecasts of the order of a week are often of remarkable quality.

The purpose of this chapter is to identify how modern meteorological techniques can improve our understanding of and quantify the elements and parameters of the atmosphere we depend on when assessing local icing conditions, in either nearby or remote environments. This is valid for case studies (related to failures), but also as an assisting tool to establish design loads in areas where the icing conditions are largely unknown. This means looking relatively deep into the state of the art of current physical and dynamic models of the atmosphere, and also at some techniques for establishing accurate descriptions of the initial atmospheric conditions in 3D. The latter is of course crucial for the reliability of short- to mid-term prognoses (hours and days).

It will certainly be much too complex to look here into all the details of the science of modern weather forecasting; we will therefore focus on some aspects concerning practical applications for atmospheric icing. In subsequent sections, the most critical weather parameters will be discussed in relation to the importance of the different icing types. For all icing types however, the temperature, wind speed and wind direction (relative to the line) are always important parameters. The required accuracy of these weather elements may vary with the icing type in question. Each of the icing types is treated separately and in details in other chapters of this book. In this chapter they are however described where appropriate in order to complete the discussion on the meteorological aspects relating to them.

This chapter is an extension of a keynote speech to the 11th International Workshop on Atmospheric Icing of Structures in Montreal (Fikke 2005b).

1.2 Atmospheric Icing – A Brief Survey of Icing Processes and their Meteorological Aspects

Atmospheric icing is a generic term for all types of accretion of frozen water substance, generally belonging to two main categories: (1) precipitation icing, and (2) in-cloud icing. Both may cause severe damage to the types of infrastructures mentioned above. Such icing is often considered as an exclusive phenomenon for circumpolar regions, or in mountains in Central Europe, Asia or North America, but it is experienced and reported any place on earth where snow occurs, or in elevated areas where temperatures can drop below the freezing point. Therefore, icing is reported from mountainous regions in such places as Spain, Algeria, South Africa, New Zealand, Latin America, etc. (Cigré 2006).

Each type of atmospheric icing is treated separately in other chapters of this book. Therefore, only some general meteorological characteristics of these icing types are briefly discussed below. Further discussion of icing processes can be found in IEC TR 1774 (1997), IEC TR 60826 (2003), ISO 12494 (2000), and Cigré TB 291 (2006).

1.2.1 Precipitation Icing

Precipitation icing may result in glaze, wet snow or dry snow, depending on how the precipitation is influenced by variations in temperature near the ground and up to a few hundred metres above ground. Such icing is experienced any place where precipitation, in combination with freezing temperatures, occurs. Probably the most severe impact on society was the well known freezing rain event in Eastern Canada and North-Eastern USA, in January 1998 (Farzaneh and Savadjiev 2001), where millions of people lost electricity for days and weeks, and industry, business and the public were paralyzed due to loss of energy, telecommunication breakdowns, inaccessible roads, etc. Wet snow accretion occurs wherever snow occurs; although it is most severe in countries where high precipitation rates near the freezing point are frequent, like Japan, Iceland, Norway and many other European countries, it has been recorded in countries around the Mediterranean Sea.

Freezing rain requires a specific temperature distribution with elevation, as shown in Fig. 1.4, where the parameters are: surface temperature ($T_{surface}$), maximum temperature and its height (T_{max} and Z_{max}), depth of melting layer ($H_{melting}$), and depth of the subfreezing layer ($H_{subfreezing}$) (Thériault et al. 2006).

A temperature inversion occurs in the lowest layer, which means that the temperature increases with height instead of the normal decrease. If the temperature nearest the ground is below freezing and the temperature at the top of the inversion layer is above freezing, a layer above the cold surface air will develop where falling snow can melt. If the temperature at the top of the inversion is high enough and/or the melting layer is deep enough, then the snowflakes can melt totally and form raindrops. When these raindrops fall into the freezing layer near the ground, they become supercooled, and may remain as liquid water drops until they hit objects in the airflow or the ground itself. As long as they are in the liquid state the droplets freeze immediately upon impact.

Depending on parameters like the depth and elevation of the freezing layer, surface temperature, thickness of the melting layer, maximum temperature of the inversion, etc., the type of precipitation that reaches the ground may be freezing rain, ice pellets, slush, refrozen wet snow or snow (Thériault et al. 2006). Also, the vertical component of air movement is of significant importance in the formation of different hydrometeors (Thériault and Stewart 2007).

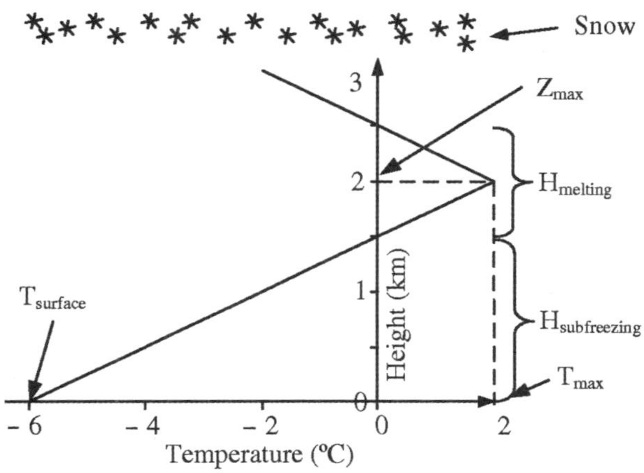

Fig. 1.4 Example of vertical temperature distribution (schematic) of the lower atmosphere for freezing rain formation. Parameters are: surface temperature ($T_{surface}$), maximum temperature and its height (T_{max} and Z_{max}), depth of meting layer ($H_{melting}$), and depth of the subfreezing layer ($H_{subfreezing}$) From Thériault et al. (2006) (Reproduced by permission of American Geophysical Union)

Several meteorological processes and topographical effects can provide conditions for such temperature inversions. In the case of significant freezing rain formation, it is necessary to have a situation where the inversion, together with the rainfall, can prevail for a long enough time to allow for the accretion build-up. Any place where there is a basin and cold air can be trapped for a certain amount of time may cause inversion, whenever a warm front (or warmer air) is passing over. As long as the wind speed is low, the inversion can persist for a long period (hours to days), but the cold air may quickly be mixed with the warmer air when the wind aloft is strong enough, and the inversion will disappear.

A more severe situation occurs when the topography channels cold air continuously near the ground from other areas, due to the combination of topography and distribution of high and low pressure systems in the atmosphere. Probably the most famous example of this kind is the St. Lawrence valley in Québec, Canada. In January 1998, such a situation was maintained for about five days. During this time, a sequence of three precipitating low-pressure systems crossed over the cold air basin that was maintained at the bottom of the valley.

The formation of snow accretions is described by Sakamoto (2000), and by Admirat in Chapter 4 herein, entitled "Wet Snow Accretion on Overhead Lines". Sakamoto also described the formation of dry snow accretions on overhead power lines (Sakamoto 2000).

Dry snow may accrete when wind speeds are sufficiently low, typically less than 2 m/s. Although this sometimes causes heavy snowfall, density never exceeds 100 kg/m^3. Hence the accreted masses are, in most cases, much lower than the loads the power lines are designed for; consequently, dry snow accretions are not discussed further in this chapter.

Wet snow is generally formed during a very narrow surface air temperature interval, slightly above $0\,^{\circ}C$. Snowflakes falling through air with increasing temperatures near the ground may eventually meet temperatures above the freezing point. The exact temperature interval for wet snow formation is not yet fully described, but it is probably within the range of $+0.5 - +2.0\,^{\circ}C$. This is supported by many observations of a rather narrow band where wet snow accretions occur, seldom more than a vertical layer of 100–150 m thickness. This is consistent with the fact that the vertical temperature gradient of the atmosphere is around $-0.6\,^{\circ}C/100$ m during precipitation.

As soon as snowflakes meet above-freezing temperatures, they start to melt. And when liquid water appears between the branches of a snowflake, it becomes sticky and can adhere to other objects. However, once a snowflake becomes very wet, like slush, the adhesion force is diminished, and most of its mass will drop off the object. It is not fully known at which liquid-to-frozen water mixing ratio the adhesive forces are strongest, but Cigré WG B2.16 (2006) states that flakes adhere readily to objects when their liquid water content (LWC) lies between 15 and 40% of the total mass of the snowflake.

From a meteorological point of view, it is clear that the most critical point is the accuracy of the estimated air temperatures. The elevation of the $0\,^{\circ}C$ isotherm may also be relevant, since this will indicate the time and total exposure to above-freezing temperatures.

Another observation about wet snow accretion that deserves further study is that it often occurs on the leeward side of gently sloped hills, where strong and relatively laminar winds may occur in slightly stable stratification. In such cases, the airflow may subside 100–200 m from the top of the ridge and be adiabatically heated by up to 1.0 or 1.5 °C, and hence the snowflakes partly melt. This is observed in Norway and Iceland; however it has not been reported with greater detail.

1.2.2 In-cloud Icing

In-cloud icing occurs only within clouds consisting of supercooled droplets, which are droplets that remain liquid at a temperature below 0 °C. Depending on the cloud LWC, the size distribution of the cloud droplets, temperature and wind speed (perpendicular to the object), soft or hard rime may occur. This type of icing can then appear only above the cloud base and also above the 0 °C isotherm. It therefore occurs most often near the top of exposed mountains, typically for constructions like telecommunication towers, ski lifts, wind turbines and other vulnerable installations and structures.

The intensity and duration of in-cloud icing depends on the flux of liquid water in the cloud, which again depends on many parameters such as temperature, wind speed, stability, depth of cloud, height above cloud base and distance from coastline.

Makkonen and Lozowski take a closer look at in-cloud icing in Chapter 3, entitled "Numerical Modelling of Icing on Power Network Equipment".

Hoar frost is a phenomenon in which water vapour is transformed directly to the solid phase (ice deposition) and forms light, nice-looking crystals. Hoar frost forms most often during cold winter nights, near open water. The low density and light weight makes it harmless for most structures; however, hoar frost on overhead power lines may cause very significant energy losses due to corona discharge. Hoar frost also causes very visible sparks and noise from the pantograph on overhead lines feeding power to trains.

All the processes and parameters mentioned above, for both precipitation icing and in-cloud icing, have to be reasonably well described in the atmospheric (weather forecasting) models in order to make them capable of illustrating and forecasting the different icing types. This will be discussed further in the next sections of this chapter.

1.3 Icing Models

The current models for all types of icing are presented and discussed in subsequent chapters and, therefore, will not be dealt with here. In this section, we shall only identify and specify the environmental parameters that most of these models depend upon for their implementation. Models are only presented to emphasise the importance of these parameters.

To describe precipitation icing (wet snow and freezing rain) the most important parameters are (Cigré 2006):

- Precipitation rate
- Surface air temperature
- Liquid water content of snow flakes
- Wind speed
- Wind direction
- Air temperature
- Relative humidity
- Visibility

For in-cloud icing the parameters are:

- Liquid water content in the cloud
- Droplet size distribution
- Air temperature
- Wind speed
- Wind direction
- Relative humidity

These parameters describe the immediate environment of the object. It is likewise important to include parameters of the accreting object itself, such as surface properties, shape, linear dimensions, torsional stiffness, etc.

Due to the very significant economic impacts of icing on manmade structures during the 20th century, numerous icing models have been developed to describe the physics of icing and how it impacts the different structures affected, especially concerning the amount of ice that can be accreted. An updated history of these efforts is given in Lozowski and Makkonen (2005).

The fundamental physics of ice accretion are demonstrated by equation (1.1) from the ISO Standard ISO 12494 (2000), also called the Makkonen model

$$\frac{dM}{dt} = \alpha_1 \alpha_2 \alpha_3 \cdot w \cdot A \cdot V \tag{1.1}$$

where

α_1 = collision efficiency (for in-cloud icing and freezing rain, $\alpha_1 = \alpha_1$(V,D,d))
α_2 = sticking efficiency (manly for wet snow)
α_3 = freezing efficiency (determines "dry" and "wet" growth for rime ice and freezing rain)
M = accreted ice mass (per unit length)
d = median volume droplet diameter
w = liquid water mass/unit volume
V = wind speed (perpendicular to accreting object)
A = cross-sectional area of object with diameter D.

Over the years, numerous studies, including both laboratory and field studies, have been performed to quantify the coefficients α_1, α_2 and α_3. There has been however, by far, much less effort expended to determine the meteorological input elements, namely air temperature, wind (speed and direction), precipitation (type and rate), cloud liquid water, and relative humidity. In particular, the vertical profile of temperature (stability of the air) is certainly of critical importance for freezing rain and wet snow formation.

It has been argued that there are no reliable data for liquid water content of clouds in a particular icing situation. This is, or at least was, quite true. But the major developments in meteorological science over the last decades have also increased our understanding of icing-related parameters very significantly, including the water cycle of the atmosphere.

Regional and small-scale topography and surface characteristics also have large influence on the local variations of most weather parameters. This chapter will describe how modern meteorology can provide significantly better input parameters to the mentioned icing models, than can be provided by classical observational data and field measurements.

1.4 Introduction to Numerical Weather Prediction Models

As mentioned above, atmospheric icing is a generic term for many very complex phenomena involving several basic processes in the atmosphere, relating to water cycle, temperature, wind speed and wind direction, vertical stability, and formation of clouds and precipitation, in addition to the micro-physical processes connected to the different phases of atmospheric water (vapour, liquid and solid). Initially, we are interested in these processes as they develop in the boundary layer of the atmosphere, at or near the surface of the earth. However, what happens with the weather at the lowest levels is indeed a result of the processes higher up.

In the higher atmosphere, the dynamic processes are governed mainly by the global pattern of high- and low-pressure systems, together with the distribution of continents, oceans, great lakes and large mountain ranges.

In the lower atmosphere, where we live and install and operate our infrastructures, also called the atmospheric boundary layer, these processes are very significantly influenced by surface properties, with its small-scale relief (hills, ridges, valleys, forests, plains, cities, rural landscape, rivers, lakes, etc.) and current conditions (wet, dry or frozen soil, cultivated land, developed areas with concrete and asphalt, open or frozen lakes, bare or snow-covered fields or forests, etc.). It is easy to understand that the earth's surface must generate the parameters for the boundary conditions of the lower atmosphere. It is also easy to understand that there must be a lot of processes along this boundary, involving the exchange of sensible heat (warming and cooling by conduction and convection), latent heat (evaporation and condensation of water vapour), and radiation (long-wave or infrared, short-wave or visible light, and ultra-violet) between the solid or liquid surface and the air above.

In addition, the wind is influenced by friction over this surface. At the surface of the earth, the wind speed must be zero, while above the planetary boundary layer (800–1 000 m), the wind is ruled by the pressure systems and the large-scale characteristics of the earth.

Modern meteorology aims at including all these physical and dynamic processes, to the extent that is practically possible, with the resources available. Modern computer technology has made it possible to model the atmosphere by solving the relevant physical equations for a set of 3D grid points throughout the atmosphere. Near the ground, the grid points are relatively dense and the vertical distance between grid layers is relatively small, while higher up, the spatial variations are on larger scales and hence the horizontal and vertical distances between grid points can be larger. What is called numerical weather prediction involves solving the equations numerically, i.e. computing the weather variables on the spatial grid, and stepping the variables forward in time to produce a forecast.

The initialization procedures for meteorological models consist of combining previous forecasts with values from observations and measurements made by manual and automatic surface weather stations, as well as other instruments such as radar, sodar, balloon soundings and satellites from all over the world, thereby creating an array of initial values at each grid point of the model. When the initial condition of the three-dimensional atmosphere is described in such a way, the governing equations of the weather model are used to integrate the variables forward in time, to produce a forecast. The governing equations are usually the Euler equations for compressible fluid flow and equations describing energy conservation, and mass conservation for important quantities such as humidity and condensed water. In addition, physical laws that describe the transfer of radiation, the formation of clouds and precipitation, chemical reactions, etc., are included in the equations.

As some weather observations and measurements are made more or less continuously (e.g. aircraft, radar and satellite measurements), these data are immediately assimilated by the models; these are then used to update and correct grid values calculated by the model.

1.4.1 Global Models

Large-scale weather systems are governed by the development of the major high- and low-pressure systems over the world; therefore, such models do not typically need the high density of grid points required to provide a detailed description of the topography and physical conditions of the surface. On the other hand, they must cover a large part of the earth's surface, and most modern models cover the entire globe. Global models now have grid sizes down to 0.25°, or about 25 km, with 90 vertical layers, as currently used by the global model at the European Centre for Medium Range Weather Forecasting (ECMWF).

1.4.2 Regional Models

Regional models, such as the widely used MM5 and WRF (Weather Research Forecasting) models, developed at the US National Centre for Atmospheric Research, will typically cover parts of a continent and its adjacent ocean. Here, more features of the region can be included and typical grid sizes are about 0.1°, or about 10 km, with some 60 layers in the vertical direction. The domain of a regional model can then be considered as a "box", where the weather parameters on the "box sides" (boundaries) must be provided by a larger global model. Inside the regional model, the same (or even more) weather variables are integrated in time, as within the global model, using similar dynamical and physical relations, but with higher resolution in space and time. This technique is called "nesting of grids".

1.4.3 Local Models

The model nesting technique can be used at even smaller time and space scales, depending on the precision and details wanted for a weather analysis or forecast. At present, there are examples of such model nesting down to a few tenths of meters in horizontal grid size, for specific purposes, where local terrain features are critical for the accuracy and reliability of the prediction of certain weather parameters. In particular, this technique can provide more accurate information on fine-scale or local parameters, than can be provided by a coarse network of observations, or indeed where such observations are missing. However, such very fine-scale models cannot yet handle the water cycle.

Some examples on the applications of nesting of such local scale models into global models will be shown in the subsequent sections.

1.5 Some Preliminary Applications of Fine-Scale Models

1.5.1 Wind Studies

As mentioned, the MM5 and WRF models are used for regular forecasting purposes. Other models may be used for limited areas, for example air quality forecasts in cities during inversion periods. Below 1-km grid spacing, super-fine models may be nested in for even more detailed studies. Figure 1.5 shows an example of domains of nested models, taken from Holstad and Lie (2006). Figure 1.5a shows the domain of a model with a 1 km resolution where a domain of 250-m resolution is nested within the black frame. The 1-km domain is approximately $80 \cdot 80\ km^2$. Figure 1.5b shows the 250-m model domain where an even more detailed model with 75-m resolution is nested inside the inner frame. The extent of the 250-m domain is approximately $22.5 \cdot 22.5\ km^2$. A runway for a proposed airport in the area is marked in red.

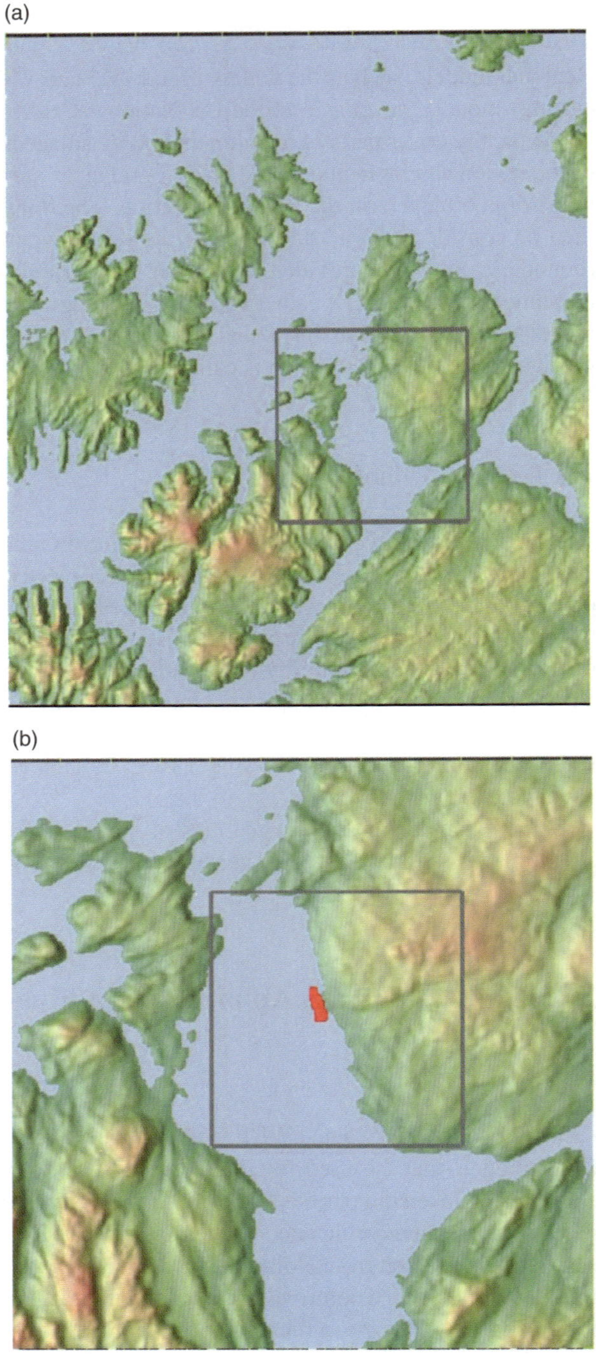

Fig. 1.5 Example of nesting of domains. (**a**) shows the domain of a model with 1 km horizontal resolution. The domain of a nested model with a resolution of 250 m is indicated within the *black square* frame. (**b**) shows this 250 m domain where a 75 m resolution model is nested within the inner *square* frame. The area of the 1 km domain is about $80 \cdot 80\ \text{km}^2$, and the area of the 250 m model is about $22.5 \cdot 22.5\ \text{km}^2$. The runway for a possible airport is indicated in *red* on the right panel. Holstad and Lie (2006) (Reproduced by permission of Storm Weather Center AS, Norway)

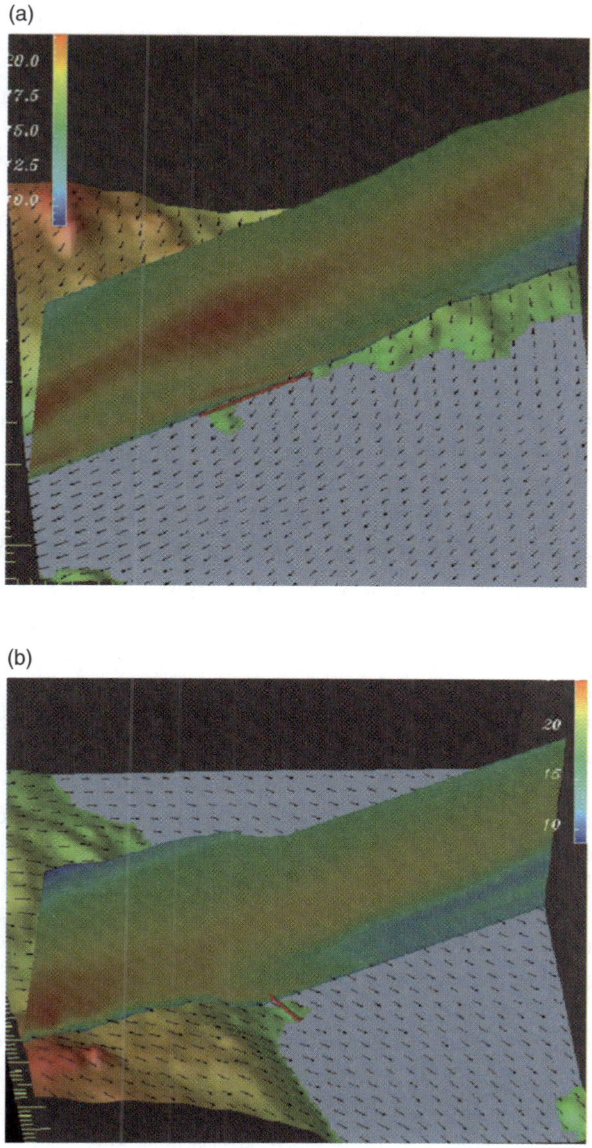

Fig. 1.6 Examples of output from the 75 m model shown in Fig. V-1 a. (**a**) shows the wind speed field in a vertical cross section along and over the runway. (**b**) shows the wind speed field in a cross section perpendicular to the runway. Colour codes for wind speeds are shown in bars. The possible runway is shown as *red lines* (Reproduced by permission of Storm Weather Center AS, Norway)

Figure 1.6 shows examples of the output from the 75-m model in Fig. 1.5b, where wind speeds are indicated with colours in the vertical cross-sections, and surface winds are indicated with arrows. Part a shows the cross-section along the runway, and part b shows the cross-section perpendicular to the runway.

Models as that shown in Figs. 1.5 and 1.6 are often used in regular computational fluid dynamics (CFD). However, conventional CFD models have significant limitations as to reliable descriptions of small-scale weather phenomena. There are several reasons why nested atmospheric models have better performance than conventional CFD models. The most important ones are:

- The boundary fields are forced by external models and thus by time-dependent weather development.
- The physical characteristics of the atmosphere are preserved consistently from global to local scales.
- Wind shear, both vertically and horizontally, is handled on all scales simultaneously.
- Stratification has a dramatic influence on the airflow pattern. This means that upstream characteristics of topography and temperature distributions must be incorporated dynamically (this is extremely important in the case of freezing rain).
- Resources with respect to both manpower and computers are much less than for conventional CFD studies. Nested fine-scale weather models run very efficiently, together with global atmospheric models.
- Historical, quality-controlled meteorological data are available for many years back, and can be used for case studies of historical events.

It should also be noted that many countries now have digital topographical maps or geographical data with a grid size down to 25 m. These may be provided by governmental mapping agencies at reasonable cost, and they can easily be implemented in the models. For such reasons, we can look forward to local weather descriptions which are representative for each span of an overhead power line, whenever appropriate and at reasonable cost and manual effort.

It must be kept in mind that, although icing is very much a function of water in the air, it is also a function of wind speed, wind direction and air temperature. The water available for icing also depends on clouds, precipitation, adiabatic cooling (condensation) or heating (evaporation), etc. Therefore, the picture of icing cannot be complete unless we take the complete weather situation into account also, and unless we can describe the relevant parameters with a spatial resolution that is comparable with the span length of an overhead line.

It is the opinion of the authors that probably the most significant progress in the understanding of atmospheric icing will be obtained through detailed studies of the meteorological processes in the atmosphere in the near future.

In order to demonstrate applications for overhead line purposes, two studies in Norway will be mentioned. The first is related to local wind in very complex terrain. Two parallel 132-kV lines (double circuit) were planned to feed a new aluminium factory in Sunndalsøra, located at the bottom of a fjord in mid-Norway (Holstad et al. 2001). This location is known to have very special wind conditions. Long-time local measurements of wind and the regular design code told to design for a gust wind speed of 75 m/s at 30 m height for these lines. The lines were to cross a flat area at the mouth of two valleys, one to the south and the other towards the ESE. The mountains around this place reach up to more than 1 800 m above sea level (asl); very strong, turbulent winds are generated from these mountains and are also forced out of the valleys. A model study, confirmed by local measurements up to

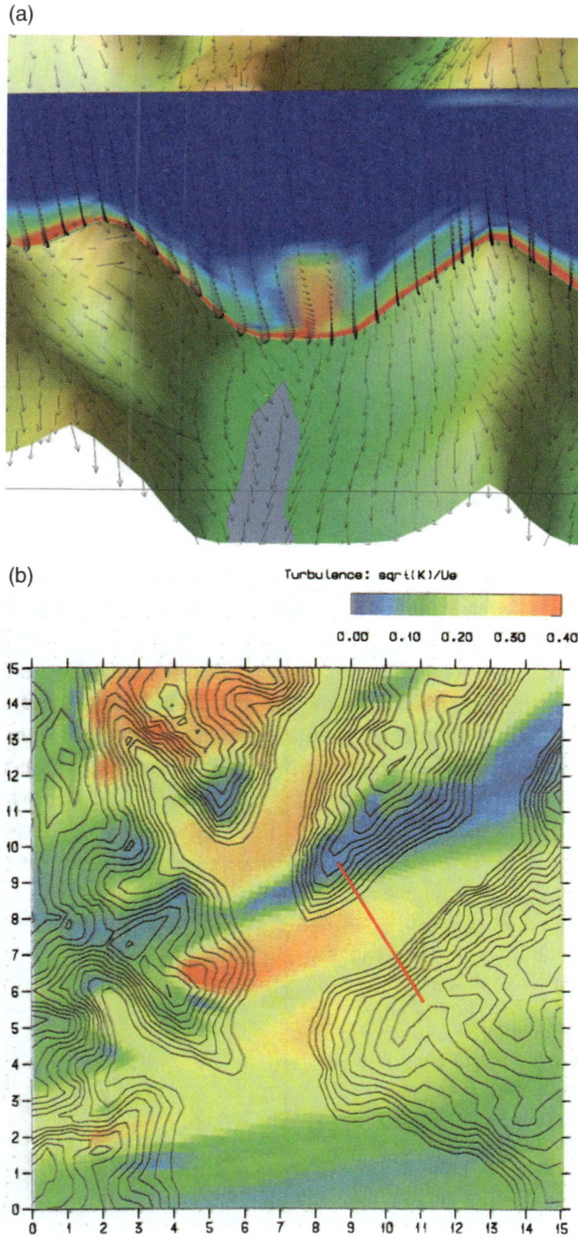

Fig. 1.7 Model studies related to transmission line projects in Norway. (**a**) is from Sunndalsøra. *Arrows* show wind speed and direction over the surface and in a vertical cross section. *Colours* on the surface represent topography (*blue* is sea level and *red* is above 1 000 m) and turbulent kinetic energy in the cross-section (Holstad et al. 2001). (**b**) shows the turbulence generated behind steep mountains (*red*: high turbulence intensity) in a fjord area. The approximate location of a fjord crossing is shown with the *red line*. Topography is shown with *contour lines*. Areas without contour lines are sea level (fjords) (Lie 2001) (Reproduced by permission of the Norwegian Meteorological Institute)

50 m above the ground, could justify that the design gust wind speed at 30 m height could be reduced to 65 m/s. Figure 1.7a shows an example from this model. The minimum grid size in this case was 250 m and the model domain was $15 \cdot 15\ km^2$. Surface wind speeds and directions are shown on the surface terrain. The vertical

plane shows wind profiles (arrows) and turbulent kinetic energy in colours (scale not shown). The cross section is approximately where new 132-kV lines were planned to cross the valley bottom.

Figure 1.7b shows another study (Lie 2001) where a similar model was used to evaluate wind conditions over a 3.5-km fjord crossing of a new 132-kV line in Western Norway. In particular, it was of interest to study the kinetic energy of a vortex street formed behind a mountain peak on the upwind side of the fjord span. The study showed that the kinetic energy had dissipated significantly at the location of the span and therefore did not affect the span.

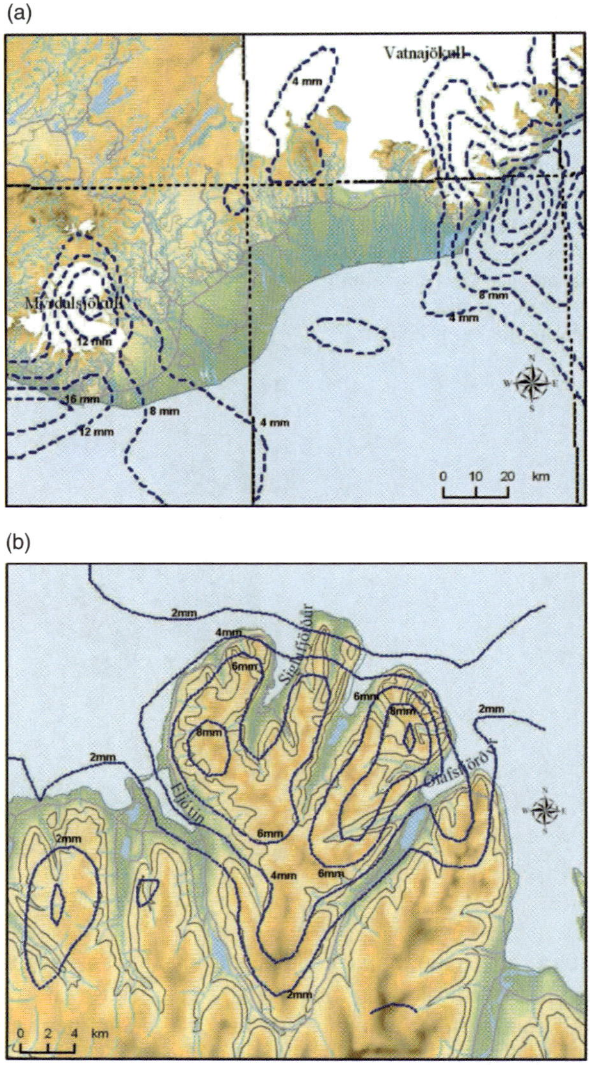

Fig. 1.8 Modelling of wet snow in Iceland with a dynamic weather forecasting model. (**a**) shows model isohyets (*contour lines* of equal precipitation) for 3 hr precipitation at temperatures near the freezing point over an area in south-eastern Iceland (*dashed curves*). From (Olafsson et al. 2002a). (**b**) shows the same for an area in northern Iceland (isohyets in *solid lines*). (Olafsson et al. 2002b). Note that the scale is different in the two examples (Reproduced by permission of Haraldur Olafsson, Iceland)

The examples shown here relate to wind only, and not specifically to atmospheric icing. However, the local wind speed and direction are always of great importance to the accretion of all types of atmospheric ice as well. Also, as will be shown later in this chapter, the water scheme of the atmosphere can also be included, in a very reasonable manner, in similar local scale atmospheric models.

1.5.2 Icelandic Wet Snow Studies

Probably the best attempts at wet snow modelling, based on dynamic weather models, were made in Iceland. Their model includes a realistic description of mountains and it is probably the first direct approach to connect wet snow modelling with regular weather forecasting models. Two of their tests are shown in Fig. 1.8a for South-Eastern Iceland (Olafsson et al. 2002a,b) for the North-Eastern part (Olafsson et al. 2002b).

The most promising aspect of the Icelandic work is that the modelled precipitation is very close to observations. In mountainous terrain like that of Iceland, this is considered to be very difficult. There is therefore reason to be more optimistic than before concerning modelling of wet snow accretions, given that temperatures can be equally well predicted. Probably the most significant constraint for further development is the current lack of good field data during wet snow accretions, with a time resolution that is adequate for comparison with weather models.

1.6 Condensation Schemes in NWP Models – Relevance for Icing Prediction

The first numerical weather prediction (NWP) models from 1950 through the 1970s had only a few simplified prognostic equations, and the emphasis was on the dynamical fields (pressure and wind). As models have evolved due to faster computers and improved knowledge, more and more emphasis has been put on a detailed treatment of the physical processes, such as phase changes of water, the evolution of precipitation, radiative processes, and energy exchange between the atmosphere and the underlying surface. The most sophisticated parameterization schemes for clouds and precipitation are typically found in non-hydrostatic, mesoscale models that are run for limited areas at high spatial resolution. The global models of the major forecast centres (e.g., ECMWF, NCEP, UK MetOffice, JMA, Météo France, etc.), which are run for 1–2 weeks ahead, often use somewhat simpler physical parameterizations in order to reduce computational cost. Below, we will briefly review the cloud physics schemes of some of the models, emphasizing the significance of the model assumptions for the ability to simulate atmospheric icing.

In today's NWP models, the following variables are typically predicted: horizontal wind components (u, v), air temperature (T), surface pressure (ps), specific humidity (q), cloud water mixing ratio (q_c). The time evolution equations for the last two can be written as follows

$$\frac{\partial q}{\partial t} = Trans_q - Cond + Evap_c + Evap_{pre} \tag{1.2}$$

$$\frac{\partial q_c}{\partial t} = Trans_c + Cond - Evap_c - \text{Precip} \tag{1.3}$$

where $Trans_q$ and $Trans_c$ denote the contributions of advective and turbulent transport, respectively, *Cond* is the rate of condensation, *Evap* the rate of evaporation, with subscript *c* referring to cloud water (liquid or ice), and subscript *pre* referring to precipitation. Finally, *Precip* denotes the local rate of conversion of cloud water into precipitation.

A critical issue is how to make the distinction between the water and ice phases in clouds and precipitation. Equation (1.3) lumps the two phases together into a common variable that we may term total cloud water (q_c). In many models, only the total cloud water is predicted, and the liquid and ice components of the total water are obtained via simple diagnostic relations that typically depend only on temperature

$$q_{c,liq} = (1 - icefraq) \cdot q_c \tag{1.4}$$

$$q_{c,ice} = icefraq \cdot q_c \tag{1.5}$$

where *icefraq* is the fraction of the condensate within the cloud, in frozen form. Often *icefraq* is simply a prescribed function of temperature, for instance increasing linearly from 0 at 0 °C to 1 at −20 °C (Rasch and Kristjánsson 1998) or −9 °C (Smith 1990). While such assumptions may broadly reflect the average of observed conditions in a region (e.g. Moss and Johnson 1994), they fail to take into account the fact that the occurrence of supercooled water is not only a function of temperature, but also other factors, such as static stability, seeding from above, vertical motions, cloud age, and the types and concentrations of atmospheric aerosols, some of which may serve as ice nuclei. In other words, it is not sufficient to know that only, say 25% of the condensate on average is in liquid form at a given temperature, e.g. −15 °C, because on a particular day, due to special weather conditions, all the condensate in an air mass may be in liquid form (supercooled) at that temperature, and the icing amounts would be greatly underestimated if only the average fraction was used.

To illustrate this, we may refer to the study by Wilson and Ballard (1999), in which results from two versions of a cloud parameterization scheme in the UK Met Office Unified Model were compared. In the first version, equations (1.4) and (1.5) were used, with *icefraq* varying linearly between a value of 1 at temperatures of −9 °C or lower and a value of 0 at and above 0 °C. In the second version, based on Rutledge and Hobbs (1983), ice water content was treated as a separate prognostic variable, meaning that ice and liquid water can coexist in a model grid point, with their relative amounts being determined from physical principles, rather than a prescribed ratio.

For the two cases presented in Wilson and Ballard (1999), major improvements were found, using the new scheme. In the first case, from January 1997, a stratocumulus cloud deck over the British Isles, consisting of supercooled cloud droplets, was investigated. The old condensation scheme, using a simple linear *icefraq* vs. temperature relationship, failed miserably, as the cloud was incorrectly diagnosed as consisting mainly of ice crystals, leading to efficient precipitation, and a spurious downward transport of humidity towards the ground, eventually leading to extensive fog occurrence. By contrast, the new scheme, which one, does not allow ice crystal nucleation at temperatures above $-10\,^{\circ}$C, and two, only predicts a slow growth of ice crystals at temperatures above $-10\,^{\circ}$C, consistent with theory, maintained the condensate within the clouds, in the form of supercooled droplets. Consequently, the spurious fog occurrence was avoided and the temperature distribution of the lowest km of the atmosphere was well reproduced, including the subsidence inversion, having its base at 900 hPa.

When cloud liquid water and cloud ice are treated as separate prognostic variables, equations (1.3)–(1.5) are replaced by the following two equations

$$\frac{\partial q_{c,liq}}{\partial t} = Trans_{c,liq} + Cond - Evap_c - \mathrm{Precip}_{liq} - Freez_c + Melt_c \qquad (1.6)$$

$$\frac{\partial q_{c,ice}}{\partial t} = Trans_{c,ice} + Dep - Subl_c - \mathrm{Precip}_{ice} + Freez_c - Melt_c \qquad (1.7)$$

where subscripts *liq* and *ice* now indicate the phase of the condensate, the terms *Freez* and *Melt* represent rates of freezing of cloud liquid water and melting of cloud ice, respectively, while *Dep* and *Subl* represent respectively the rates of vapour deposition and sublimation of cloud ice.

The global NWP models of the European Centre for Medium-range Weather Forecasts (ECMWF), the National Center for Environmental Prediction (NCEP) and Météo France (the ARPEGE model) all use cloud microphysical schemes of a similar complexity to that of the Unified Model. On the other hand, some regional models, run for limited areas at very high spatial resolution, adopt much more detailed and computationally costly microphysics schemes, with a significantly larger number of prognostic variables. Examples of this type of models are the MM5 model and its successor, the WRF model.

Figure 1.9 illustrates the basic features of the MM5/WRF scheme. Note that in its most elaborate form it contains 8 prognostic variables for condensate. This is because the scheme carries 5 prognostic species for the mass or mixing ratio (mass relative to the mass of dry air) of condensate (the Q terms), and for 3 of these the particle number is also predicted (the N terms).

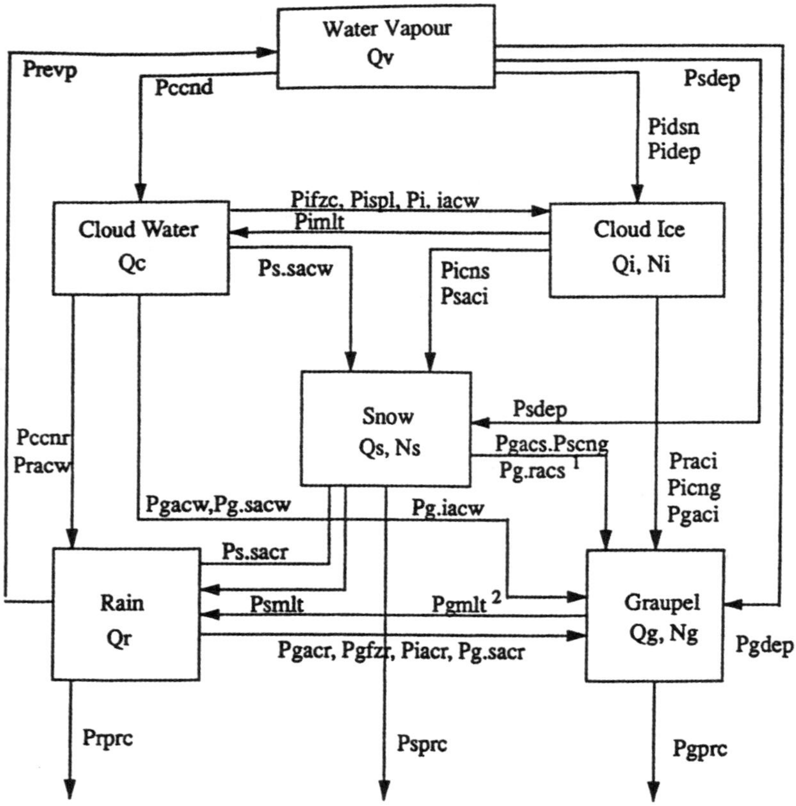

Fig. 1.9 A schematic depiction of the cloud microphysics scheme of the MM5/WRF models (Reisner et al. 1998). The *boxes* indicate the prognostic variables in the model, while the *arrows* represent cloud microphysical processes (Permission is granted by John Wiley & Sons Ltd on behalf of Royal Meteorological Society)

Reisner et al. (1998) investigated two wintertime cases of in-cloud icing over Colorado, both of which represented a potential problem for air traffic. In both cases, a warm, humid air mass was lifted along a warm front. The air was stable and there was no precipitation entering the cloud from above. As a result, a cloud consisting of supercooled droplets persisted for several hours at temperatures between −5 °C and −15 °C. Liquid water concentrations as high as 0.35 g/m^3 were observed at a temperature of −10 °C, 3 km above sea level. This feature was well simulated by the MM5 model, using the cloud microphysics scheme depicted in Fig. 1.9. Interestingly, the ability to simulate the supercooled cloud water was crucially dependent upon the parameterization of the size distribution for snow crystals.

Other examples of nesting models from global or regional scale to local scale can be found in Doyle and Durran (2003), Czyzyk and Runk (2007) and Hauge et al. (2008).

1.7 A Case Study: Using Numerical Weather Prediction Models to Forecast In-cloud Atmospheric Icing Episodes

The parameterization of the microphysical processes in NWP models has improved significantly over the past decades. The increase in computing power has made it possible to incorporate several microphysical processes into NWP models, including explicit calculation of both mixing ratio, and for some moisture variables, also the number concentration, such that more accurate predictions of precipitation and clouds can be made. One should expect that there is a potential to use NWP models with such detailed microphysics to predict atmospheric icing.

In this study, a numerical weather prediction model (Weather Research and Forecasting modelling system (WRF) was used to carry out high resolution simulations, with a focus on in-cloud icing. WRF is a non-hydrostatic mesoscale NWP model, developed jointly by several institutions in the United States (Skamarock et al. 2005). The model is meant to gradually replace its predecessor, the MM5, which has been widely used both for weather forecasting and various weather-related applications. Simulations were carried out for selected cases and the model results are quite good compared to measurements of supercooled cloud water and in-cloud icing.

To study the effect of the microphysics in the model, two different parameterization schemes were tested. Most attention was paid to the control simulations, where the Thompson scheme for cloud microphysics was used, including a prognostic calculation of ice particle number concentrations (Thompson et al. 2004). In addition, simulations were carried out with a simpler, more economical cloud microphysics scheme; WSM6 (WRF Single-Moment 6-class).

Two different mountains were used as validation sites: Ylläs in Northern Finland and Gamlemsveten on the northwest coast of Southern Norway. Both are frequently exposed to moist air masses in winter, and atmospheric icing is a well known problem at both sites.

1.7.1 First Experiment - Ylläs

Measurements were made at the top of Mt. Ylläs (67.6°N, 24.3°E) in Northern Finland, at an elevation of 706 m. It is a rounded peak and is the highest mountain in a large region. At Mt. Ylläs, accurate in-situ measurements of in-cloud icing have been carried out for several years, using a rotating multi-cylinder instrument (Makkonen 1992). The simulations were configured with a triple nested domain. Starting with a 120×120 grid with spacing, $\Delta x = 13.2$ km, and stepwise nested down to a 130×130 grid with $\Delta x = 0.825$ km. Such high horizontal resolution gives a detailed representation of the terrain in the model, which is very important when phenomena forced by the terrain itself are studied, for example the orographic production of cloud water, when moist air masses are lifted along a hillside.

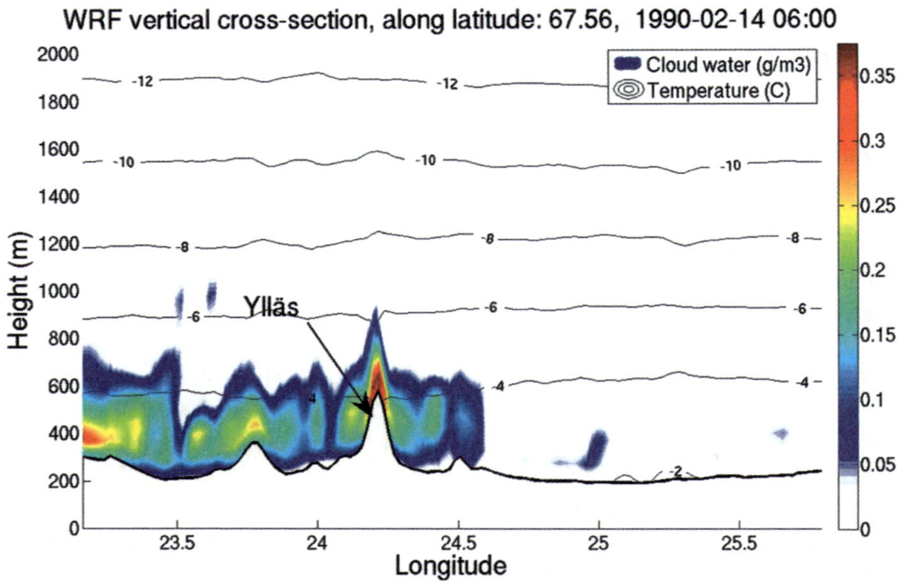

Fig. 1.10 Mt. Ylläs, Finland. Vertical cross section of cloud liquid water content (LWC) and temperature at 06 UTC 14 February 1990. Measured LWC $= 0.27\ g/m^3$, modelled LWC $= 0.37\ g/m^3$, wind direction: SE

Simulations for three icing events at Ylläs were carried out. The simulated amount of supercooled cloud water at the lowest level of the model was compared to the measured amount at the same juncture. Figures 1.10, 1.11 and 1.12 show vertical cross-sections of the simulated supercooled cloud water at the same times as the measurements were carried out.

The figures show that the top of Ylläs is well above the cloud base for all three situations, and that the model slightly overestimates the amount of cloud water. A maximum in the LWC just above the mountaintop indicates a significant production of cloud water by orographic lifting of the air masses along the mountainside.

The table below summarizes the results from the experiments at Ylläs. In the same table we have included the results from the same icing events obtained by Vassbø et al. (1998) using the HIRLAM model, with a grid spacing of 5.0 km in the highest resolution domain.

Table 1.1 shows that HIRLAM consistently failed to simulate any cloud water at all in the lowest level (31), so LWC values from levels 29 and 30 were studied instead. The WRF simulations show a strong dependence between LWC and horizontal grid resolution. This emphasizes the importance of high horizontal resolution and a good representation of the model terrain. When the simpler cloud microphysics scheme is used (WSM6), a large portion of the condensate is in the form of ice (not shown), and the supercooled liquid water is partly removed by precipitation, resulting in an underestimation of the supercooled liquid water.

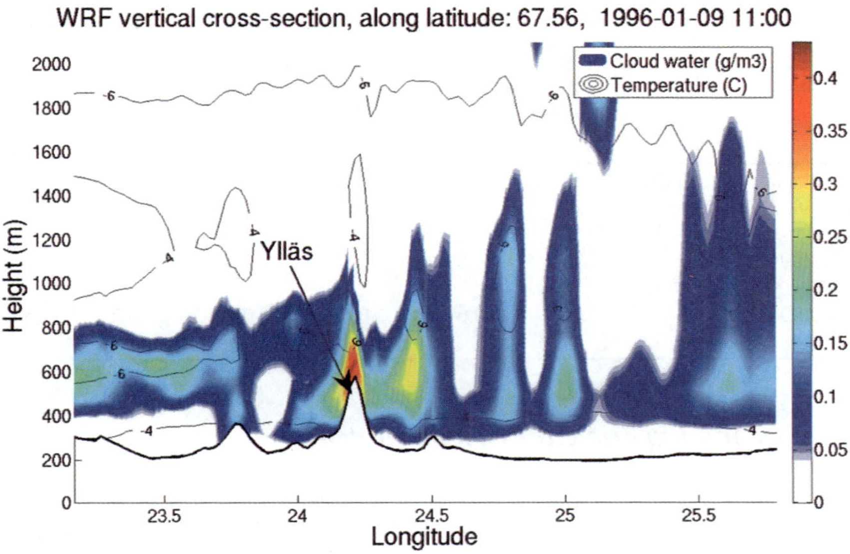

Fig. 1.11 Mt. Ylläs, Finland. Vertical cross section of cloud liquid water content and temperature at 11 UTC 9 January 1996. Measured LWC = 0.30 g/m^3, modelled LWC = 0.39 g/m^3., wind direction: SW

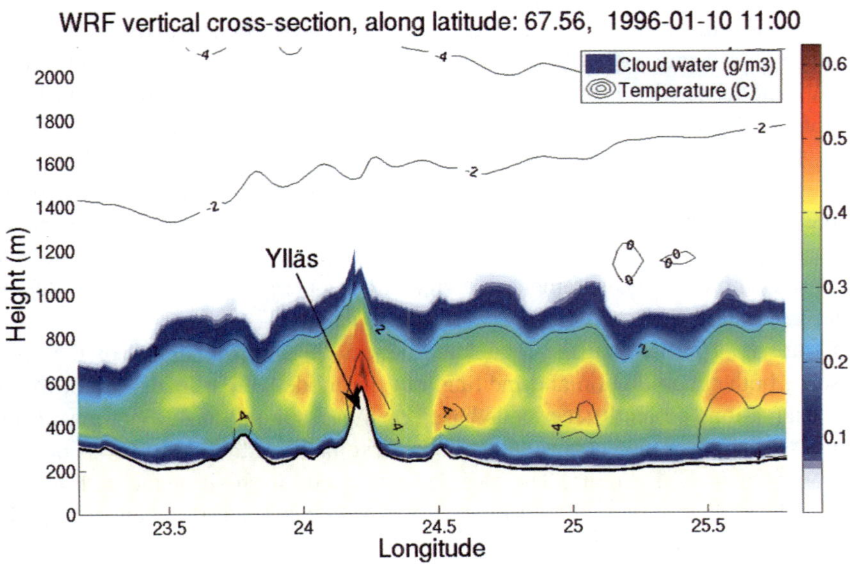

Fig. 1.12 Mt. Mt. Ylläs, Finland. Vertical cross section of cloud liquid water content and temperature at 11 UTC 10 January 1996. Measured LWC = 0.43 g/m^3, modelled LWC = 0.51 g/m^3, wind direction: SSW

Table 1.1 Ratio between modelled and measured amount of cloud water. The WRF results are all from the lowest level

Simulation	WRF Control	WRF Control	WRF Control	WRF WSM6	HIRLAM		
	Domain 3	Domain 2	Domain 1	Domain 3	Lev 31	Lev 30	Lev 29
Grid spacing	0.83 km	3.30 km	13.20 km	0.83 km	5.0 km	5.0 km	5.0 km
14.02.1990 06 UTC	1.37	0.80	0.13	0.80	0.0	0.44	0.48
09.01.1996 11 UTC	1.30	0.83	0.08	0.57	0.0	0.33	0.67
10.01.1996 11 UTC	1.19	0.76	0.0	0.67	0.0	0.23	0.49

1.7.2 Second Experiment - Gamlemsveten

The second experimental site is situated at Gamlemsveten (62.58°N, 61.32°E), 10 km northeast of Ålesund in Norway. Mt. Gamlemsveten has an elevation of 800 m and is exposed to moist air masses from the sea. Typically, the mountaintop is well above cloud base when such air masses enter the area.

For this site, direct validating measurements of cloud water were not available, but measurements of accumulated ice on a steel wire were carried out by Kjeller Vindteknikk AS. The measurements were retrieved from web camera pictures of an iced guy wire of a 10 m meteorological mast at the top of Gamlemsveten (Fig. 1.13). Further, ice loads (kg/m) were then calculated based on an assumption of ice density.

The WRF simulations were produced with a triple-nested domain, with a horizontal grid spacing of 0.9 km in the finest grid. Comparing model results from a NWP model to measurements of accumulated ice loads is not an easy task. To be able to do such a comparison at all, further processing of the modelled data through an ice accumulation model needed to be done (Tallhaug et al. 2005 and ISO 12494). Time series of cloud water, temperature and wind speed from WRF simulations were used as input into the accumulation model, resulting in a time series of accumulated ice (kg/m) on an ISO reference object. Time series of accumulated ice based on on-site measurements of wind and temperature, combined with cloud water from the WRF simulations were also calculated.

Figure 1.14 shows that the model reproduces the main tendency of the icing episode, both accumulation and melting, but the modelled ice loads do not match the observations perfectly. Ice loads based on WRF simulations at 794 masl (blue curve) seem to give the best results, with the best match to the observations on both 7 February and 10 February. The calculations based on on-site wind and temperature measurements (yellow and pink curves) underestimate the icing rate in the period between 8 February and 10 February, partly because the quality of the wind measurements is poor due to ice accretion on the wind sensor.

The observations show a period between 8 February and 10 February with heavy ice accumulation, which all the modelled curves underestimate. By studying the

Fig. 1.13 Mt. Mt. Gamlemsveten, Norway. Typical web camera picture used for the measurements (Reproduced by permission of Kjeller Vindteknikk AS, Norway)

general weather situation (not shown) for the same period, we find a situation with deep convective clouds in cold air masses (snow showers) from the northwest. In this particular case, the WRF model underestimates the amount of supercooled cloud water in situations where clouds consist of a mixture of frozen and liquid condensates.

The results from the Ylläs experiment suggest that there is good potential for quantitative forecasts of in-cloud icing episodes, using current NWP models at high spatial resolution, with sophisticated cloud microphysics parameterizations. The icing episodes studied in this experiment were dominated by stratified continental air, with clouds consisting mostly of supercooled water. The model seems to perform well under such conditions. A dramatic improvement is found when comparing to the HIRLAM model results from 10 years ago.

The results from the experiment at Gamlemsveten are a bit more intricate to analyze because the modelled ice loads were calculated from temperature and wind speed, in addition to cloud water. There are also several uncertainties regarding the comparison between the calculated and the observed ice loads. The agreement between the modelled and the observed ice loads seems to be best in weather situations with low stratus clouds, containing mostly liquid cloud water. The model seems to overestimate the degree of glaciation in a period with convective clouds in a cold air mass, when cloud ice is present in the cloud.

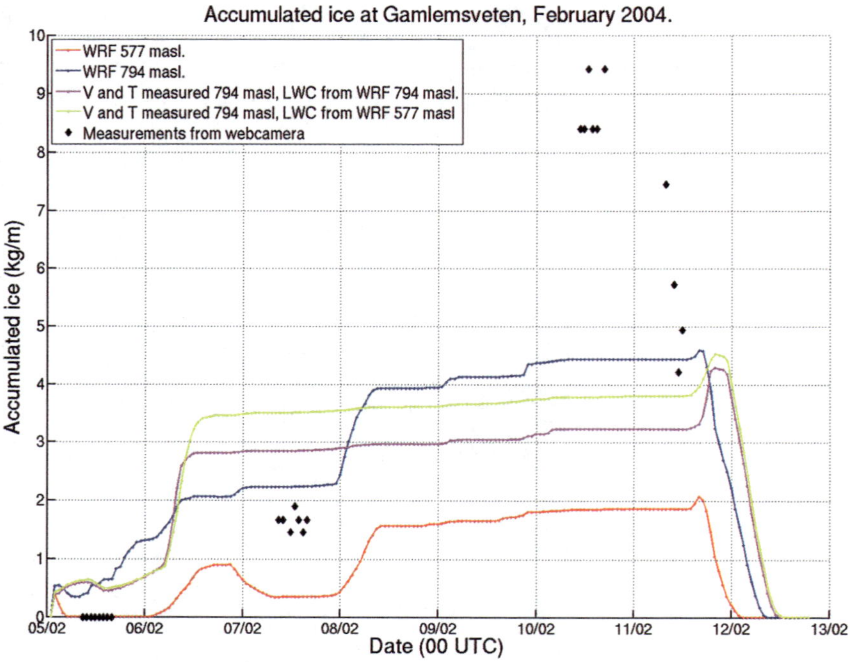

Fig. 1.14 Mt. Gamlemsveten, Norway. Accumulated ice loads, modelled and observed. 577 masl (meters above sea level) refers to the elevation of the mountain in the smoothed model terrain, while 794 masl refers to the real height of Gamlemsveten. Observations are only available in clear weather conditions during daytime

1.8 Concluding Comments

It has been demonstrated in this chapter that the modern science of meteorology has developed tools that are able to deal also with many adverse weather impacts on vulnerable structures, such as overhead power lines, radio and TV towers, wind turbines, ski lifts, etc. In particular, they are able to handle topographical features down to scales comparable with one span of an overhead line. They are also rapidly enhancing their ability to describe, physically, the H_2O cycle in the atmosphere, especially relating to precipitation type and amount, as well as the liquid water content of clouds, as a function of height above cloud base.

It will never be possible to use atmospheric models alone, without any local measurements for control and calibration. However, the models should be tested further, in combination with field measurement programs, in order to continually improve them for future practical applications. In particular, the currently (2007) ongoing COST Action 727 "Measuring and Forecasting Atmospheric Icing on Structures", involving 12 European countries, will provide such new achievements in this field during the next few years (see Fikke et al. 2007a).

The overall gain, with respect to atmospheric icing, from all the efforts mentioned in this chapter, will be to improve the accuracy, details and reliability of all critical weather-related parameters that are discussed in the subsequent chapters of this book. Consequently, the capacity of the atmospheric models will constantly be improved with respect to analyses of icing events in the past and forecasting them within time scales of 3–48 hrs, with reasonable reliability for both statistical and operating purposes. Therefore, a close collaboration with meteorological scientific communities is strongly recommended in order to develop dedicated services for all industries depending on adequate atmospheric icing information.

Acknowledgments The authors are grateful to Dr Ivar Lie, Storm Weather Center AS, for valuable discussions and input to this chapter.

References

Admirat P, Sakamoto Y (1988) Calibration of a wet snow model on real cases in Japan and France. In: Proc 4th International Workshop on Atmospheric Icing of Structures (IWAIS 1988), Paris, France, September

Cigré (2001) Guidelines for field measurement of ice loadings on overhead power line conductors. Cigré Task Force 22.06.01, TB179

Cigré (2006) Guidelines for meteorological icing models, statistical methods and topographical effects. Cigré Working Group B2.16, TB 291

Czyzyk S, Runk K (2007) Operational Forecast Support by National Weather Service Forecast Office in Las Vegas during the Terrain-Induced Rotor Experiment. In: Proc 12th Annual Conf on Mountain Meteorology, American Meteorological Soc

Doyle JD, Durran DR (2003) High-resolution simulations of wave-induced turbulence and rotors using NRL's COAMPS. In: Proc 10th Annual Conf on Mountain Meteorology, American Meteorological Soc

Ervik M, Fikke SM (1982) Development of a mathematical model to estimate ice loading on transmission lines by use of general climatological data. IEEE Transactions on Power Apparatus and Systems, PAS-101, No. 6 June 1982: 1497–1503

Farzaneh M, Savadjiev K (2001) Icing Events Occurrence in Québec: Statistical analysis of field data. Int J of Offshore Polar Eng, 11, no 1 March: 9–15

Fikke SM (2005a) COST Action 727: Measuring and forecasting atmospheric icing on structures. In: Proc 11th International Workshop on Atmospheric Icing of Structures (IWAIS 2005), Montreal, Canada, June 2005, Paper IW64

Fikke SM (2005b) Modern meteorology and atmospheric icing. In: Proc 11th International Workshop on Atmospheric Icing of Structures (IWAIS 2005), Montreal, Canada, June 2005, Paper IW73

Fikke SM, Johansen OS (1987) Earlier Norwegian iceload research. A review of investigations and results. In: Proc 2nd International Workshop on Atmospheric Icing of Structures (IWAIS 1984), Trondheim Norway, June 1984. EFI TR 3439, June: 11–18

Fikke SM et al. (2007a) COST Action 727 Atmospheric icing on structures. Measurements and data collection on icing. State of the art. Veröffentlichung MeteoSchweiz Nr 75

Fikke SM, Heimo A, Säntti K (2007b) COST 727 – Report from Phase 1. In: Proc 12th International Workshop on Atmospheric Icing of Structures (IWAIS 2007), Yokohama, Japan, October

Goia ML (2000) Damages caused by icing and wind to the Romanian OEL. In: Proc 9th International Workshop on Atmospheric Icing of Structures (IWAIS 2000), Chester, June

Golikova TN, Toporkava GD, Nikitina LG (1989) Ascertaining ice-load maps of the USSR territory. Trans Improving the reliability of high voltage lines. Moscow, Energoatomizdat, 1989: 107–122

Hauge G, Holstad A, Lie I (2008) The use of ultra high resolution weather prediction models for real-time forecasting in complex terrain. Weather and Forecasting, in press.

Holstad A, Lie I (2006) Simulation of wind conditions for the proposed new Hammerfest airport using a fine scale atmosphere model. Storm Technical Report 2006/7. Storm Weather Center, Bergen, Norway

Holstad A Lie I, Utnes T, Ødegaard V (2001) Wind conditions in Sunndalsøra: A study using fine-scale models, Res. Rep. no 125, The Norwegian Meteorological Institute, July

IEC (1997) Overhead lines – Meteorological data for assessing climatic loads. International Electrotechnical Commission Technical Report 61774, first edition: 1997–2008

IEC (2003) Design criteria of overhead transmission lines. International Electrotechnical Commission (IEC) Technical Report 60826, Ed. 3.0

ISO (2000) Atmospheric icing of structures. International Standardization Organisation (ISO) International Standard 12494

Krómer I (1993) Hungarian icing activity survey. In: Proc 6th International Workshop on Atmospheric Icing of Structures (IWAIS 1993), Budapest, September 1993: ix–x

Lehtonen P, Ahti K, Makkonen L (1986) The growth and disappearance of ice loads on a tall mast. In: Proc 3rd International Workshop on Atmospheric Icing of Structures (IWAIS 1996), Vancouver, Canada, May

Lie I (2001) Wind conditions in Fjærlandsfjorden, Res. Rep. no 129, The Norwegian Meteorological Institute, November

Lozowski EP, Makkonen L (2005) Fifty years of progress in modelling the accumulation of atmospheric ice on power network equipment. In: Proc. Eleventh International Workshop on Atmospheric Icing on Structures, Montreal, CD-ROM

Makkonen L (1992) Analysis of rotating multicylinder data in measuring cloud-droplet size and liquid water content. J Atmos Oceanic Technol, 9: 258–263

Moss SJ, Johnson DW (1994) Aircraft measurements to validate and improve numerical model parameterizations of ice to water ratios in cloud. Atmos Res, 34: 1–25

Olafsson H, Eliasson AJ, Thorsteins E (2002a) Orographic influence on wet snow icing. Part I: Upstream of mountains. In: Proc 10th International Workshop on Atmospheric Icing of Structures (IWAIS 2002), Brno, Czech Republic, June 2002, Paper 2–2

Olafsson H, Eliasson AJ, Thorsteins E (2002b) Orographic influence on wet snow icing. Part II: Downstream of mountains. In: Proc 10th International Workshop on Atmospheric Icing of Structures (IWAIS 2002), Brno, Check Republic, June 2002, Paper 2–3

Poots G (ed) (2000) Ice and snow accretions on structures. Philosophical Transactions, vol 358, no 1776, The Royal Society London, November

Popolanský F (2000) Economical aspects of ice failures caused in power transmission on the territory of former Czechoslovakia. In: Proc 9th International Workshop on Atmospheric Icing of Structures (IWAIS 2000), Chester, June

Rasch PJ, Kristjánsson JE (1998) A comparison of the CCM3 model climate using diagnosed and predicted condensate parameterizations. J Climate, 11: 1587–1614

Reisner J, Rasmussen RM, Bruintjes RT (1998) Explicit forecasting of supercooled liquid water in winter storms using the MM5 mesoscale model. Q J R Meteorol. Soc, 124: 1071–1107

Rutledge SA, Hobbs PV (1983) The mesoscale and microscale structure and organization of clouds and precipitation in midlatitude cyclones. VIII: A model for the 'seeder-feeder' process in warm-frontal rainbands. J Atmos Sci, 40: 1185–1206

Sakamoto Y (2000) Snow accretion on overhead wires. Roy Soc Phil Trans, 358, no 1776, November: 2941–2970

Skamarock WC, Klemp JB, Dudhia J, Gill DO, Barker DM, Wang W, Power JG (2005) A description of the Advanced Research WRF Version 2. NCAR Technical Note, NCAR/TN-468+STR

Smith RNB (1990) A scheme for predicting layer clouds and their water content in a general circulation model. Q J R Meteorol Soc, vol 116: 435–460

Tallhaug L, Harstveit K, Fidje A (2005) Ice accumulation observed by use of web camera and modelled from meteorological parameters. IN: Proc Boreas VII, Wind energy production in cold climates, Saariselkä, Finland

Thériault J, Stewart RE, Milbrandt JA, Yau MK (2006) On the simulation of winter precipitation types. J Geophysical Research, vol 111: D18202, doi:10.1029/2005JD006665

Thériault J, Stewart RE (2007) On the vertical effects of air velocity on winter precipitation types. Natural Hazards and Earth System Sceinces, vol 7: 231–242

Thompson G, Rasmussen RM, Manning K (2004) Explicit forecasting of winter precipitation using an improved bulk microphysics scheme. Part I: Description and sensitivity analysis. Mon Wea Rev, vol 132: 519–542

Thorsteins E, Elíasson AJ (1998) Iceload measurements in test spans in Iceland – statistical analysis of data. In: Proc 8th International Workshop on Atmospheric Icing of Structures (IWAIS 1998), Reykjavik, June: 285–289

Vassbø T, Kristjánsson JE, Fikke SM, Makkonen L (1998) An investigation of the feasibility of predicting icing episodes using numerical weather prediction model output. In: Proc 8th Int. Workshop on Atmospheric Icing on Structures (IWAIS 1998), Reykjavik, June: 343–347

Wareing BJ, Chetwood P (2000) Ice load data from Deadwater Fell. In: Proc 9th International Workshop on Atmospheric Icing of Structures (IWAIS 2000), Chester, June

Wilson DR, Ballard SP (1999) A microphysically based precipitation scheme for the UK Meteorological Office Unified Model. Q J R Meteorol Soc, 125: 1607–1636

Chapter 2
Statistical Analysis of Icing Event Data for Transmission Line Design Purposes

Masoud Farzaneh and Konstantin Savadjiev

2.1 Introduction

Climatic loads on overhead transmission lines are random variables. Wind, icing, and combined loads of wind and ice vary greatly in time and space. Therefore, a statistical approach is most appropriate for their study and forecast. For this reason, statistical analysis of field data for icing events is very important at all stages of planning: design, construction, and operation of overhead transmission line networks, especially for lines that cross regions with severe icing conditions. Results from this analysis allow for forecasting climatic loadings; this determines the economic characteristics and operational reliability of the line and its principal components. Analysis of large-scale databases covering vast areas and long time periods also allows for a better understanding of the spatial and temporal evolution of icing events and the extent of their impact on the power network.

The need for long-time series of available, reliable and pertinent field records of meteorological data containing historical information for icing events is increasing continuously, especially since probabilistic methods have been recommended in transmission line design (see IEC/CEI-60826 2003). While such databases for wind are, in general, relatively readily available, collecting data for icing on line equipment is a difficult, costly, and time-consuming process.

In this chapter, we aim, first, at discussing issues pertaining to icing event data measurement and collection, and then at reviewing statistical analyses and modelling of icing using the ice data. This chapter is therefore subdivided into two main parts. The first part, 2.1 – Measurements and Database, gives some basic definitions pertaining to icing events (2.1.1), describes the sources for obtaining meteorological field data (2.1.2), and the principal measurement instruments: Passive Ice Meter (PIM) (2.1.3), Icing Rate Meter (IRM) (2.1.4), Load Cell (LC) (2.1.5), tubular-steel-rod racks in Norway (2.1.6), and horizontal metallic rods at stand Studnice, Czech Republic (2.1.7). Section 2.1.8 is devoted to databases containing statistical information on icing events. The second part, 2.2 – Statistical Analysis and Modelling Ice

M. Farzaneh
University of Québec at Chicoutimi, 555 Boulevard de l'Université, Chicoutimi, Canada G7H 2B1
e-mail: farzaneh@uqac.ca

M. Farzaneh (ed.), *Atmospheric Icing of Power Networks*,

Loads on Overhead Transmission Lines, is a brief review of the principal statistical distributions, models, and methods in use in transmission line engineering related to atmospheric icing: modelling freezing-rain data from the PIM network (2.2.1), extreme value analysis (2.2.2), combined wind-on-ice loads (2.2.3), modelling hourly icing rates on transmission line conductors (2.2.4), joint distribution of wind speed and air temperature (2.2.5), and modelling spatial and temporal evolution of icing events (2.2.6).

2.2 Measurements and Database

2.2.1 Basic Definitions

In this chapter, each observed occurrence of frozen water on transmission line equipment is referred to as *atmospheric icing event*, or simply *icing event.* Evolution in time of a typical icing event can be subdivided into three more or less distinct phases: *accumulation*, *persistence*, and *shedding* (dropping off). Figure 2.1 illustrates the evolution of the icing build-up on conductors of an actual 315-kV transmission line during a typical icing event. In this figure, the accumulation phase between 2:00 h and 14:00 h is followed directly by ice shedding phase between 14:00 h and 17:00 h. Due to the relatively short total duration, there is no persistence phase.

Atmospheric icing is a very complex phenomenon, which may present various types, shapes, and physical properties; it is customary in engineering to consider two main processes for ice formation determining the type of ice accretion: *precipitation*

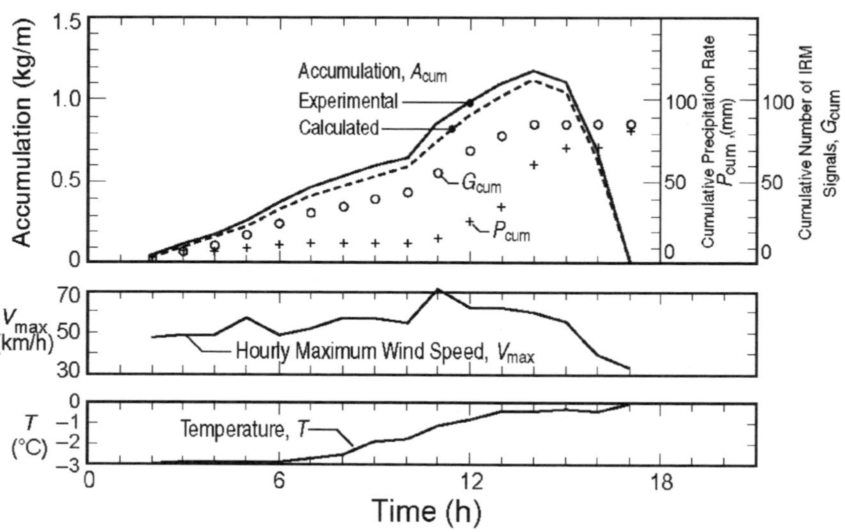

Fig. 2.1 Evolution in time of a typical, precipitation-icing event (Reproduced by permission of IEEE)

icing and *in-cloud icing*. Principal properties of ice accretion build-up and related atmospheric conditions corresponding to these processes may be summarized as follows:

Precipitation icing: Occurs as *freezing rain* (glaze), density 700 to 900 kg/m^3, temperature $-10\,^\circ\text{C} < t < 0\,^\circ\text{C}$; *wet snow*, 400 to 600 kg/m^3, $0\,^\circ\text{C} < t < 3\,^\circ\text{C}$.

In-cloud icing: Occurs as *hard rime*, density 600 to 800 kg/m^3, temperature $-10\,^\circ\text{C} < t < 1\,^\circ\text{C}$, or *soft rime*, 200 to 600 kg/m^3, $-20\,^\circ\text{C} < t < 1\,^\circ\text{C}$.

Usually, icing events are difficult to refer to as belonging to one or another of these categories because they represent a mixture of both types. Savadjiev and Farzaneh (2005) proposed a simplified classification, based on the presence (or absence) of precipitation during the accumulation phase of an icing event, which provides statistically more reliable basis for the analysis of these events. For example, the curve in Fig. 2.1 depicting the precipitation rate and the cumulative quantity of precipitation (in mm/h), P_{cum}, shows that this specific icing event is of precipitation-ice type. Another possible distinction between both types of icing, recommended by IEC/CEI, is considered in Section 2.2.5.

In general, in transmission line engineering only freezing rain (glaze) and hard rime are considered to be dangerous, and as having great influence on line reliability. Wet snow and soft rime are rarely critical for line components and structures. In some rare cases wet snow may accrete, forming build-ups with great diameter which, after freezing at low temperatures, may produce significant combined wind-on-ice forces.

A process closely related to icing, *sublimation*, characterized by direct transformation of solid phase into vapour, or inversely, *hoar frost*, is not considered in this chapter because of its negligible impact on transmission line reliability and design.

2.2.2 Sources of Field Data for Icing Events

In overhead transmission line engineering, different means exist for obtaining field records containing information for icing events. These differ mainly by the sources of data, methods of observation, measuring devices and instrumentation, type, content, and length of time series in the database. Recognizing the importance of sources of such data, IEC/CEI currently recommends the creation and implementation of new programs for collecting pertinent meteorological and icing field data, or further development of existing icing measurement programs.

The main sources and methods for obtaining statistical field data for icing events and their impact on overhead transmission lines can be summarized as follows:

2.2.2.1 National Weather Services

This basic source of climatic information is widely used because of its multiple advantages, such as the abundance, availability, and diversity of data covering relatively long periods of observation across an entire country. A disadvantage of using this information for transmission line design purposes is the relatively sparse distribution of meteorological stations supplying data usually recorded far from transmission line routes.

2.2.2.2 Electrical Facilities

This is the largest source of icing event information, indispensable for planning, design, and operation of newly installed lines, or upgrading existing ones as a consequence of re-evaluating design loadings. It covers:

- Information from previous experience during operation of overhead line networks in areas exposed to icing.
- Direct measurements on samples fallen to the ground, or samples collected from accessible non-energized line conductors or ground wires. Samples of glaze collected by this relatively easy and natural way, especially those fallen from conductors or ground wires, are usually broken and sometimes do not conserve their original shapes or dimensions.
- Field observations of icing on vegetation, or neighbouring distribution and telecommunication lines.
- From specially designed icing measurement devices, such as the Passive Ice Meter (PIM) used in Quebec for freezing-rain observations (Fig. 2.2), and tubular steel-rod racks used in Norway (Fig. 2.4) and the Czech Republic (see Section 2.1.7).
- Using test-spans specially designed, installed, and instrumented for icing test purposes near to transmission line routes, or on the top of hills in areas with increased icing event recurrence frequency. These experimental spans simulate the behaviour of phase conductors and ground wires on actual transmission lines under the impact of climatic loadings, icing, wind, wind on iced conductors, and great variations of the ambient temperature.
- From permanently instrumented actual (energized) transmission lines in service. Test facilities of this kind presently exist and are operated is several countries. In the province of Quebec, Canada, such ice monitoring and warning systems, SYGIVRE, have been developed for studying atmospheric icing and its impact on power networks. Most of the test stations, installed on actual transmission lines in service, are instrumented with modern measurement equipment supplying real-time icing event information available to users via satellite: Icing Rate Meter (IRM), load cell (force sensor), hygrometric/ambient-temperature probe, electrically heated anemometers, rain gauge (Laflamme 1996; McComber et al. 1996; Druez et al. 1999; Savadjiev and Farzaneh 1999, 2001, 2004).

2.2.3 *Passive Ice Meter (PIM)*

A large number of varying constructions and ice-measurement devices have been used in several countries in Europe since the first half of the 20th Century (Burgsdorff

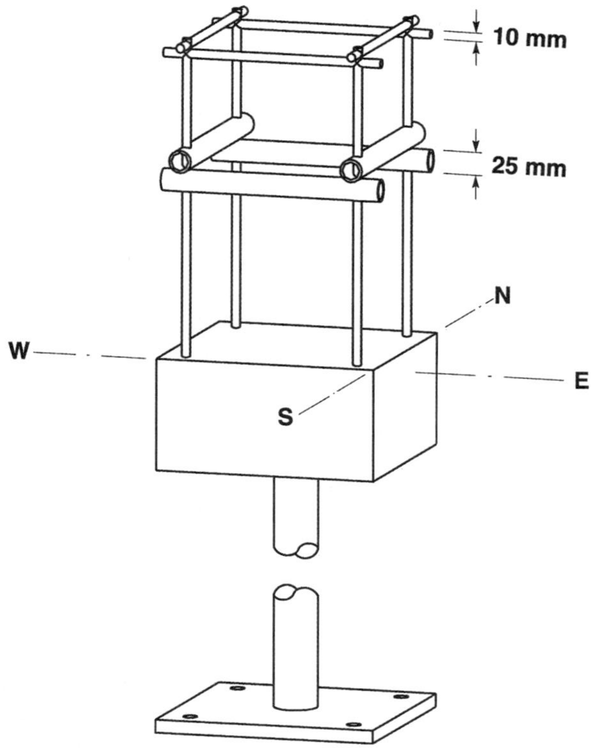

Fig. 2.2 Passive Ice Meter (PIM) (Reproduced by permission of IJOPE)

1947). The Passive Ice Meter (PIM), sketched in Fig. 2.2, is designed for measuring ice accumulation during freezing-precipitation icing events in Quebec. Its metallic receiving cylinders – 4 × 10 mm simulating ground wires, and 4 × 25 mm simulating phase conductors – as well metallic plates simulating surfaces of structure members are installed 1.2 m above the ground, and oriented N-S and E-W.

A great advantage of the PIMs is that, due to their low cost and ease of installation and operation, they can be used to form vast networks of measurement sites covering the entire territory or country. In Quebec, over the years after 1974, about 170 PIMs were installed throughout the province, spaced on average 50 km apart. A Passive Ice Meter is a simple and reliable instrument for measuring equivalent radial ice thickness, shape, and physical characteristics of the ice deposit on its receiving cylinders and plates. In addition, measurements made on the four faces of the instrument supply information for the influence of wind direction.

Operation of the PIM network gives very useful and abundant field data records for precipitation icing events and thus allows for establishing better knowledge of the freezing rain climate in the province. For instance, Laflamme and Périard (1996) used such information to create 100-year return period freezing-rain forecasting maps covering the St. Lawrence Valley in Quebec. Some distributions of freezing-rain icing events were also established from the analysis of the PIM network database: (i) Annual Number of Icing Event Recurrence (ANIER), (ii) Icing Event

Residency Period (IERP), and (iii) Total Annual Duration of Icing Events (Farzaneh and Savadjiev 2005). Modelling these distributions, as discussed in Section 2.2.1, is important for assessing combined wind-on-ice loadings, i.e. determining statistics-based return periods for wind-speed design associated with icing events (see Section 2.2.3).

2.2.4 Icing Rate Meter (IRM) and Load Cell (LC)

The IRM, developed for icing detection, consists of a vertical 25.4-mm long and 6.2-mm diameter cylindrical probe situated on the top of the instrument (Fig. 2.3). When ice accretes on the probe, the natural frequency at which the probe vibrates (40 kHz) decreases at a rate of about 2 Hz for each milligram of accreted mass (Laforte et al. 1995). After an accretion of 60–65 mg, an electronic controller de-ices the probe by heating and shaking, emits a signal recorded by a cumulative counter and thus completes an icing/de-icing cycle. The hourly number of cycles (IRM signals), G, is presently widely used as a warning device signalling the beginning of an icing event. The hourly number of signals may also be used for an indirect estimation of ice accretion rate on neighbouring transmission lines or other

Fig. 2.3 Icing Rate Meter (IRM) (Reproduced by permission of IEEE)

objects or equipment. However, for this purpose, a correlation must be established between the hourly number of IRM signals and the actual icing accumulation rate on transmission line conductors (see Section 2.2.4).

In wet, precipitation-icing conditions – large supercooled water droplets falling vertically at temperatures slightly below the freezing point – the receiving efficiency of the IRM probe is very low. Sometimes this leads to malfunctioning (breakdowns) during the operation: (i) positive icing accumulation rates are observed without emission of IRM signals ($G = 0$), or (ii) some extreme icing rates greater than 0.1–0.2 kg/m/h leading to complete blockage of the probe by a dense cover of glaze.

2.2.5 Load Cell

The Load Cell (LC) is a force sensor incorporated between a suspension insulator string and the cross arm of a tower of the instrumented transmission line used as an icing-test site (station). In Quebec, such force sensors are installed at several SYGIVRE stations, e.g. at tower No 50 of the 315-kV line at Mont Bélair, near Quebec City. Load cell readings are proportional to the weight of the supported conductor (iced or not) within the weight span, plus the weight of the suspension insulator itself. Hourly variations of this force are used to estimate the actual icing accumulation per unit length of the conductor, A_{cum}. The conversion from load cell readings, C, in volts, into icing rate, in kg/m/h, can be made by using computer algorithms (Savadjiev and Farzaneh 1999).

Load cells are now installed by several electrical utilities at each icing test site where neighbouring IRMs are to be calibrated by direct measurement of the actual icing rate. However, the use of LCs is not as widespread as that of IRMs, because it is limited by difficulties inherent to their installation, maintenance, and operation on energized HV transmission lines in service.

2.2.6 Tubular Steel-Rod Racks

Tubular steel-rod racks, used for icing measurements in Norway, are shown in Fig. 2.4. They consist of a 10.5-m long horizontal rod, 55 mm in diameter, suspended 5 m above ground and connected to a dynamometer which mechanically records the maximum load since the previous resetting. A set of dampers is installed for absorbing vibrations due to wind. In addition, some of the racks are equipped with electronic load cells that record the loads every three hours on a local logger. Such devices have been permanently installed in icing test stations situated in the vicinity of existing or to-be-built high-voltage (HV) lines, especially those crossing areas exposed to severe icing. A database recorded from the operation of the Norwegian rack network contains time series of annual maxima of ice accretion amounts on tubular steel-rods in kg/m. Unit ice weight data for ice accumulation on 55-mm cylinders, converted into accumulation on actual line conductors, or into equivalent

Fig. 2.4 Tubular steel-rod rack used for icing-measurements in Norway (Reproduced by permission of CIGRÉ)

radial ice thickness, may be used for the extreme value analysis (see Section 2.2.2) for forecasting design loads with a given return period (Ervik and Fikke 1982a, 1982b).

2.2.7 Horizontal Metallic Rods at Stand Studnice, Czech Republic

Another example of specially installed devices for long-period field measurements of icing is the stand made up of horizontal metallic rods installed at Studnice, Czech Republic. Popolansky et al. (1998) reported about a long-period ice measurement experiment, now known worldwide, carried out continuously since the 1940–1941 winter season. Ice unit weight was measured on 30-mm diameter horizontal rods installed 5 m above the ground. Results for the annual maxima in kg/m are shown graphically in Fig. 2.5.

Example 1 Numerical values given in Table 2.1 are for annual maxima of ice unit weight (kg/m) measured graphically from Fig. 2.5.

The data in Table 2.1 is presently the longest homogenous and uninterrupted time series of icing annual maxima known in the world. These data will be used in Section 2.2.2 to illustrate extreme value analysis.

2.2.8 Icing Event Database Notations

The notations used in this chapter for designating the climatic loadings and meteorological variables are as follows:

A = Icing accumulation in kg/m
G = Hourly number of IRM signals
H_r = Relative humidity of air in %
I_R = Hourly icing rate [+ accumulation, − shedding (disappearance), in kg/m/h]:

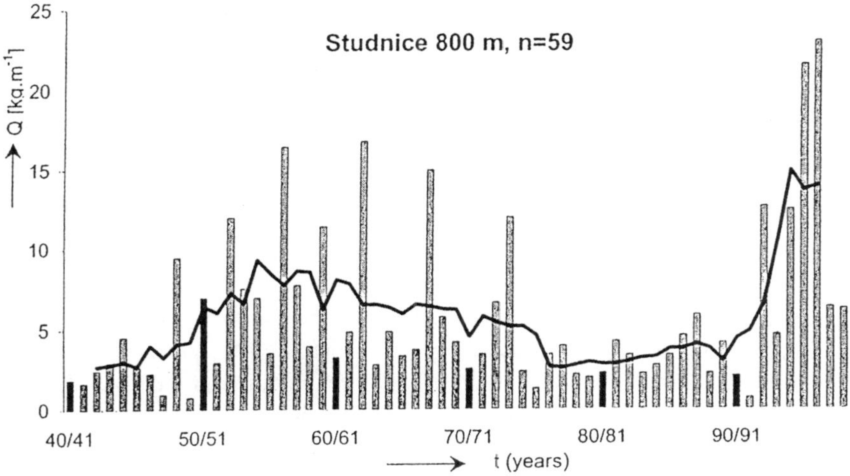

Fig. 2.5 Annual maxima of ice unit weight measured at Studnice, Czech Republic, 1940–1999 (Reproduced by permission of CIGRÉ)

k_{ie} = Type of icing:
k_{ie} = 1 – in-cloud icing
k_{ie} = 2 – precipitation icing
T = Ambient temperature in °C
V_{av} = Mean wind speed in km/h
V_{max} = Maximum wind speed in km/h
Z = Wind direction in degrees
P = Precipitation rate in mm/h

Table 2.1 Annual maxima of ice unit weight from Studnice, Czech Republic, 1940–1999

Winter season	Ice unit weight (kg/m)	Winter season	Ice unit weight (kg/m)	Winter season	Ice unit weight (kg/m)	Winter season	Ice unit weight (kg/m)
40–41	1.6	55–56	3.4	70–71	2.3	85–86	3.3
41.42	1.5	56–57	16.4	71–72	3.2	86–87	4.4
42–43	2.2	57–5/	7.6	72–73	6.6	87–88	5.7
43–44	2.8	58–59	3.8	73–74	12.0	88–89	2.1
44–45	4.4	59–60	11.3	74–75	2.2	89–90	4.0
45–46	2.6	60–61	3.1	75–76	1.1	90–91	2.0
46–47	2.1	61–62	4.7	76–77	3.3	91–92	0.5
47–48	0.8	62–63	16.7	77–78	3.8	92–93	12.5
48–49	9.3	63–64	2.6	78–79	2.0	93–94	4.5
49–50	0.7	64–65	4.6	79–80	1.8	94–95	12.4
50–51	6.9	65–66	3.3	80–81	2.1	95–96	21.3
51–52	2.8	66–67	3.5	81–82	4.0	96–97	23.1
52–53	12.0	67–68	14.9	82–83	3.2	97–98	6.3
53–54	7.5	68–69	5.5	83–84	2.1	98–99	6.2
54–55	6.8	69–70	4.0	84–85	2.5	–	–

Table 2.2 Classification of 1,739 hours of icing data recorded at the Mont Bélair test site

			Precipitation icing	In-cloud icing
Accumulation	Normal	$(G > 0)$	$(P > 0)$	$(P = 0)$
	Abnormal	$(G = 0)$	$(P > 0)$	$(P = 0)$
Persistence	Normal	$(G = 0)$	$(P > 0)$	$(P = 0)$
	Abnormal	$(G > 0)$	$(P > 0)$	$(P = 0)$
Shedding	Normal	$(G = 0)$	$(P > 0)$	$(P = 0)$
	Abnormal	$(G > 0)$	$(P > 0)$	$(P = 0)$

Following the classification proposed by Savadjiev and Farzaneh (2005), the information contained in the databases for icing events can be classified as per the type, precipitation icing or in-cloud icing, as a function of the presence or absence of precipitation ($P > 0$ or $P = 0$) during the accumulation phase. The phases of an event can be considered as "normal" or "abnormal", depending on the hourly number of IRM signals, G, as illustrated in Table 2.2.

2.3 Statistical Analysis and Modelling Ice Loads on Overhead Transmission Lines

Presently, a large quantity of models for icing on transmission line conductors and structures is available in the specialized literature. However, in this section, only some statistics-based models and techniques for data analysis are considered, due to their importance for overhead line design and reliability evaluation.

2.3.1 Modelling Freezing-Rain Data from PIM Measurements

2.3.1.1 Freezing Rain Maps

As mentioned previously, the wide spread of PIMs throughout Quebec, and the extreme value analysis of annual maxima for ice thicknesses with a 100-year return period allow for plotting freezing rain maps for the St. Lawrence Valley (Laflamme and Périard 1996). Maps of this kind could be very useful for transmission line design purposes. However, two essential remarks concerning the construction method for these maps can be made.

First, forecasting climatic loads for a 100-year return period is not reliable if obtained from the analysis of time series shorter than $N = 30$–50 years of observation over the entire territory subject to mapping. Unfortunately, such long period series are not available in the majority of countries concerned by atmospheric icing and its impact on the power networks. To be truly useful, such maps should be constantly corrected and updated using continuous input of data from consecutive observations and extreme value analyses.

Second, all models for freezing-rain accretion on transmission line conductors underline the large influence of the wind direction relatively to the line axis (see for example, Chouinard et al. 2005; Jones 2003). This is because two mutually perpendicular transmission lines constructed at the same point of the map will be exposed to different ice accretion during the same icing event. For a correct forecasting of icing, *wind rose* information for the entire territory covered by the map is necessary (Section 2.2.3).

2.3.1.2 Annual Number of Icing Event Recurrence (ANIER)

The empirical histogram for ANIER for the territory of Quebec is shown in Fig. 2.6. Over 25% of ANIER data are between 0 and 2 occurrences per year, per site. The great variability of observed ANIER can be explained by the major influence of local topography. For instance, on the Mont Bélair icing site, the average ANIER observed is more than 25 events per year, which is four times greater than the average for the St. Lawrence Valley.

Statistical analysis shows that the empirical distribution of ANIER can be modelled as a discrete variable using the *negative binomial distribution* (Farzaneh et al. 2001)

$$\text{Cumulative distribution function, } F(N_E) = \sum_{i=1}^{N_E} \binom{m+i-1}{i} p^m (1-p)^i \quad (2.1)$$

$$\text{Probability density function, } f(N_E) = \binom{m+N_E-1}{N_E} p^m (1-p)^{N_E} \quad (2.2)$$

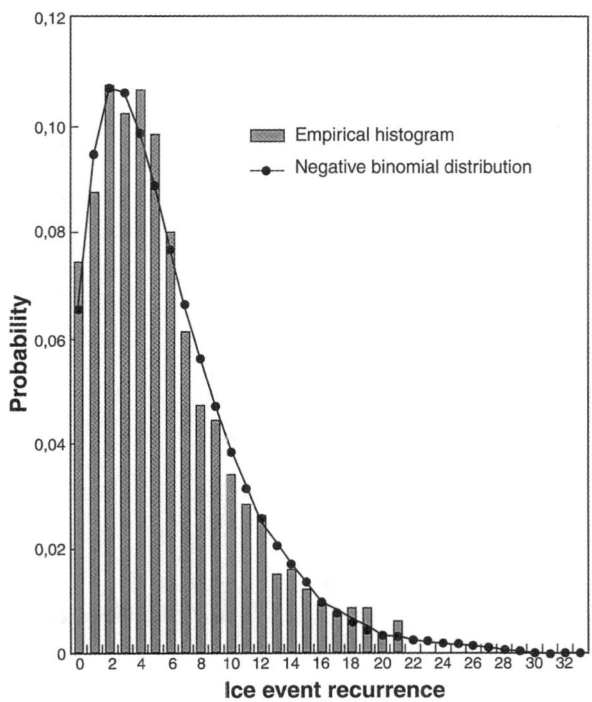

Fig. 2.6 Annual number of occurrences of precipitation-ice events in Quebec (Reproduced by permission of IJOPE)

where N_E is the annual number of icing events taken as an integer variable; $m = 2$ (a positive integer) and $p = 0.25$ are parameters determined numerically by the matching moments method. The theoretical density function (2.2) is shown in Fig. 2.6, together with the empirical histogram.

2.3.1.3 Icing Event Residency (Duration) Period (IERP)

The distribution of the residency period, t_{rp}, of icing events in Quebec ranges between a few hours and 19 days. It is largely asymmetrical towards the right and about 95% of t_{rp} are shorter than 4 days. The persistence of icing events in the northern regions of Quebec is much longer compared to that found in the southern regions. However, this effect cannot be observed from the empirical distribution in Fig. 2.7 because the majority of PIM stations are situated between 45° and 50° latitude, which corresponds to the southern part of Quebec below Lake Mistassini. It can be shown that, in the range of 2 to 17 days, including more than 92% of all data, the Weibull distribution law best fits the empirical histogram

$$\text{Cumulative distribution function, } F\left(t_{rp}\right) = 1 - \exp\left[-\left(\frac{t_{rp}}{c}\right)^{k}\right] \quad (2.3)$$

$$\text{Probability density function, } \quad f\left(t_{rp}\right) = \frac{k}{c}\left(\frac{t_{rp}}{c}\right)^{k-1} \exp\left[-\left(\frac{t_{rp}}{c}\right)^{k}\right] \quad (2.4)$$

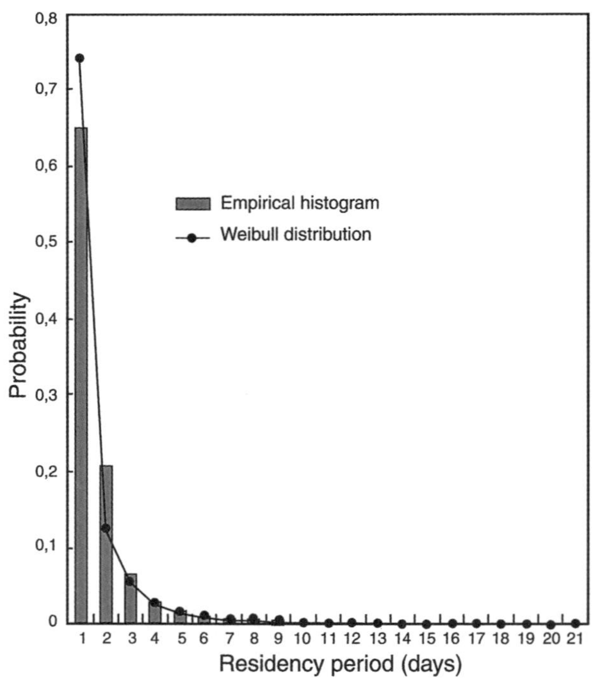

Fig. 2.7 Residency period (duration) of precipitation-ice events in Quebec (Reproduced by permission of IJOPE)

where $c = 0.85$ is the scale parameter, and $k = 0.74$ is the shape parameter of the best-fitting function, determined numerically by the *least squares* method (Johnson 1978).

2.3.1.4 Total Annual Residency Period of Icing Events in Quebec

The empirical distribution of total annual residency period (Fig. 2.8) is the combined distribution of ANIER and IERP using the following algorithm

$$t_{\mathrm{a.rp},i} = \sum_{j=1}^{N_{\mathrm{E},i}} t_{\mathrm{rp},j} \tag{2.5}$$

where $N_{\mathrm{E},i}$ is the annual number of icing events in the ith year, and $t_{\mathrm{rp},j}$ is the residency period of the jth icing event during the ith year.

The equality of the mean, $\bar{t}_{\mathrm{a.rp}}$, and the standard deviation, $\sigma_{\mathrm{a.rp}} = 150$ h per year (Table 2.2), are typical for an exponential distribution

$$F\left(t_{\mathrm{a.rp}}\right) = 1 - \exp\left(-\frac{t_{\mathrm{a.rp}}}{b}\right) \tag{2.6}$$

where $b = \bar{t}_{\mathrm{a.rp}} = \sigma_{\mathrm{a.rp}} = 150$ hours per year ($\approx$ 6 days per year). Savadjiev and Farzaneh (2003) proposed to use this parameter as an estimation of the return period

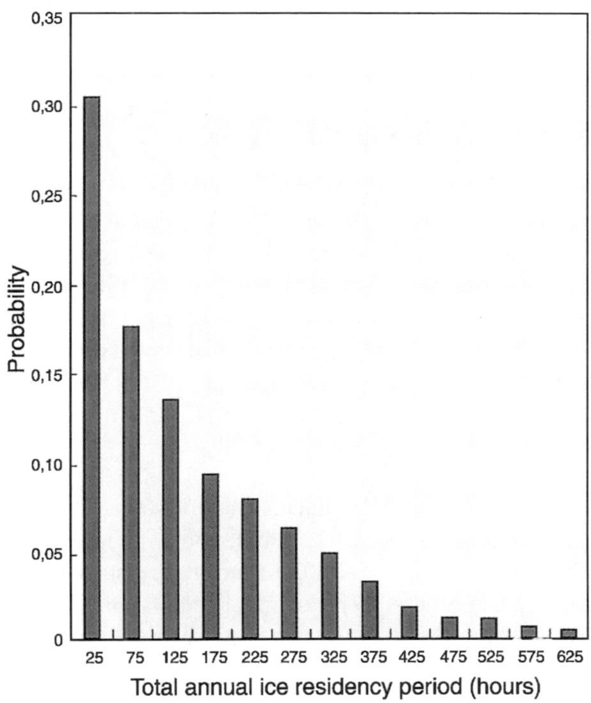

Fig. 2.8 Total annual duration of precipitation-ice events in Quebec (Reproduced by permission of IJOPE)

Table 2.3 Statistical parameters of freezing-rain icing events in Quebec

Statistical parameter	Annual number of icing event recurrence	Ice residency period	Total annual ice residency period
Mean	6	1 day	150 h ($\approx$ 6 days)
Standard deviation	5.1	1.5 days	150 h ($\approx$ 6 days)
Coefficient of variation	0.85	1.5	1
Coefficient of skewness	1.48	4.75	–

of the wind speed associated with icing events for the climatic conditions in Quebec (see Section 2.2.3).

A summary of the principal statistical parameters of freezing-rain icing events in Quebec is given in Table 2.3.

2.3.2 Extreme Value Analysis

Statistical analysis of data for extremes observed in the past allows for assessing and forecasting potential future extremes. Indeed, overhead transmission lines are exposed to climatic loadings, such as wind, icing, and combined loads of wind-on-ice, which are random values. Matching largest extreme of loading with smallest extreme line component resistance is the basis for reliability-based design in modern design methods (IEC/CEI-60826 2003). Extreme values, taken from random populations, are also random values with their own distribution. Establishing these distributions and assessing their parameters is an important part of transmission line engineering for the determination of critical weather loads.

2.3.2.1 Exact Distribution of Extremes

Statistical analyses of extremes and underlying probabilistic theories have been developed since the beginning of the last century. However, their introduction, popularization, and wide application in engineering are mainly due to Gumbel (1958).

If a series of N annual samples, each of size n, are taken from the same population - *initial* or *parent* distribution, $F(x)$, the probability $\Phi_n(x)$ that the largest among n independent and equally distributed observations is less than or equal to x is given by

$$\Phi_n(x) = F^n(x) \tag{2.7}$$

For instance, if the initial distribution studied represents the distribution of hourly maxima of wind speed, V_h, samples determining the distribution of annual maxima, V_{max}, are of size $n = 8,760$ h. If the initial distribution and its parameters are known, Eq. 2.7 can be used for numerical calculation of $\Phi_n(x)$. While calculations of extremes from the initial exponential distribution are easy to perform, some initial distributions, such as the normal distribution, require complicated formulae solved by powerful computers.

2.3.2.2 Return Period of Repeated Occurrences

If an event A occurs repeatedly, and the random variable t is the time between two consecutive occurrences of A with a prescribed intensity X, the mean value T is the return period of the event A

$$\bar{t}\,(X) \equiv T\,(X) = \frac{1}{1 - F\,(X)} \tag{2.8}$$

where $F(X)$ is the cumulative distribution function (CDF) of the random variable X. When the extremes are observed annually, the return period is time expressed in years.

2.3.2.3 Asymptotic Distribution of Extremes

If the initial distribution $F(X)$, or the sample size n are unknown, or n goes to infinity, the exact distribution of largest extremes, $F(X_{\max})$, cannot be determined directly by Eq. 2.7. However, it is known from the statistical theory of extremes that, as $n \to \infty$, there exist asymptotic forms of extreme distributions known as *asymptotic distributions of extremes*. The form of these is independent from the exact form of the initial distribution, but depends only on the behaviour of their tail toward the largest (or smallest) values. There exist three asymptotic-distribution forms toward which converge large classes of initial distributions. Precisely for this reason, asymptotical theory of extremes is very useful in engineering and greatly facilitates the analysis. Gumbel (1958) had classified the asymptotic distributions as the following three distinct types:

Type I: Asymptotic distribution for initial distributions with unlimited and exponentially decaying tail in the direction of extremes. Type I asymptotic distribution holds for largest extremes coming from a wide class of initial distributions of exponential type – normal, lognormal, exponential, largest extremes of type I, Weibull, gamma, etc. This is the major asymptotic distribution of extreme values used largely in engineering for studying extreme wind and icing loads (IEC/CEI-60826 2003).

Type II: From initial distributions of Cauchy/Pareto type with unlimited, polynomially decaying tail in the direction of extremes. Type II asymptotic distribution is known also as Frechet distribution. In general, long-tailed initial distributions do not possess moments, and lead to much larger extremes than those of type I.

Type III: From bounded initial distributions in the direction of extremes. Type III asymptotic distribution for smallest extremes is also known as Weibull distribution. Some initial distributions, such as the exponential, Weibull, gamma, and lognormal, which belong to type I for largest extremes (wind, icing, and combined loads), belong to type III for smallest extremes (minimal ambient temperatures, breaking strength of materials) because these distributions are limited toward the left.

2.3.2.4 Type I Asymptotic Distribution

Type I asymptotic distribution, also called Gumbel distribution of largest extremes, has a cumulative distribution function

$$\Phi(X) = \exp\{-\exp[-\alpha(X-u)]\} = \exp\left\{-\exp\left[-\frac{c_1}{\sigma_X}\left(X-\bar{X}+\frac{c_2}{c_1}\sigma_X\right)\right]\right\} \tag{2.9}$$

where $\bar{X}$ and $\sigma_X$are the mean and the standard deviations of the annual maxima, respectively; u is the location parameter and α is the inverse of the scale parameter; c_1 and c_2 are constants, depending on the number of years of observation N. For $N \to \infty$, $c_1 \to 1.28255\ldots$, $c_2 \to 0.5772\ldots$.

Design, reference values for largest extremes with a prescribed return period, T, can be calculated by inversing Eq. 2.9

$$X_{\max}(T) = \bar{X} - \frac{c_2}{c_1}\sigma_X + \frac{\sigma_X}{c_1}\left\{-\mathrm{Ln}\left[-\mathrm{Ln}\left(1-\frac{1}{T}\right)\right]\right\} \tag{2.10}$$

Mathematical expressions for types II (Frechet distribution) and III asymptotic distributions are omitted here because they are less important for the extreme value analysis than those of type I (Gumbel). For more details on them one may consult IEC/CEI-60826 (2003), Gumbel (1958), and Castillo (1988). On the other hand, type III asymptotic distribution for smaller extreme values is the well-known Weibull distribution (2.3).

2.3.2.5 Initial (Parent) Distributions

Johnson (1978) showed that the Weibull distribution (2.3) should be preferably used as initial distribution for wind speed. He used the least-squares method to estimate the scale and shape parameters from samples of field data. The shape of the initial wind-speed distribution can be very different as a function of the shape parameter k (Fig. 2.9).

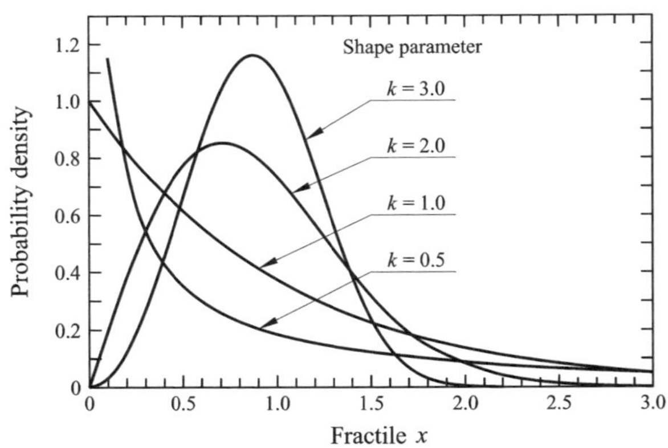

Fig. 2.9 Weibull distribution with scale parameter c = 1, and different shape parameters, k (Reproduced by permission of NRC)

In general, statistical analysis of extreme wind speed is easier to perform, compared to icing events. This is mainly due to annual samples for hourly wind speed, from which the annual maximum is selected, contain $n = 8,760$ measurements during long periods of years of observation, N. Icing events being rare phenomena, even in countries with severe icing conditions (Canada), the annual number of icing events is typically 1–5 per year (see the distribution of ANIER in Quebec, Fig. 2.6). What is more, the available time series for icing records are typically in the range of only 10–25 years.

Weibull distribution is largely used as initial (parent) distribution, not only for wind, but also for icing and combined loadings. This is because the Weibull distribution may have different shapes, its right tail is of exponential type with type I asymptotic behaviour, while its left tail is bounded by zero, exactly as the meteorological variables are (IEC/CEI-60826). For illustration of this assertion, the initial distribution of maximum ice unit weight per icing event is analyzed in Example 2, using data records from the Mont Bélair icing test site in Quebec.

Example 2 The histogram of maximum ice accumulation per icing event (in kg/m), during 378 icing events observed between 1992 and 2006 at the Mont Bélair icing test site in Quebec, is shown in Fig. 2.10. Statistical parameters of this distribution are as follows:

Mean = 0.492 kg/m
Standard deviation = 0.734 kg/m
Coefficient of variation = 1.492
Coefficient of skewness = 4.151

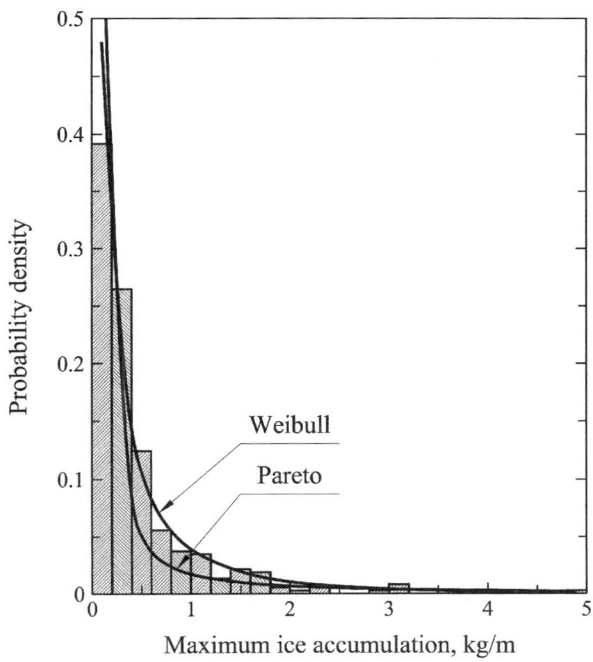

Fig. 2.10 Maximum ice accumulation during 378 icing events at Mont Bélair, Quebec, Canada, between 1992 and 2006. Theoretical fitting functions: Weibull and Pareto

On the same figure are plotted two theoretical fitting curves representing the following probability functions

$$\text{Weibull: } F(x) = 1 - \exp\left(-\left(\frac{x}{c}\right)^k\right) = 1 - \exp\left(-\left(\frac{x}{0.3749}\right)^{0.6757}\right) \quad (2.11)$$

where the numerical values of the shape coefficient $k = 0.6757$, and the scale coefficient $c = 0.3749$ in (2.11) are estimated using the least-squares method, and

$$\text{Pareto: } F(x) = 1 - \left(\frac{a}{x}\right)^b = 1 - \left(\frac{0.018}{x}\right)^{0.3845} \quad (2.12)$$

where $a = 0.018\,\text{kg/m}$ is the location parameter (minimum value of the variate), and the numerical value of the shape parameter $k = 0.3845$ are determined using the method of maximum likelihood.

Visual inspection of the fitting in Fig. 2.10 seems to show that the Weibull law fits better the experimental distribution, though the Chi-Square Goodness-of-Fit test rejects both Weibul and Pareto hypotheses. In spite of the exceptionally large annual number of icing events observed at Mont Bélair (on average 27 icing events per year), due to the relatively small number of years of observation (14 years), it is impossible to determine if the right tail of the experimental distribution is of exponential or Pareto type. Establishing the type of the initial icing distribution is still more difficult for other sites with much lower mean annual number of icing event occurrences. However, large-scale statistical studies carried out in several countries (and the province of Québec) show that in the majority of cases largest extremes of icing-amount follow type I asymptotical distribution, Gumbel distribution (IEC/CEI-60826 2003).

In some rare cases of wet snow, type I asymptotic distribution is not applicable for analysis of extremes. Some recent studies (Le Du and Laurent 2005; Jones 2003) recommend for this special case the application of a threshold method with fitting of icing extremes by a generalized Pareto distribution. The Peak-Over-Threshold (POT) method will be analyzed in a later section.

2.3.2.6 Probability Papers

Graphical methods, such as the probability-paper techniques, for analysis, estimation, and extrapolation of extremes in engineering are a useful tool and are used extensively by engineers. The principle of the probability paper, constructed for a given family of distributions, is transforming the scales in such a manner that the plotted curve $F(X)$ against X appears as a straight line. This simple method makes it possible to draw the fitting line manually and to check the goodness-of-fit by simple inspection. There exist probability papers constructed for different distributions: normal, lognormal, Weibul, Gumbel, Frechet, etc. The ordinary, semi-logarithmic paper may also be used for the exponential distribution. In the extreme value analysis of climatic loadings, the Gumbel probability paper is widely used. A specimen of Gumbel probability paper is shown in Fig. 2.11. Probability paper for extremes of

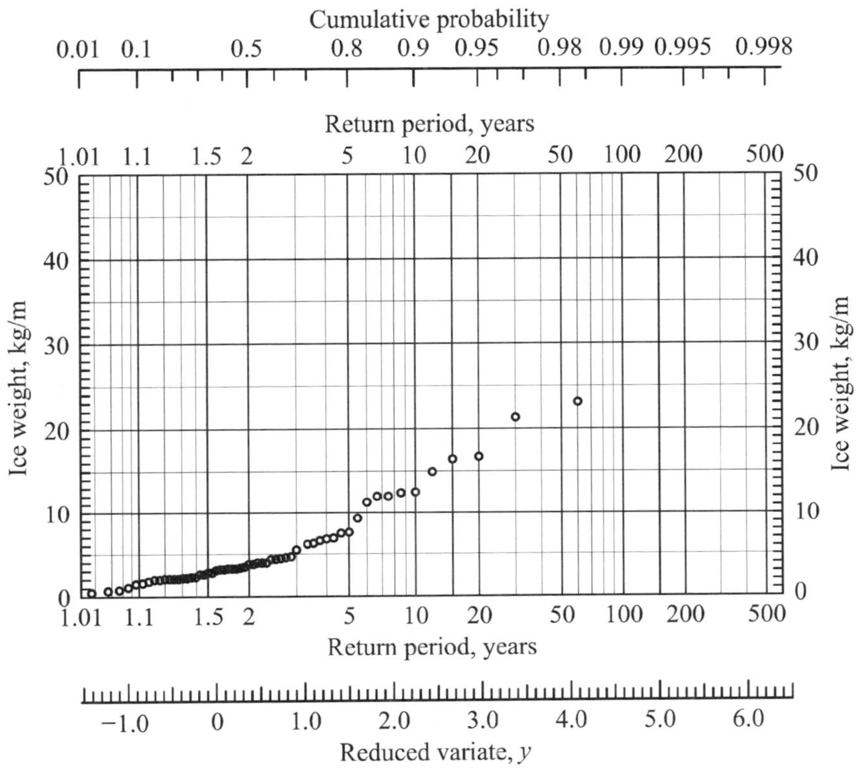

Fig. 2.11 Data for annual maxima of unit ice weight from Studnice, Czech Republic, on Gumbel probability paper for largest extremes

asymptotical type II, Frechet distribution (see Castillo 1988) is shown in Fig. 2.12. Figure 2.13 shows another specimen of Gumbel probability paper where the ordinate represents radial ice thickness.

In established practice, the return period T is plotted as abscissa in transformed, double logarithmic scale (the same for Gumbel and Frechet probability paper)

$$y = -\mathrm{Ln}\,[-\mathrm{Ln}\,(F\,(X))] = -\mathrm{Ln}\left[-\mathrm{Ln}\left(1 - \frac{1}{T}\right)\right] \tag{2.13}$$

The random variate, ice unit weight or radial ice thickness, is plotted as ordinate in linear scale for Gumbel probability paper, and in logarithmic scale for Frechet probability paper.

Transformation (2.13) changes the Gumbel distribution (2.9) into a straight line

$$y = \alpha\,(X - u) \tag{2.14}$$

or

$$x = u + \frac{y}{\alpha} = a + by \tag{2.15}$$

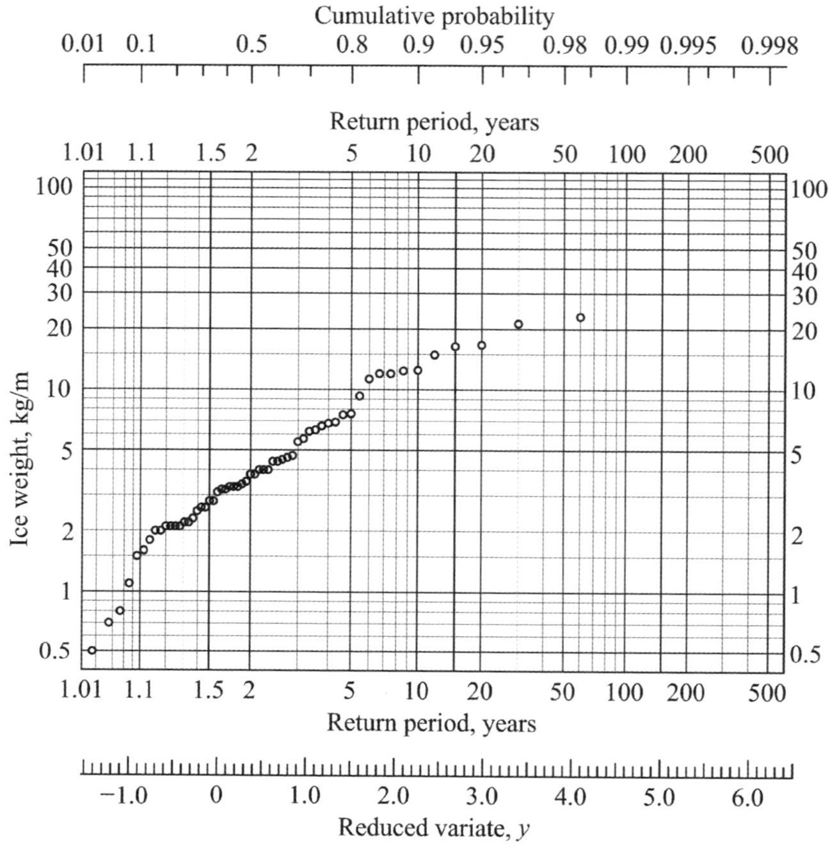

Fig. 2.12 Data for annual maxima of unit ice weight from Studnice, Czech Republic, on Frechet probability paper for largest extremes

Similarly, the Frechet distribution is transformed in a straight line by

$$y = \alpha\,(\mathrm{Ln}X - u) \tag{2.16}$$

or

$$x = \exp\,(a + by) \tag{2.17}$$

2.3.2.7 The Plotting-Position Problem

If a sample of annual maxima from N years of observations are arranged in increasing order

$$X_1 \leq X_2 \leq X_3 \leq \ldots \leq X_m \leq \ldots \leq X_{N-1} \leq X_N \tag{2.18}$$

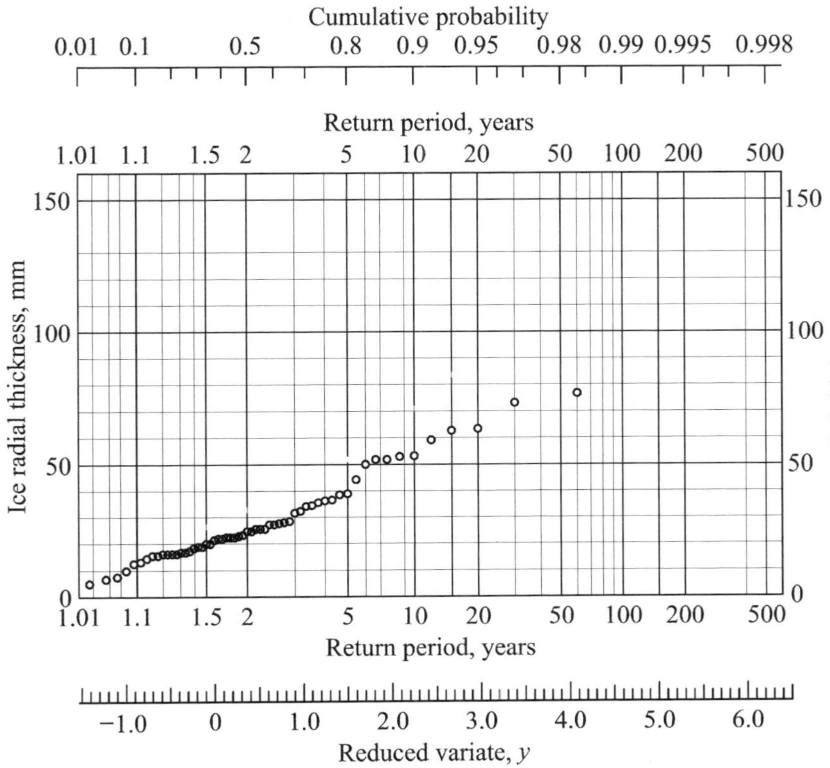

Fig. 2.13 Data for annual maxima of the equivalent radial ice thickness, from Studnice, Czech Republic, on Gumbel probability paper for largest extremes

they can be plotted on probability paper using the well known formula, first proposed by Weibul, and strongly recommended by Gumbel (1958)

$$\bar{F}\,(X_m) = \frac{m}{1+N} \tag{2.19}$$

representing the mean frequency of the mth value, that is of rank m. Several other formulae are proposed in the specialized statistical literature, such as

$$\bar{F}\,(X_m) = \frac{m-0.5}{N} \tag{2.20}$$

or

$$\bar{F}\,(X_m) = \frac{(m-0.3)}{(N+0.4)} \tag{2.21}$$

based on the mth value median.

Recently, the problem concerning the *best* plotting position on probability paper was discussed by Makkonen (2006). One of the principal arguments against formulae for plotting positions, as in Eqs. 2.20 or 2.21, is the return period of the largest extreme ($m = N$) which, from Eq. 2.8, becomes $T = 2N$ for Eq. 2.20, and $T \approx 1.43N$ for Eq. 2.21, both much larger than the sample size N. For comparison, the classical Weibul formula, Eq. 2.19, gives $T = N + 1$.

Example 3 Plot the data from Table 2.1 on probability paper for largest extremes of type I, Gumbel.

Solution: The members of the time series in Table 2.7 are arranged into increasing order and are given rank m from 1 (smallest) to $N = 59$ (largest). Mean CDF corresponding to each member of the series is calculated from Eq. 2.19

$$\bar{F}(w_m) = \frac{m}{60}$$

The return period of the member of rank m is calculated from Eq. 2.8

$$T(w_m) = \frac{1}{1 - \bar{F}(w_m)}$$

The reduced variable y_m, corresponding to Gumbel's distribution of type I, is calculated from Eq. 2.13

$$y_m = -\mathrm{Ln}\left[-\mathrm{Ln}\left(\bar{F}(w_m)\right)\right] = -\mathrm{Ln}\left[-\mathrm{Ln}\left(1 - \frac{1}{T(w_m)}\right)\right]$$

The points with coordinates $[w_m, T(w_m)]$ or $[w_m, y_m]$ are plotted in Fig. 2.14 on Gumbel (type I) probability paper.

Example 4 Plot the same data from Table 2.1 on Frechet probability paper for largest extremes of type II.

Solution: Using the procedure in Example 3, the points with coordinates [Ln(w_m), $T(w_m)$] or [Ln(w_m), y_m] are plotted in Fig. 2.12 on Frechet (type II) probability paper for largest extremes.

Example 5 Transform the data in Table 2.1 from annual maxima of ice unit weight w in kg/m, into equivalent radial ice thickness t (mm) on the 30-mm measurement rod, with the standard density of 900 kg/m^3.

Solution: Unit ice weight w can be calculated as a function of the diameter of a cylindrical rod D (mm) and the radial ice thickness t (mm) from the following formula

$$w_m = \frac{0.02773 t_m (t_m + D)}{9.81}, \text{ kg/m} \tag{2.22}$$

Solving (2.22) for the ice thickness t yields

$$t_m = \frac{-D + \sqrt{D^2 + 1414.6 w_m}}{2}, \text{ mm} \tag{2.23}$$

Using the procedure in Example 3, the points with coordinates $[t_m, T(t_m)]$ or $[t_m, y_m]$ are plotted in Fig. 2.13 on Gumbel (type I) probability paper for largest extremes.

Example 3 shows that, in this specific case, the unit-weight data plotted on Gumbel probability paper in Fig. 2.11 present a non-linear trend (concavity toward the top). The same data, when plotted on Frechet probability paper (Example 4, Fig. 2.12) show a still more pronounced non-linear trend (convexity toward the top). Only the data transformed into radial ice thickness in Example 5, Fig. 2.13, show a quasi-perfect overall linear trend.

2.3.2.8 Fitting Straight Lines on Probability Paper

If the scatter of the plotted observations is small, the fitting straight line (2.15) or (2.17) can be drawn manually, by visual inspection and using a ruler. However, in some cases, the manual method is not precise enough and the estimation of the two parameters of the fitting straight line, a and b, has to be made using the least-squares method (Gumbel 1958). It consists in calculating successively the sample mean and standard deviations for the random variate

$$\bar{X} = \frac{1}{N} \sum_{m=1}^{N} X_m \tag{2.24}$$

$$s_X = \sqrt{\frac{\sum_{m=1}^{N} X_m^2 - \frac{\left(\sum_{m=1}^{N} X_m\right)^2}{N}}{N-1}} \tag{2.25}$$

the reduced variable for each observation ($1 \leq m \leq N$)

$$y_m = -\mathrm{Ln}\left[-\mathrm{Ln}\left(\frac{m}{N+1}\right)\right] \tag{2.26}$$

and the fitting straight line $x = a + by$, where

$$a = \bar{X}_m - \frac{s_X \bar{y}_m(N)}{\sigma_y(N)} \tag{2.27}$$

$$b = \frac{s_X}{\sigma_y(N)} \tag{2.28}$$

$$\bar{y}_m (N) \equiv c_2 = \frac{1}{N} \sum_{m=1}^{N} y_m = \frac{1}{N} \sum_{m=1}^{N} \left\{ -\mathrm{Ln} \left[-\mathrm{Ln} \left(\frac{m}{N+1} \right) \right] \right\} \tag{2.29}$$

$$\sigma_y (N) \equiv c_1 = \sqrt{\frac{\sum_{m=1}^{N} y_m^2}{N} - \left(\frac{\sum_{m=1}^{N} y_m}{N} \right)^2} \tag{2.30}$$

2.3.2.9 Construction of Confidence (Control) Interval Band

The construction of confidence-interval band (or control band) covering the entire range can be made using the methodology proposed by Gumbel (1958). Usually, these control bands are constructed for $p = 0.68269$, corresponding to $\pm\sigma$ (one standard deviation) of the normal distribution, and $p = 0.95450$, corresponding to $\pm 2\sigma$. The entire range of the observed annual maxima is subdivided into three zones:

$0.15 \leq F(X) \leq 0.85$, where the mth central values are normally distributed. In this central zone, the values $\pm\Delta_m$ for $p = 0.68$ (or $\pm 2 \times \Delta_m$ for $p = 0.95$) are added to or subtracted from the values X_m situated on the fitting straight line at abscises $m/(N+1)$, with

$$\Delta_m = \sqrt{\left\{ \frac{\pi^2}{6} + 1.1396 \, [y\,(T) - 0.5772] \, \frac{\pi}{\sqrt{6}} + +1.1 \, [y\,(T) - 0.5772]^2 \right\}} \times \frac{\sqrt{6}}{\pi} \frac{s_X}{\sqrt{N}} \tag{2.31}$$

where $y(T)$ is the reduced variable.

The zone including the largest observed extremes with ranks:

For $N \geq 100$: $m = N, N-1, N-2$
For $N \geq 50$: $m = N, N-1$
For $N \geq 10$: $m = N$

The lengths Δ_m to be added (subtracted) from the theoretical straight line, calculated from the exact distribution of the largest order statistics, N, $(N-1)$, $(N-2)\ldots$, are given in Table 2.4 (source: Gumbel 1958, p. 218).

Table 2.4 Control band width for largest extremes

Interval	Probability	
	$p = 0.68269$	$p = 0.95450$
Δ_N	$1.14078/\alpha$	$3.0669/\alpha$
Δ_{N-1}	$0.75409/\alpha$	$1.7820/\alpha$
Δ_{N-2}	$0.589/\alpha$	$1.35/\alpha$

where α is the inverse of the scale parameter b of the fitting straight line (2.15)

- Zone of extrapolation. In this zone, the control band limits represent straight lines parallel to the theoretical straight line drawn at distance $\pm\Delta_N$ or $\pm 2\Delta_N$.
- If the ends of these intervals are joined by straight lines they form a continuous band covering the entire range of observations.

Example 6 Estimate the parameters of the fitting straight line and construct 68 and 95% control bands for the data from Table 2.1 when plotted on Gumbel, type I, probability paper (Fig. 2.11).

Solution: The best fitting straight line is calculated as follows:

$$w = a + by = a + b\left\{-\text{Ln}\left[-\text{Ln}\left(1 - \frac{1}{T}\right)\right]\right\} = 3.225 + 4.289\left(1 - \frac{1}{T}\right)$$

where, from Eq. 2.27, $a = 3.225$, from Eq. 2.28, $b = 4.289$, from Eq. 2.30, for $N = 59$ years $c_1 = 1.1734$, from Eq. 2.29 for $N = 59$ years $c_2 = 0.5518$. The fitting straight line and control band, 68 and 95%, are shown in Fig. 2.14.

Figure 2.14 illustrates some principles concerning fitting straight lines on probability paper:

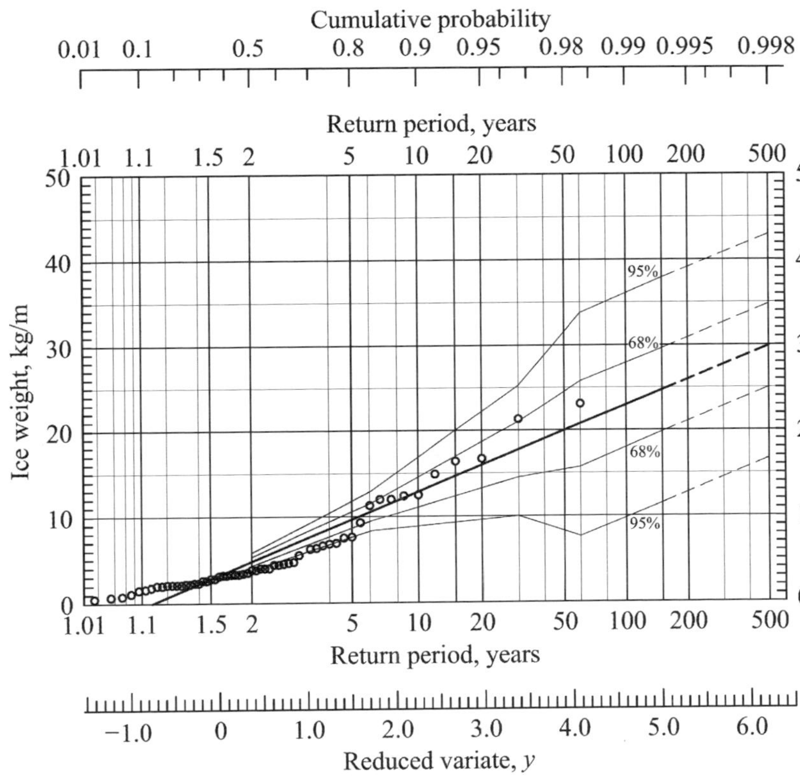

Fig. 2.14 Gumbel probability chart for icing unit-weight data in Example 5, with fitting straight line and confidence-interval limits

- The best fitting straight line and control band limits are extrapolated for illustration to return a period of 500 years without any statistical significance. A basic rule for applying statistics in meteorology states that an extrapolation should not go beyond 2 times of the size N of the studied sample. For this specific example, the extrapolation must not be made beyond 100–150 years. The expected ice loads on 30-mm diameter cylindrical rods installed 5 m above ground can be found in Table 2.5.
- For application to conductors of actual transmission lines, the reference ice-weight data from Table 2.1 should be recalculated for different heights above ground, conductor diameter, or span length in accordance with international standards (IEC/CEI-60826 2003), or national standards, such as SN-40.1 (1993) in force at Hydro Quebec.
- It should be noted that the fit of the observed data in Fig. 2.14 is not good enough – some of the observations are out of the control band limits, and the 95% control band becomes uselessly wide. Gumbel (1958) stated that, in the asymptotical theory, the good fit is based on three assumptions (related here for icing events): (1) the distribution of the icing-event maximum loads is of the exponential type; (2) the annual number of icing events is sufficiently large; (3) maximum ice accumulation observed during icing events occurring within the same winter season are independent. For the data in Example 3, it is impossible to verify the first two assumptions because no data for the initial distribution are published. The third assumption is more or less met if the consecutive icing events are separated by periods without icing (see also Le Du and Laurent 2005; Jones 2003).
- It can be seen in Fig. 2.14 that the observation points with ordinate $w > 5\,\mathrm{kg/m}$ show an approximate linear trend with slope greater than that of the remaining points. This suggests a Gumbel, type I, asymptotic distribution toward the right tail (largest extremes).

Table 2.5 Forecasting ice loads in Example 6

Return period, years	Ice load expected, kg/m	Confidence interval limits, kg/m			
		Upper 95%	Upper 68%	Lower 68%	Lower 95%
50	20	33	25	15	7
100	23	36	28	18	10
150	25	38	30	20	12

2.3.2.10 Graphical Estimation of the Fitting Straight Line Parameters

As mentioned previously, in several practical cases, the observational data when plotted on Gumbel probability paper (Fig. 2.11 or 2.14) show a non-linear trend, mainly due to the median and smallest values. Nevertheless, the largest extremes (of interest in this analysis) show a linear trend and can be fitted by a straight line covering only the right tail of the asymptotic distribution (Castillo 1988). This procedure is illustrated in Example 7, Fig. 2.15.

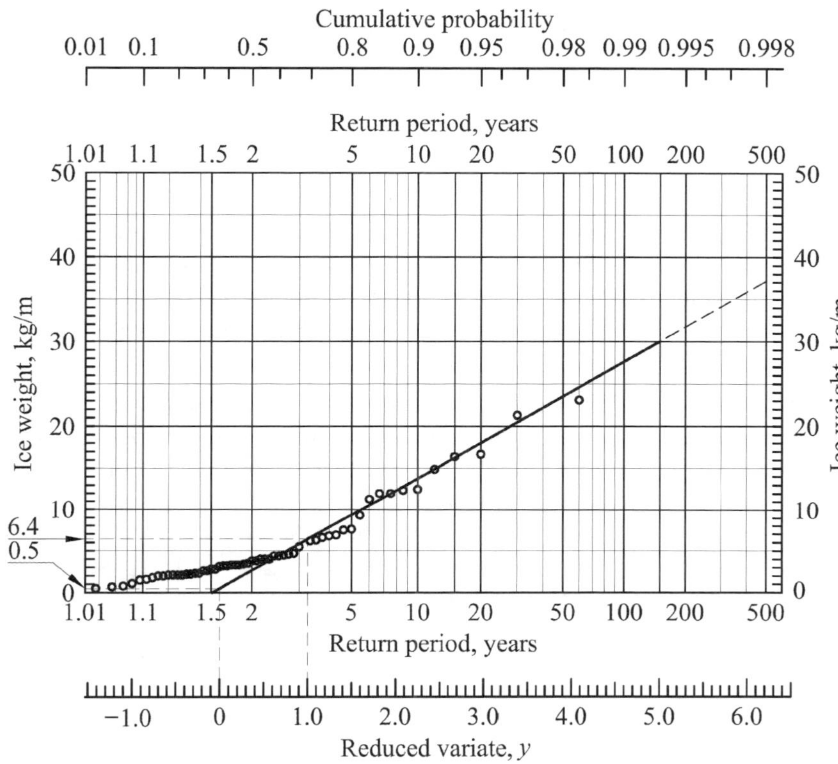

Fig. 2.15 Fitting the right tail of the ice-weight data in Example 6 on Gumbel probability paper

Example 7 Determine the parameters of the straight line fitting the right tail of the observational data from Example 3, Table 2.1, when plotted on Gumbel probability paper (Fig. 2.11).

Solution: The fitting straight line in Fig. 2.15 (drawn manually by ruler) is presented by Eq. 2.15

$$\mathrm{w} = \mathrm{a} + \mathrm{by}$$

The parameters a and b are determined graphically as follows:

For $y = 0$, $w(0) = a = 0.5$; for $y = 1$, $w(1) = 6.4\,\text{kg/m} = 0.5 + b$, and hence $b = 5.9$. Using the fitting straight line

$$w = 0.5 + 5.9y = 0.5 + 5.9\left\{-\mathrm{Ln}\left[-\mathrm{Ln}\left(1 - \frac{1}{T}\right)\right]\right\}$$

The following reference ice loads, on 30-mm diameter cylindrical rods installed 5 m above ground, can be expected, as shown in Table 2.6.

Table 2.6 Forecasting ice loads in Example 7

Return period, years	Ice load expected, kg/m
50	23.5
100	27.6
150	30.0

Example 8 Fit the empirical data from Example 3, Table 2.1, when plotted on Frechet probability paper (Fig. 2.12). Determine the parameters of the straight line fitting the right tail (toward the largest extremes) of the observational data. Forecast the expected values of the reference unit ice weight for return periods of 50, 100, and 150 years.

Solution: Data for unit ice weight from Example 3 when plotted on Frechet probability paper (Fig. 2.12) show a non-linear trend (convexity), which indicates that the asymptotical distribution is not of type II. Nevertheless, the right tail of the experimental data (with ordinate $w > 9\,\mathrm{kg/m}$) shows a linear trend and thus suggests a Frechet, type II, asymptotic distribution in the right tail. The straight line fitting the right tail of the experimental distribution is shown in Fig. 2.16. The parameters of the straight line, $\mathrm{Ln}(w) = a + by$, are determined as follows:

For $y = 0$, $w(0) = 6\,\mathrm{kg/m}$; $a = \mathrm{Ln}[w(0)] = \mathrm{Ln}(6)$, $a = 1.7918$

For $y = 1.5$, $w(1.5) = 10\,\mathrm{kg/m}$; $\mathrm{Ln}10 = 1.7918 + 1.5b$. Solving for b yields $b = 0.3405$

The fitting straight line is presented by equation

$$w = \exp(a + by) = \exp\left(1.7918 + 0.3405\left\{-\mathrm{Ln}\left[-\mathrm{Ln}\left(1 - \frac{1}{T}\right)\right]\right\}\right)$$

Estimated values for reference unit ice load as a function of the return period are given in Table 2.7.

It can be seen from the table that type II asymptotical distribution, Frechet distribution, forecasts somewhat larger extremes than does the Gumbel, type I asymptotic distribution. This is a general rule due to the long right tail of the initial distribution of Pareto (Cauchy) type.

2.3.2.11 Transforming Data for Unit Ice Weight into Equivalent Radial Ice Thickness

Usually, probability distributions of ice unit weight w and equivalent radial ice thickness t are different, because these random variables are linked by Eq. 2.22, a

Table 2.7 Forecasting ice loads in Example 8

Return period, years	Ice load expected, kg/m
50	22.7
100	28.7
150	33.0

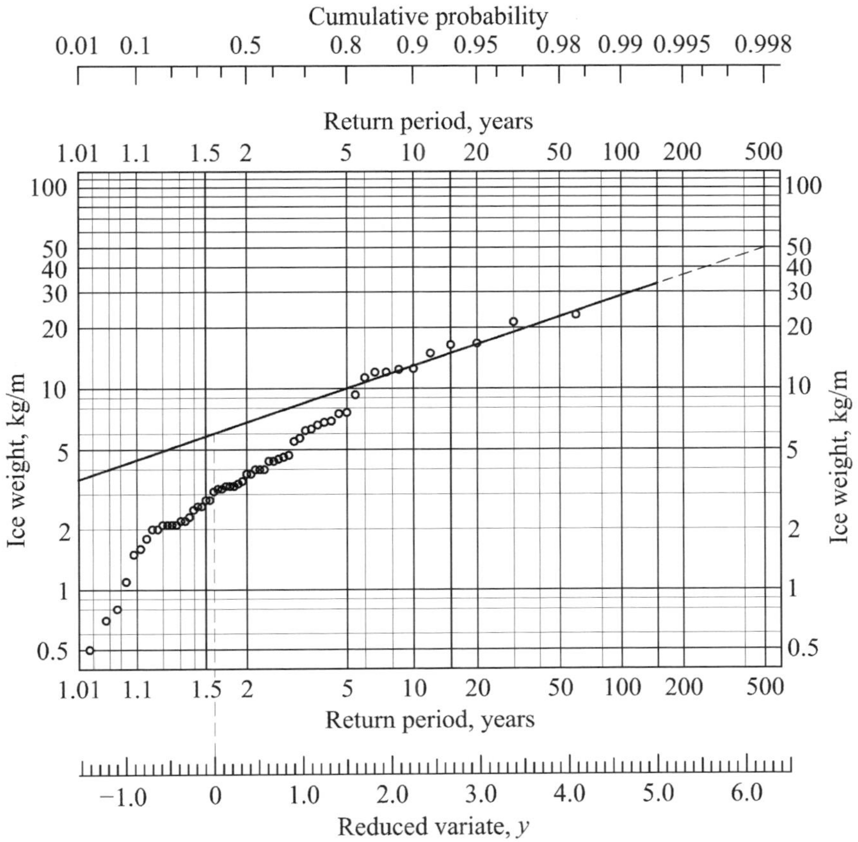

Fig. 2.16 Fitting the right tail of the ice-weight data in Example 7 on Frechet probability paper

quadratic function (Savadjiev et al. 1998). Sometimes, the transformation, Eq. 2.23, of the icing data allows for improving the fitting on probability paper. That is the case for the radial ice thickness data in Fig. 2.13, Example 5, the linear trend of which, over the entire range, suggests clearly a Gumbel, type I, asymptotical distribution.

Example 9 Fit the data for the equivalent radial ice thickness t when plotted on Gumbel probability paper (Fig. 2.13) and construct 68 and 95% control bands. Forecast the expected reference values of t, for density $900\,\text{kg/m}^3$, on 30-mm cylindrical rods installed 5 m above ground, for return periods of 50, 100, and 150 years.

Solution: The parameters of the fitting straight line in Fig. 2.17 are

$$t = a + by = 21.23 + 14.053\left\{-\text{Ln}\left[-\text{Ln}\left(1-\frac{1}{T}\right)\right]\right\}$$

The 68 and 95% control band limits calculated from Table 2.7 and Eq. 2.31 are also plotted in Fig. 2.17.

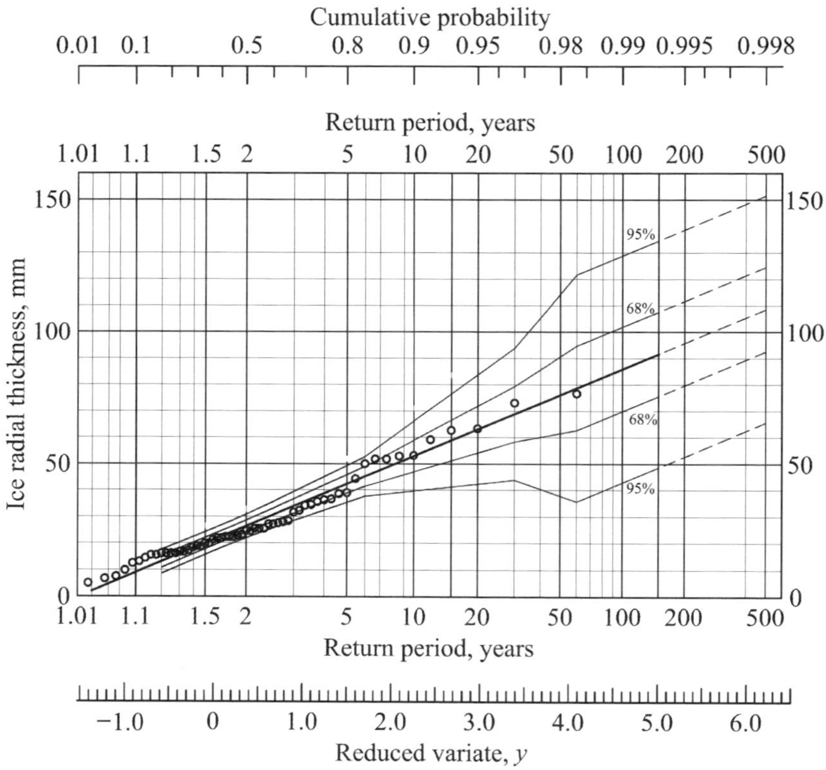

Fig. 2.17 Gumbel probability chart for radial ice thickness data in Example 8, with fitting straight line and confidence-interval limits

The following reference equivalent radial-ice thickness may be expected on 30-mm diameter cylindrical rods installed 5 m above ground, with density 900 kg/m^3, as shown in Table 2.8.

Table 2.8 Forecasting radial-ice thickness in Example 9

Return period, years	Expected ice thickness, mm	Confidence interval limits, mm			
		Upper 95%	Upper 68%	Lower 68%	Lower 95%
50	76	119	92	60	33
100	86	129	102	70	43
150	92	135	108	76	49

2.3.2.12 Generalized Extreme Value Distribution (GEV)

The three asymptotic distributions: type I (Gumbel), type II (Frechet), and type III (Weibull), may be represented by a single analytical expression called *generalized extreme value distribution* (Le Du and Laurent 2005), or *Von-Mises* form (Castillo 1988). The GEV distribution for largest extremes is

$$F(X) = \exp\left\{-\left[1 + k\left(\frac{X-u}{\alpha}\right)\right]^{-\frac{1}{k}}\right\} \tag{2.32}$$

For $k = 0$, at the limit, Eq. 2.32 becomes a type I (Gumbel) asymptotic distribution

$$F(X) = \exp\left\{-\exp\left[-\frac{X-u}{\alpha}\right]\right\}$$

For $k > 0$, Eq. 2.32 becomes a type II asymptotic distribution – Frechet distribution, and for $k < 0$ (2.32) becomes a type III asymptotic distribution – Weibull distribution.

As mentioned above, the most used asymptotic distribution in extreme value analysis is the type I Gumbel distribution. Because the shape parameter k for the Gumbel distribution is taken to be equal to zero, only two parameters, the location parameter u and the scale parameter α have to be estimated using the methods discussed in the previous sections.

In the general case, when the shape parameter k is unknown, three parameters should be estimated, k, u, and α. There exist several statistical methods for their estimation, some of which are discussed in Le Du and Laurent (2005). However, as recommended in Gumbel (1958) and Castillo (1988), the simplest method for choosing the most convenient asymptotical distribution is the use of different types of probability paper; the decision can be made by a simple inspection of the observation data and the goodness of straight line fit.

2.3.2.13 Peak-Over-Threshold Method (POT)

In some regions, wet snow may fall in considerable quantities during a few rare years separated by long periods without any accretion at all. Such snowfall events may severely affect buildings and airports, but usually are not dangerous to transmission line equipment. In these cases, the initial distribution of wet-snow events is not of exponential type, and the asymptotical distribution of largest extremes cannot be described conveniently by type I Gumbel distribution. This is mainly due to the sparse time series containing several zeroes and only a few members of great magnitude. Recently, the Peak-Over-Threshold (POT) method was recommended for statistical analysis of wet snow episodes (Le Du and Laurent 2005), while Jones (2003) proposes using this method in all situations of freezing-rain events.

The POT method consists in selecting from the initial distribution all extremes which exceed a previously fixed threshold. The selected number of extremes, depending on the chosen threshold, may be large in some years with frequent and intensive icing events, but may be zero in many others with insignificant icing occurrence or without any ice observed at all. In other words, the POT method proposes compensating the information for no icing observed in several years, with more abundant information from a few years with intensive and frequent icing, in order to constitute a sufficiently long-term series for the purposes of the subsequent extreme value analysis. The empirical distribution of the selected sample of extremes is then fitted with the CDF of a *generalized Pareto distribution*

$$F(X) = 1 - \left[1 - \frac{k}{\alpha}(X - S)\right], \; k \neq 0 \qquad (2.33)$$

where k is the shape parameter, and S is the amount of the selected threshold.

Caution should be exercised when the POT method is used to forecast icing loads on overhead transmission lines, for the following reasons:

- To ensure statistical independence of the selected icing amount extremes required by the asymptotic theory, only one maximum should be selected per icing event from one winter season. The usually small number of significant icing events per winter season limits considerably the possibility of selecting more than one or two icing extremes per year, and thus limits the length of the needed time series, even in countries with as severe icing conditions as found in Quebec (Jones 2003).
- On the other hand, the information for no-icing in some years of observation is characteristic of icing climate in a given region and should not be neglected for the purposes of the analysis.
- What should be done with some winter season maxima, which are different from zero, but are smaller than the selected threshold? Should they be neglected and thus valuable information lost? Or, in these cases, should the chosen threshold be lowered in order to select all available icing-event maxima (except the zeroes) from the initial distribution?
- What will be the statistical significance of results obtained by extreme value analysis of data from only a few years, no matter how many extremes exceeding the threshold have been selected?
- It is impossible to specify the return period to a forecasted amount or intensity of the studied variable. Using the POT method, selected maxima are not equally spaced in time – some may be only few days apart, others may be separated by several years. The clear specification of return period for extreme ice, wind, and combined wind and ice loadings is very important for the modern, reliability-based design of transmission lines, where many design criteria are specified in terms of return period.
- Wet-snow icing events, for which the POT method seems to be applicable, are rarely critical for transmission lines and are not determinants for their reliability.

The above remarks concern not only the POT method, but are pertinent for all practical statistical methods used in extreme value analysis. Icing being a rare phenomenon, difficult to study, analyze, and forecast, it is obvious that initiating ice measurement programs or further developing existing ones is the best way of improving the statistical significance and reliability of the analysis of climatic loadings for transmission line design purposes.

2.3.3 Modelling Combined Wind-on-Ice Loads

Simultaneous action of wind and icing on overhead transmission lines is often critical for structures, insulators, and even conductors. Usually, they determine the design and operational reliability of the entire line. The probability of simultaneous occurrence of extreme wind speed and extreme icing is negligibly small. The best way for assessing design values for wind speed associated with icing events, V_{ice}, is the statistical analysis of field-data records from observations of actual transmission lines exposed to wind during icing events. Unfortunately, such databases are rarely available and establishing systems or devices for the measurement of combined wind-on-ice loads is, in general, very much limited by the costs and time involved.

Taking into account the very low probability of simultaneous occurrence of extreme wind and extreme ice, several standards and design codes recommend a simplified methodology for determining combined wind-on-ice loads (IEC/CEI-60826 2003). It consists of the combination of low-probability icing (long return period) with high-probability wind (short return period) in such a way as to obtain the same specified reliability level as for extreme icing and wind loads taken separately. In some standards such as SN-40.1 (1993), another combination of involved variables is also allowed for design purposes - reduced wind combined with reduced icing. The last combination, which is usually not critical for the mechanical or electrical behaviour of overhead transmission lines, will not be discussed further in this section.

2.3.3.1 Methods for Specifying Design Values for Wind Speed Associated with Icing Events

Several approaches exist for specifying wind speed associated with icing events, V_{ice}, necessary for the calculation of combined wind-on-ice loads on transmission line conductors.

Statistical Analysis of Field Data: Krishnasamy and Kulendran (1996) reported an interesting application of the statistical approach for assessing combined wind-on-ice loads. Data records for annual extreme wind speed, annual extreme ice thickness, and wind-on-ice loads were analyzed at the time of a statistical experiment carried out in Ontario, Canada, covering data from 30 Atmospheric Environment Service weather stations. The purpose of the experiment was assessing statistics-based values for a reduction factor for wind speed associated with icing events

$$\text{Reduction factor, } k_r = \frac{\text{Wind speed used in wind-on-ice loads calculation, } V_{ice}}{\text{Extreme annual wind speed, } V_{max}} \times 100\%, \tag{2.34}$$

As a result of this analysis, the following values for k_r were obtained: mean = 56%, standard deviation = 7%, minimum = 45%, and maximum = 75%. These results are in very good agreement with the established practice in several countries (ex-Soviet Union and other East-European countries), where the reduction factor for wind speed used for wind-on-ice load calculations is taken equal to 50–56%.

Joint Probability Distribution of Wind and Ice: Chouinard (2000) used a theoretical, probabilistic approach to assess the joint probability distribution function of wind and icing during icing events, $f_{V,t}$, as the product of the marginal probability distribution function of equivalent radial ice thickness and the conditional probability distribution of wind speed, given a radial ice thickness t. Including the effect of the drag coefficient C, and integrating the joint probability distribution $f_{V,t,C}$, allows for establishing the cumulative distribution function of wind-induced loads on iced conductors

$$F_F = 1 - \iiint\limits_{\Omega} f_V(V) \cdot f_t(t) \cdot f_C(C)\, dV\, dt\, dC \tag{2.35}$$

where Ω is the domain of variation of the wind speed, ice thickness, and drag coefficient

$$\Omega = \{0 < V < \infty; 0 < t < \infty; 0 < C < \infty;\} \tag{2.36}$$

Using the CDF Eq. 2.36, wind-induced load on iced conductors can be selected for design purposes with a given return period. Unfortunately, the practical application of the theoretical method for estimating combined wind-on-ice loads on conductors is very limited. Combined loads depend not only on wind speed, icing amount, and drag coefficient, but also on several other variables, such as shape, density, and annual residency period of the ice build-up, direction of the coincident wind speed, non-uniformity of wind and icing in the line spans, height of conductors above ground level, altitude above sea level, air pressure and temperature, etc. Evaluation of integral (2.36) is very difficult, and can be made only by computer-performed Monte Carlo simulations. Some of the variables involved are often considered as fixed parameters. For instance, wind direction is always taken to be perpendicular to the line axis, while the drag coefficient and density of the ice build-up are assumed equal to 1 and 900 kg/m^3, respectively.

2.3.3.2 Specifying Return Period for Wind Speed Associated with Icing Events

IEC/CEI-60826 (2003) specifies combining extreme icing with the annual maximum of wind speed recorded during icing events and the consecutive periods where the air temperature is below 0 °C. When actual data from field observations are not available, a total period of 72 h is recommended.

ASCE (1991) recommends estimating the expected value of maximum wind speed during a seven-day (168-h) ice persistence period.

Savadjiev and Farzaneh (2003) proposed combining extreme icing with the expected value of maximum wind speed during a 6-day (144-h) annual ice residency

Table 2.9 Reference wind loading for wind-zone 1 in Quebec

Reliability level	Normal	High
Return period, T	150 years	500 years
Largest extreme, V_{max}	110 km/h	120 km/h
Mean, $\bar{V}_{max}$	75 km/h	75 km/h
Standard deviation, σ_V	10.24 km/h	10.24 km/h
Combined, V_{ice}	80 km/h	85 km/h

Note: All values, except for the standard deviation σ_V, are from SN-40.1 (1993)

period, T_{IRP}, averaged on all areas of Quebec where transmission lines are installed (see Section 2.2.1).

If the asymptotic distribution of annual maxima, $F(V_{max})$, is known or specified, the distribution $F(V_{ice})$ of the wind speed associated with icing events may be calculated using the following relation

$$F(V_{ice}) = \frac{\text{Ln}\,[F(V_{max})]}{\frac{8760}{T_{IRP}}} + 1 = \frac{\text{Ln}\,[F(V_{max})]}{\frac{8760}{144}} + 1 = \frac{\text{Ln}\,[F(V_{max})]}{60.33} + 1 \quad (2.37)$$

Example 10 Determine the reduction factor for wind speed associated with icing events (combined loads) for the climatic conditions in Quebec specified in SN-40.1 (1993): wind-zone 1: Compare the results with the specified V_{ice} in Table 2.9.

Solution: The CDFs, $F(V_{max})$, and $F(V_{ice})$, assumed to be type I Gumbel distributions, are drawn on Gumbel probability paper in Fig. 2.18. The initial wind speed distribution, $F(V_h)$, in this example, is also assumed to be type I Gumbel distribution for convenience, but as underlined previously, the actual initial wind speed distribution is probably Weibull or exponential (both bounded to the left tail), with coefficient of variation $1.8 > \text{COV} > 0$.

The statistical parameters of these wind-speed distributions are:

For $F(V_{max})$: mean 75 km/h, standard deviation 10.2 km/h
For $F(V_{ice})$: mean 42.2 km/h, standard deviation 10.2 km/h
For $F(V_h)$: mean 2.52 km/h, standard deviation 10.2 km/h

Calculated values for V_{ice}, and the reduction coefficient k_r, from curve 2 in Fig. 2.18 are:

Normal reliability level: $T = 150$ years: $V_{ice} = 77.6$ km/h, $k_r = 77.6/110 = 0.705$ (70.5%)
High reliability level: $T = 500$ years: $V_{ice} = 87.2$ km/h, $k_r = 87.2/120 = 0.727$ (72.7%)

As may be seen in Table 2.9, the calculated values for V_{ice} are very close to those specified in SN-40.1 (1993).

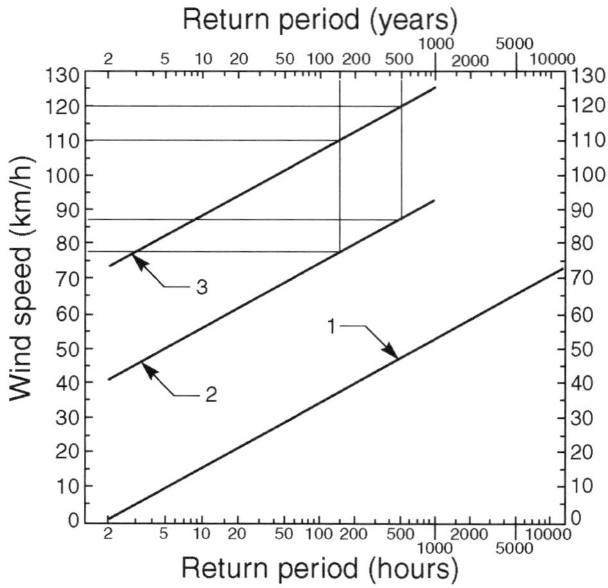

Fig. 2.18 Reference distributions for wind-load Zone 1 in Quebec: 1 – extreme wind speed, V_{max}; 2 – wind speed during icing events, V_{ice}; 3 – initial, hourly wind speed, V_h (Reproduced by permission of NRC)

2.3.3.3 Numerical Algorithm for Modelling Combined Wind-on-Ice Loads

Computer-performed Monte Carlo technique may be used for simulating different wind and icing distributions in order to establish values for the reduction factor of wind speed associated with icing events. These calculations are summarized by the following algorithm:

Step 1 Numerical simulation of initial Weibull wind speed distributions with various mean, COV, shape, and scale parameters.
Step 2 Selecting the annual maxima of wind speed from annual samples of size $n = 8,760$ h.
Step 3 Selecting the annual maxima of wind speed during icing events from subsets of size $n = 144$ h taken from the same annual samples.
Step 4 Repetition of steps 2 and 3 for at least $N = 500$ years.
Step 5 Estimating the parameters of the asymptotic Gumbel distribution for V_{max} and V_{ice}.
Step 6 Calculating $V_{max}(T)$ and $V_{ice}(T)$ for different return periods T ranging from 20 to 500 years.
Step 7 Estimating the reduction factor k_r from the ratio $V_{ice}(T)/V_{max}(T)$.

Results from the repeated application of this algorithm are shown graphically in Fig. 2.19.

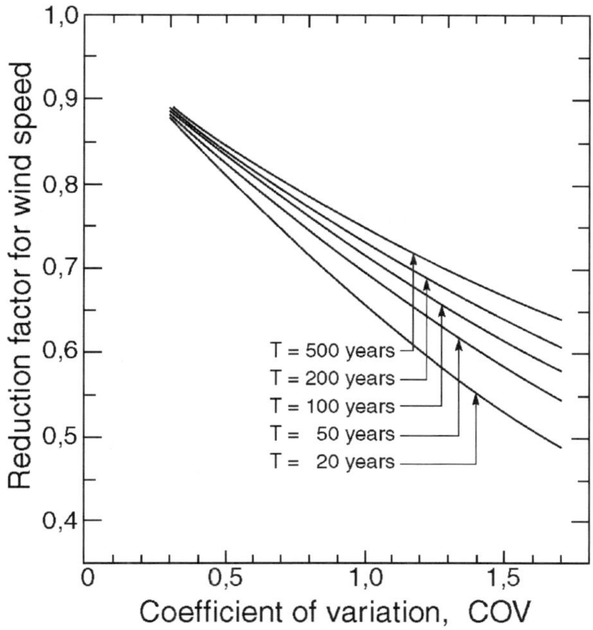

Fig. 2.19 Reduction factor k_r for wind speed, for different return periods T, and coefficient of variation of the initial wind speed distribution (Reproduced by permission of NRC)

2.3.4 Modelling the Icing Rate on Transmission Line Conductors

The random character of the climatic loadings determines the statistical approach as more appropriate for creating a working and usable model for icing on overhead transmission lines. This model should be based on the analysis of observational data covering the icing itself, but also all pertinent meteorological variables acting simultaneously and influencing the icing process. The creation of a statistics-based (probabilistic) model is important as an instrument for a natural validation of the multitude of physical, mathematical and laboratory models of atmospheric icing used in engineering practice.

In this section, real-time records from measurements of precipitation icing on the instrumented 315-kV line at Mont Bélair, Quebec, are analyzed. The following scatter-plots, Figs. 2.20–2.26, illustrate the trend observed between icing rate I_R, and climatic variables such as temperature, T, wind speed, V, wind direction, Z, precipitation rate, P, and hourly number of IRM signals, G, taken separately. The smoothed lines in these figures, obtained using the *Gaussian-kernel smoother*, illustrate the shape of the corresponding functions f_1–f_4 in the additive model (Hastie and Tibshirani 1990)

$$\mathrm{E}(I_R \mid T, G, V, P) = f_1(T) + f_2(G) + f_3(V) + f_4(P) \qquad (2.38)$$

where the symbol E means *expected value*, and f_1 through f_4 are arbitrary functions representing the influence of the considered climatic variable X (predictor) taken separately on the icing rate I_R (response). Once the shape of the smoothed

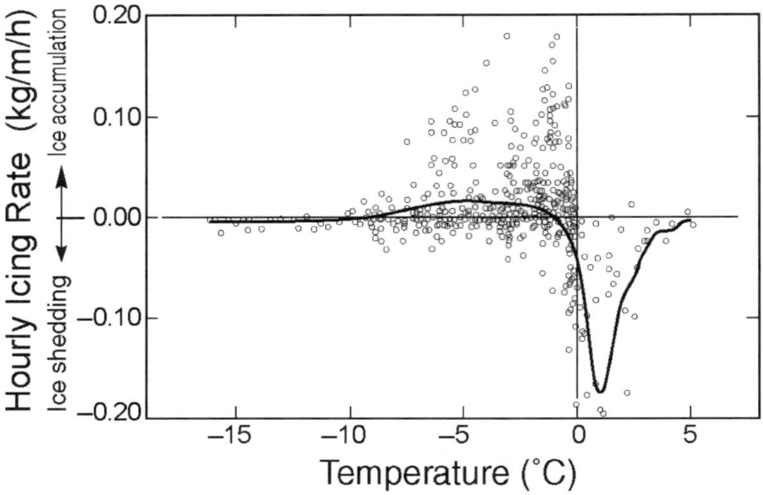

Fig. 2.20 Hourly rate of precipitation icing as a function of the ambient temperature (Reproduced by permission of IEEE)

curves is established, their numerical parameters may be estimated by multivariable regression analysis

$$I_R = a_0 + a_1 T + a_2 G + a_3 V + a_4 P \tag{2.39}$$

where a_0–a_4 are constants whose numerical values are obtained using the method of least squares.

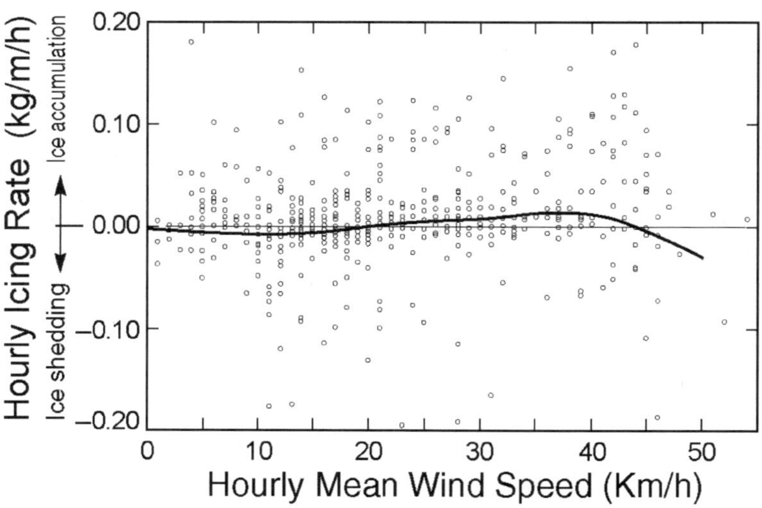

Fig. 2.21 Hourly rate of precipitation icing as a function of the hourly mean wind speed (Reproduced by permission of IEEE)

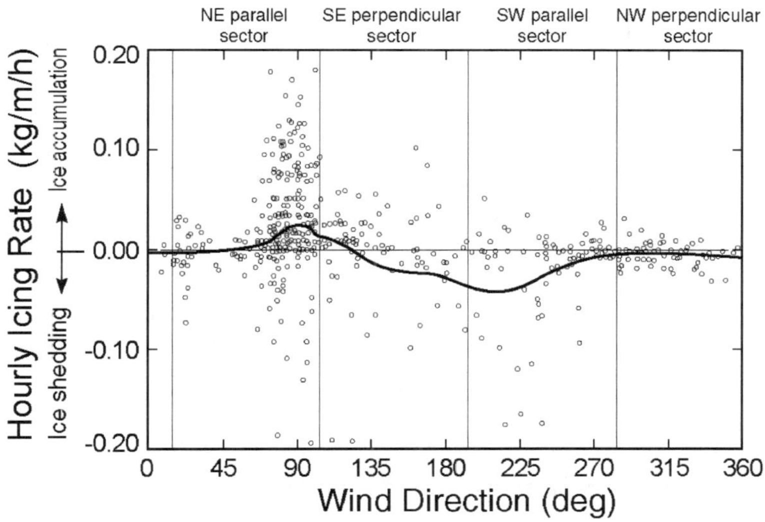

Fig. 2.22 Hourly rate of precipitation icing as a function of wind direction (Reproduced by permission of IEEE)

2.3.4.1 Influence of Temperature

In general, precipitation-icing accumulation is observed in the interval $-10\,°C < T < 0\,°C$, i.e. mainly in the second quadrant of the scatter-plot in Fig. 2.20, with a maximum at around $-5\,°C$. Positive rates are rarely observed below $-10\,°C$, because of the rapidly decreasing water content in ambient air. Of course, no accumulation is observed at positive temperatures. Two points, representing small

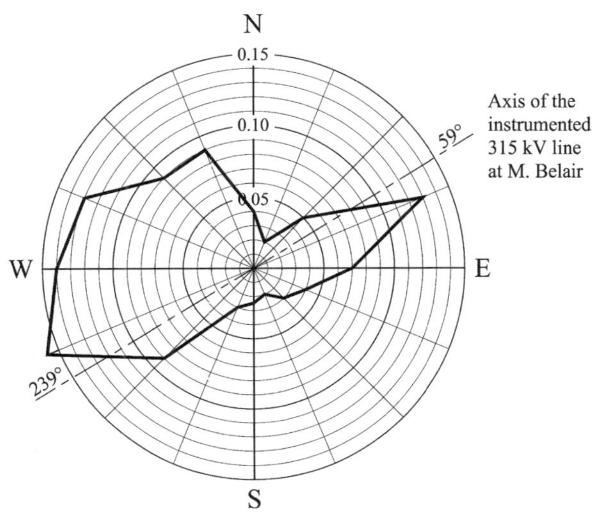

Fig. 2.23 Wind rose for Mont Bélair icing-test site during precipitation icing events

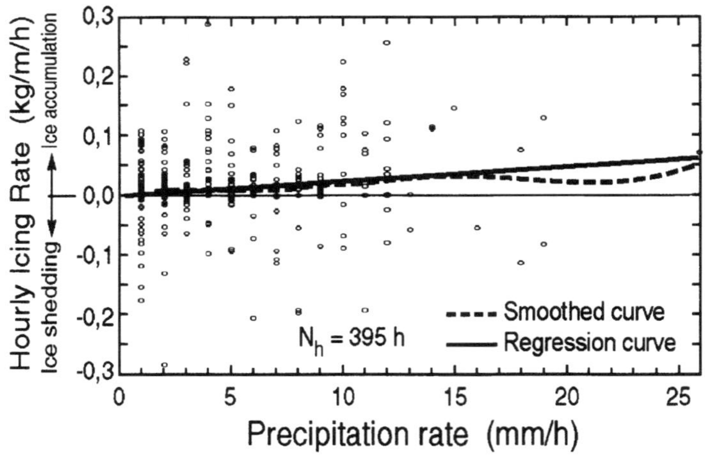

Fig. 2.24 Hourly rate of precipitation icing as a function of the hourly precipitation rate (Reproduced by permission of IEEE)

accumulations at positive temperatures and visible in Fig. 2.20, are exceptional and probably due to wet snow.

Ice shedding (negative hourly icing rate) by melting or breaking is observed at temperatures above −1 °C. Ice shedding by sublimation is predominant for temperatures below −10 °C.

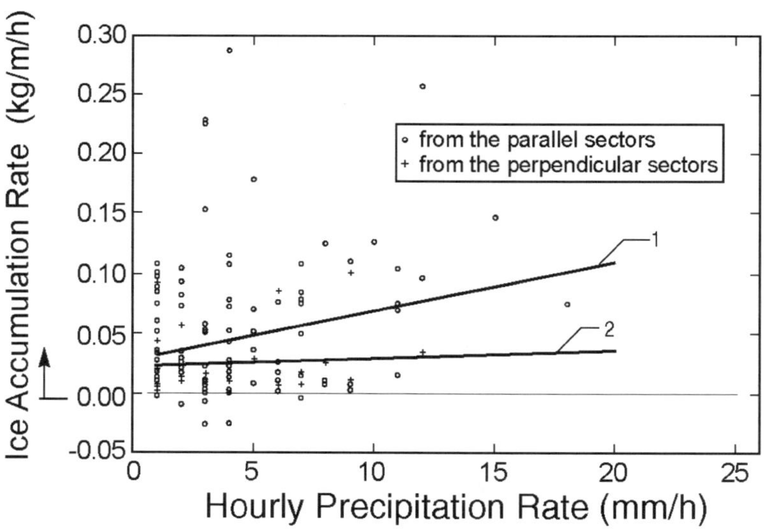

Fig. 2.25 Hourly rate of precipitation icing as a function of the hourly precipitation rate 1 – wind from the parallel sectors; 2 – wind from the perpendicular sectors (Reproduced by permission of IEEE)

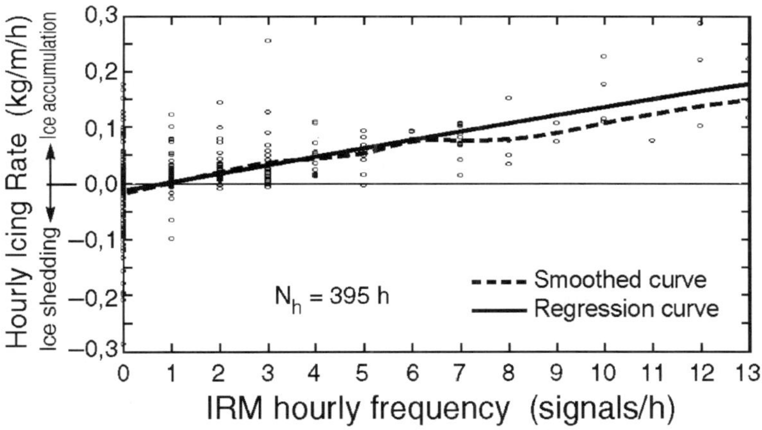

Fig. 2.26 Hourly rate of precipitation icing as a function of the hourly number of IRM signals (Reproduced by permission of IEEE)

2.3.4.2 Influence of Wind Speed and Direction

The smoothed line in Fig. 2.21 demonstrates the very small influence of the mean wind speed on the icing rate. In contrast, the wind direction has a much greater influence as illustrated in Fig. 2.22. The axis of the instrumented 315-kV line at Mont Bélair is oriented 59°(NE) – 239° (SW). It roughly coincides with the direction of the predominant winds at Mont Bélair, given by the wind rose shown in Fig. 2.23.

To study the influence of wind speed and direction on the hourly precipitation-icing rate, the whole horizon is subdivided into four 90° sectors centered along and perpendicular to the 315-kV line axis. Analysis of data in Fig. 2.22 shows that the predominant (parallel) winds blowing mainly from the NE parallel sector may be associated with the greatest icing rates observed.

This observation contradicts the traditional belief that, for precipitation icing, perpendicular winds are responsible for the greatest icing accumulations on line conductors (IEC/CEI-60826 2003; Chouinard et al. 2005; Jones 2003). This problem of great scientific interest should be further studied by analyzing field data from consecutive winter seasons at Mont Bélair, but also data from lines located near other icing-test sites or with different geographical orientation, or data from wind-tunnel simulations.

2.3.4.3 Influence of Freezing Precipitation Rate

The influence of freezing-precipitation rate, P, is illustrated in Fig. 2.24. The small positive slope of the least-squares-regression straight line is evaluated at approximately 0.0025 kg/m/h per 1 mm/h precipitation.

However, the slope of the regression curve increases significantly if this correlation is studied separately for winds from the parallel sectors ($\bar{I}_R = 0.0305\,\text{kg/m/h}$), and is much smaller for winds from the perpendicular sectors (Fig. 2.25).

2.3.4.4 Correlation between Hourly Icing Rate and the Hourly Number of IRM Signals

This correlation is illustrated in Fig. 2.26. The slope of the straight line, ≈ 0.015 kg/m/h per 1 IRM signal, is about 3 times greater than that for in-cloud icing (0.005 kg/ m/h per 1 IRM signal). These numbers may be used as calibration data for the IRM (Savadjiev and Farzaneh 2004). Nevertheless, these values should be subject to further validations. Figure 2.26 confirms the prior belief about the shape of these regression lines: (i) linear trend, and (ii) negative predominance (shedding) for $G = 0$ signals per hour. However, some "aberrant" points can be observed:

- Positive (accumulations) rates without IRM signals ($G = 0$). These measurement points are probably due to IRM blockage (breakdowns) in intensive wet conditions.
- Negative (shedding) values for $G = 1$ signal per hour.

2.3.5 Joint Distribution of Wind Speed and Air Temperature

This distribution gives the approximate limits for the transition soft rime, hard rime, and glaze shown in Fig. 2.27. The same distribution, obtained from the analysis of experimental data records at the Mont Bélair icing test site is shown graphically in Fig. 2.28 for precipitation-icing events and in Fig. 2.29 for in-cloud icing events. It can be used as a rough indication for determining the type of icing accumulation on transmission line conductors. It is interesting to note that the limits between the different types of icing shown in Fig. 2.27 shift to the right for wet-precipitation icing conditions (compare Figs. 2.28 and 2.29). These limits, if expressed by mathematical equations, can be successfully applied to training of artificial-neuron networks for recognizing the types of icing during the processing of icing field data.

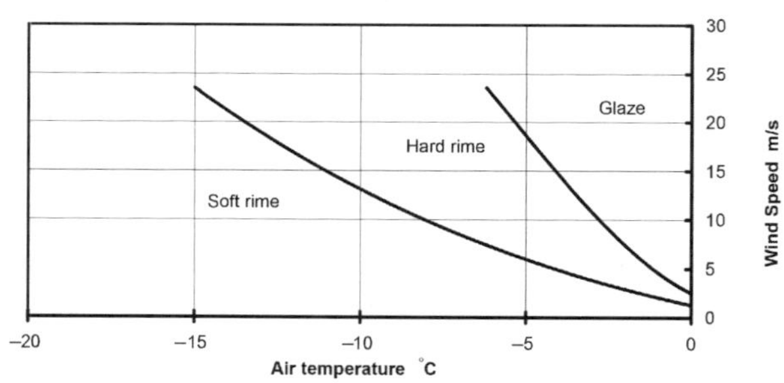

Fig. 2.27 Type of icing as a function of wind speed and temperature (Reproduced by permission of the International Electrotechnical Commission (IEC) from its International Standard IEC 60826 ed. 3.0 (2003))

Fig. 2.28 Joint distribution wind speed-air temperature during precipitation-icing (glaze) events

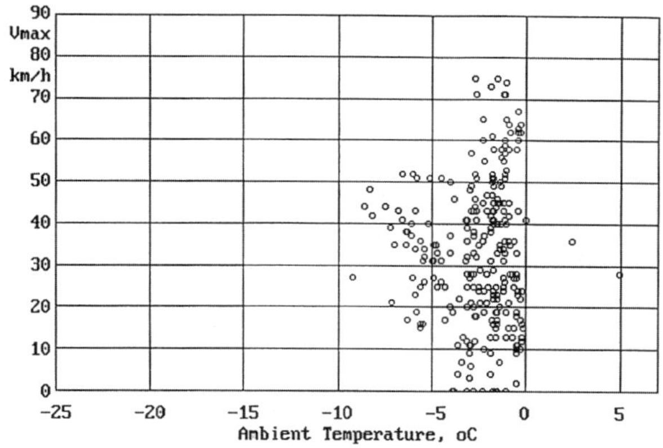

Fig. 2.29 Joint distribution wind speed-air temperature during in-cloud-icing (rime) events

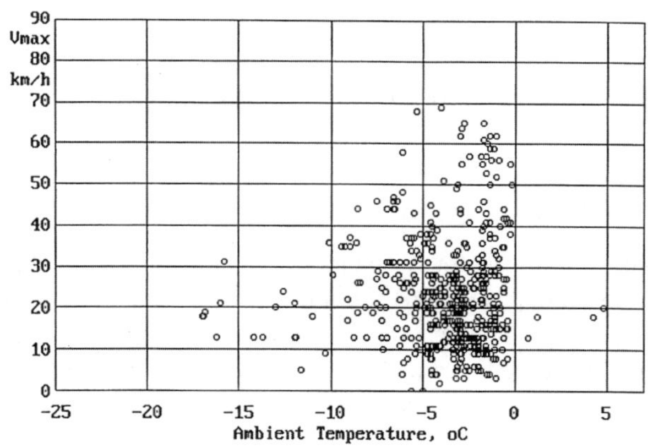

2.3.6 Modelling Temporal and Spatial Evolution of Icing Events

Overhead transmission lines are elements of large power transmission systems, and cross long distances through regions characterized by different topographic and climatic conditions. During their lifetime, transmission lines are affected not only by icing events occurring punctually in time (a few hours or days) and at the vicinity of the line route, but statistics show that ice storms with the most severe consequences are those which cover large areas in space and extend over one or more weeks. Statistical study of the spatial and temporal evolution of icing events is possible by analysis of icing-data records collected at a large number of measurement stations throughout the regions of interest, during a sufficiently long period of time. In Quebec, Guesdon (2000) and Houde et al. (2000) studied icing-event temporal and spatial evolution using data records from the PIM network, as well as data from 28 SYGIVRE stations in service at the beginning of 2000 (Fig. 2.30).

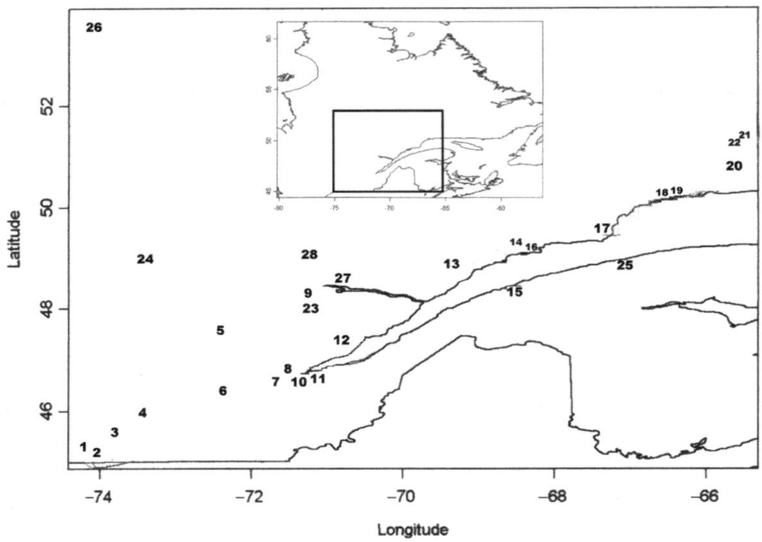

Fig. 2.30 Geographical location of 28 SYGIVRE icing-test stations in service in 2000. [Reproduced by permission of Guesdon, 2000]

2.3.6.1 Intensity Function

An icing measurement station is characterized by the temporal repartition of the icing events which are not uniformly distributed along a year. If the time coordinate of an event is measured in days from July 1st to June 30th, and the icing events are considered as a non-homogenous Poisson point process, the intensity function, $\lambda(t)$, characterizes the repartition in time of the icing events at a given location. An estimation of $\hat{\lambda}(t)$ can be obtained from the experimental data records using the Gaussian-kernel smoother technique

$$\hat{\lambda}(t)\frac{1}{nh}\sum_{i=1}^{n}\varphi\left(\frac{t-t_i}{h}\right) \tag{2.40}$$

where n is the number of icing events observed, $\varphi(x)$ is the standard normal (Gaussian) density function

$$\varphi(x)=\frac{1}{\sqrt{2\pi}}\exp\left(-\frac{x^2}{2}\right)$$

and h is the smoothing parameter determining the degree of smoothness of the curve. For example, using data from the SYGIVRE database, the intensity function (seasonal repartition) of precipitation-icing events in Quebec was obtained. This curve is illustrated in Fig. 2.31 and that for in-cloud-icing events in Fig. 2.32. Based on a relatively large number of icing events, the intensity function for the two different types of icing is noticeably different. Figure 2.31 shows two distinct

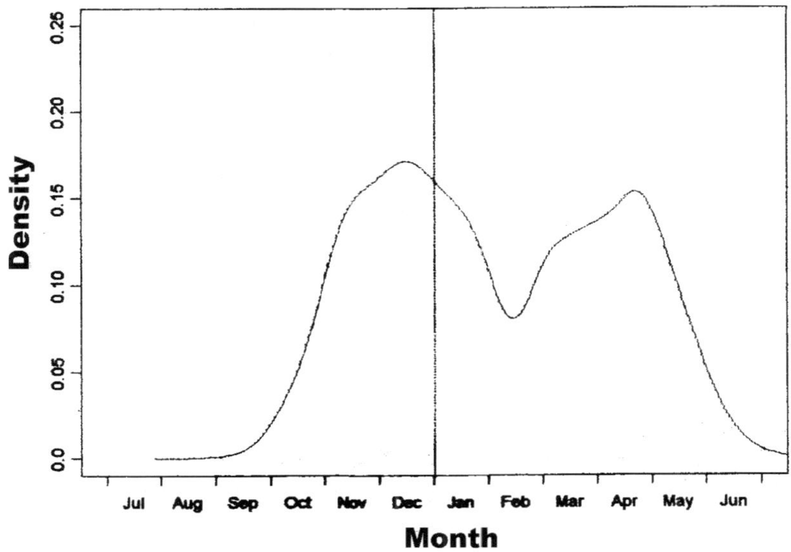

Fig. 2.31 Seasonal density of icing events of type glaze (precipitation icing)

maxima of the intensity function, in December and in April. Probably due to the low temperatures in January and February, there are much fewer freezing-rain events. In contrast, the intensity function of in-cloud icing (rime) type (Fig. 2.32) shows a maximum at the beginning of winter and decreases slowly afterwards. Similar curves may be constructed separately for all of the SYGIVRE stations in service. The

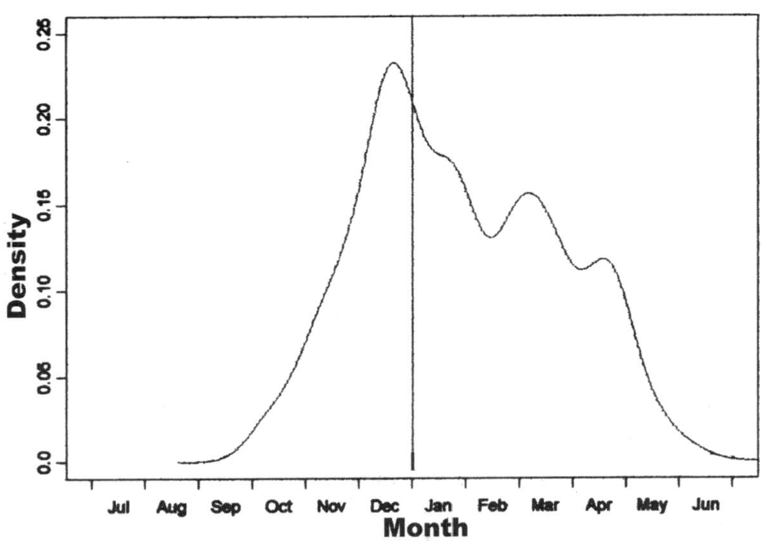

Fig. 2.32 Seasonal density of icing events of type rime (in-cloud icing)

results obtained from this characterization can be used for establishing if a station belongs to type *glaze* or *rime*, i.e. to establish the *icing type predominance* of the site.

2.3.6.2 Temporal Characteristics of Icing Event Seasons

The repartition of icing events over the year, represented by the intensity function $\hat{\lambda}(t)$, makes it possible to describe similarities and differences between winter seasons, and thus look on the temporal evolution of these seasons. For this purpose, a *distance* between consecutive years in terms of repartition and intensity of icing events is defined as follows

$$d_{ij} = \int \left|\lambda_i(x) - \lambda_j(x)\right| \tag{2.41}$$

where $\lambda_i(x)$ and $\lambda_j(x)$ are the intensity functions for years i and j. It is clear from Eq. 2.41 that if two years have exactly the same repartition of icing events (are characterized by the same intensity function), the distance defined by Eq. 2.41 will be equal to zero. In the contrary case, if two years are totally dissimilar from the point of view of repartition, number and intensity of icing events over the year, $d_{ij} = 2$. Using the proposed methodology, a temporal clustering can be made for years of SYGIVRE observation with the average distance as a grouping criterion (Fig. 2.33).

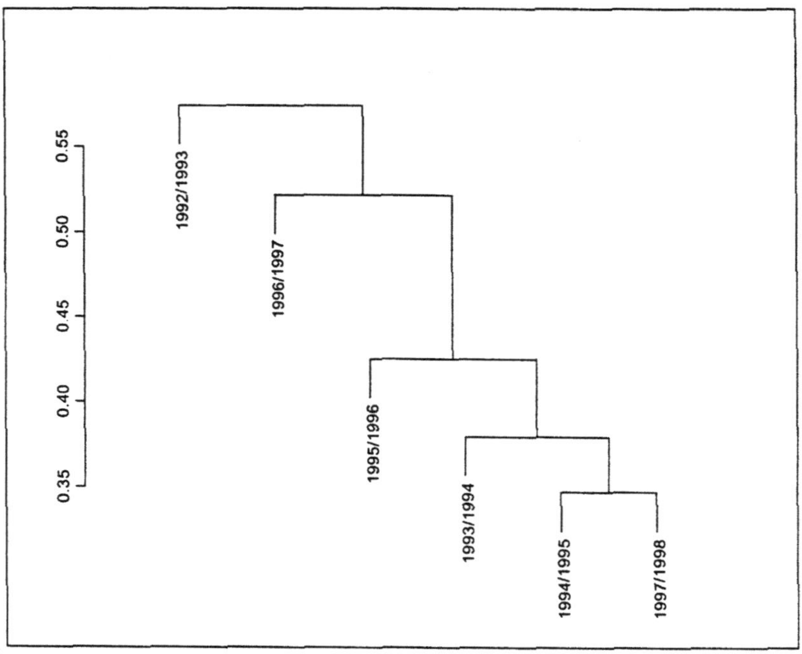

Fig. 2.33 Hierarchical temporal clustering tree for SYGIVRE stations

2.3.6.3 Spatial Characteristics of Ice-Measurement Stations

To study spatial evolution of icing events, it is important to characterize icing measurement stations by the *distance* from other stations sharing the same icing events. This distance, which is different from geographical distance, reflects the spatial grouping (association) of stations. It is given by the probability, p_{ij}, that if an icing event occurs at the vicinity of station i, it is shared by station j. This probability can be estimated by

$$\hat{p}_{ij} = \frac{n_{ij}}{n_i} \tag{2.42}$$

where n_i is the number of events observed at station i, and n_{ij} is the number of events shared by the ith and jth stations. A *distance* matrix $\mathbf{D}$ can be constructed for all SYGIVRE stations

$$\mathbf{D} = \begin{bmatrix} d_{11} & d_{12} & \cdots & d_{1N} \\ d_{21} & d_{22} & \cdots & d_{2N} \\ \cdots & \cdots & \ddots & \cdots \\ d_{N1} & d_{N2} & \cdots & d_{NN} \end{bmatrix} \tag{2.43}$$

where N is the number of stations in service. The distances d_{ij} are calculated using the relation

$$d_{ij} = 1 - \frac{n_i p_{ij} + n_j p_{ij}}{n_i + n_j} \tag{2.44}$$

and vary between 0 for two stations which share all icing events, and 1 for stations which do not share at all icing events. The precision of the evaluation of d_{ij} can be very small if the number of icing events observed is small. For this reason, only probabilities p_{ij} significantly different from 0 (at significance level 5%) are considered. This methodology allows for constructing a hierarchical clustering tree for the spatial *distances* between the measurement stations (Fig. 2.34).

In spite of the relatively small number of years of observation, the grouping of stations in Fig. 2.33 reflects the particularities of icing climate on the territory of Quebec. Some natural spatial grouping of stations can be observed in Fig. 2.34:

- Stations 1, 2, 3, 4, and 6 are situated in the region of Montréal near the St. Lawrence River and show glaze predominance (precipitation icing).
- Stations 8, (Mont Bélair), 12, and 23 situated near Québec City in hilly and mountainous terrain have rime (in-cloud) icing predominance.
- Stations 13, 14, and 17 situated north of the St. Lawrence River show a rime predominance.
- Stations 20, 21, and 26 situated in the North of the province of Quebec show a rime predominance.

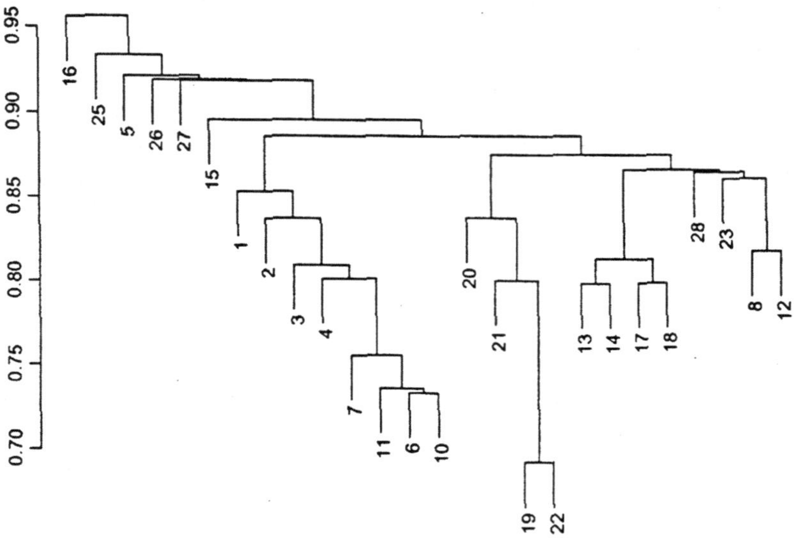

Fig. 2.34 Hierarchical spatial clustering tree for SYGIVRE stations

- Hierarchical temporal and spatial clustering trees for icing-test stations shown in Figs. 2.33 and 2.34 are a very useful tool for studying temporal and spatial evolution of icing events. This methodology allows for understanding and predicting the future propagation in time and space of an important icing event happening at a particular point of the territory at a given time. For instance, the catastrophic freezing-rain ice storm of January 1998 in Quebec affected mainly a limited area around Montreal, covered by the cluster of stations 1, 2, 3, 4, and 6 in Fig. 2.34.

2.4 Conclusions

This chapter reviews briefly the principal sources of field data for obtaining records containing information for icing events, such as national weather services, electrical facilities, direct measurements on samples fallen to the ground, field observations of icing on vegetation, neighbouring distribution and telecommunication lines, from specially designed icing measurement devices, test-spans, and permanently instrumented (energized) transmission lines in service. Some measuring devices and instruments are described, including the Passive Ice Meter, Icing Rate Meter, Load Cell, tubular-steel-rod racks used in Norway, and horizontal metallic rods at stand Studnice, Czech Republic. The methodology of observation, the type, content, and length of the time series in the constituted database are also discussed. The importance and pertinence of creating and implementing new programs for collecting meteorological and icing field data are emphasised. Further development of existing icing measurement programs is recommended.

Modelling freezing-rain data from observations made within the PIM measurement network supplies abundant and very useful information for the icing climate in Quebec: annual number of icing event recurrence, icing event residency period, total annual icing event residency period, duration of the winter season, and establishing freezing-rain mapping of the St. Lawrence Valley, spatial and temporal evolution of icing events.

Extreme Value Analysis is one of the most important procedures in transmission-line engineering. Using multiple numerical examples, the importance of the distribution of largest extremes of type I (Gumbel) is underlined. Construction of confidence-interval limits for the regression straight line on probability paper leads to an improved fitting of the observational data and forecasting icing loads with a prescribed return period. The minimum number of years of observation, and the influence of the type and shape of the initial distribution on the rapidity of convergence toward asymptotic extreme value distributions are also discussed.

Probabilistic modelling of combined wind-on-ice loads on transmission line conductors is developed. Based on PIM observations, a 144-h total annual ice residency period is proposed as a return period for the wind speed associated with icing events. Reduction factors for wind speed during ice storms ranging between 0.4 and 0.7 are proposed as a function of the coefficient of variation of the initial wind speed distribution and the prescribed return period of the combined wind-on-ice loads.

Analysis of icing data from the Mont Bélair icing test site in Quebec, Canada, equipped with Icing Rate Meter and Load Cells for direct measurement of precipitation and in-cloud icing on transmission line conductors, makes it possible to obtain correlations between hourly icing rate and the ambient temperature, wind speed and direction, hourly precipitation rate, and hourly number of IRM signals. These correlations are the basis for developing probabilistic models for icing events.

The joint distribution of wind speed and air temperature determines the approximate limits between the three principal types of icing on transmission line conductors: soft rime, hard rime, and glaze. A successful application of the mathematical expression of these limits is the training of artificial-neuron networks for recognizing the types of icing during the processing of field data.

Modelling spatial and temporal evolution of icing events is a very promising branch of the statistical analysis of atmospheric icing. Using the newly introduced notations: *intensity function*, characterizing the severity of icing events at a given site during winter seasons, and icing event transmission *distance*, it becomes possible to study ice storm propagation through space and their evolution in time. This methodology, based on the analysis of enormous quantities of icing field data records covering the entire territory of a country and long periods of observations, will contribute to a deeper understanding of atmospheric icing and its impact on large scale modern power transmission systems.

References

ASCE (1991) Guidelines for Transmission Line Structural Loading. Committee on Electric Transmission Structures, American Society of Civil Engineers, New York

Burgsdorff BB (1947) Construction and Operation of Transmission Lines in Regions with Severe Icing Conditions. Publ. by GOSENERGOIZDAT, Moscow (in Russian)

Castillo E (1988) Extreme Value Theory in Engineering. Academic Press, New York

Chouinard L, Feknous N, Sabourin G (2005) Design Winds during Ice Storm as a Function of Direction for Transmission Lines. In: Proc of the 11th International Workshop on Atmospheric Icing of Structures, Montréal: 271–274

Chouinard L (2000) Combination of Wind and Ice Loads for the Reliability Analysis of Electric Distribution and Transmission Networks. In: Proc of the 9th International Workshop on Atmospheric Icing of Structures, Chester, Session 1, 6 p

CIGRÉ (2006) Guidelines for Meteorological Icing Models, Statistical Methods and Topographical Effects. Working Group B2.16, Brochure 291, 116 p

Druez J, McComber P, Farzaneh M (1999) Correlation between Measurement of an Ice Detector and the Mass of Ice Accreted on Two Different Sized Conductors. Canadian Journal of Civil Engineering, vol 26, no 6: 869–875

Ervik M, Fikke SM (1982a) Development of a Mathematical Model to Estimate Ice Loading on Transmission Lines by Use of General Climatological Data. IEEE Trans on PAS, vol PAS-101, no 6

Ervik M, Fikke SM (1982b) Attempts Toward Estimating Ice Loadings Based on General Climatological Data. In: Proc of the 1st International Workshop on Atmospheric Icing of Structures, Hanover: 33–40

Farzaneh M, Savadjiev K (2005) Statistical Analysis of Field Data for Precipitation Icing Accretion on Overhead Lines. IEEE Transaction on Power Delivery, vol 20, no 2: 1080–1087

Farzaneh M, Savadjiev K, Druez J (2001) Icing Event Occurrence in Quebec: Statistical Analysis of Field Data. International Journal of Offshore and Polar Engineering, vol 11, no 1: 9–15

Guesdon C (2000) Étude des repartitions des évenements de verglas et de givre à travers le Québec. Master's of Engineering Thesis, Université du Québec à Chicoutimi

Gumbel E (1958) Statistic of extremes. Columbia University Press, New York

Hastie T, Tibshirani R (1990) Generalized Additive Models. Publ. by Chapman & Hall/CRC, London

Houde L, Guesdon C, Farzaneh M, Chouinard L (2000) Analysis of Spatial Patterns for Icing Events in Quebec. In: Proc of the 9th International Workshop on Atmospheric Icing of Structures, Chester, Session 2, 8 p

IEC/CEI-60826 (International Electrotechnical Commission) (2003) Design Criteria of Overhead Transmission Lines. Technical Report, Third edition, 2003–10, Geneva

Jones K (2003) Ice Storms in the St. Lawrence Valley Region. Technical Report ERDC/CRREL TR-03–1, US Army Corps of Engineers, 129 p

Johnson GL (1978) Economic Design of Wind Electric Systems. IEEE Trans on PAS, vol PAS-97, no 2: 554–561

Krishnasamy S, Kulendran S (1996) Combined Wind and Ice Loads from Historical Extreme Wind and Ice Data. In: Proc of the 7th International Workshop on Atmospheric Icing of Structures, Chicoutimi: 119–124

Laflamme J (1996) Icing Rate Measurements: A Key Way of Estimating Ice Loads on Structures. In: Proc of the 7th International Workshop on Atmospheric Icing of Structures, Chicoutimi: 175–180

Laflamme J, Périard G (1996) The Climate of Freezing Rain over the Province of Québec in Canada: A Preliminary Analysis. In: Proc of the 7th International Workshop on Atmospheric Icing of Structures, Chicoutimi: 19–24

Laforte JL, Alaire MA, Laflamme J (1995) Wind Tunnel Evaluation of a Rime Metering Device Using a Magnetostrictive Sensor. Atmospheric Research 36, Elsevier Science: 287–301

Le Du M, Laurent C (2005) T-Return Period Values of Meteorological Design Parameters. In: Proc of the 11th International Workshop on Atmospheric Icing of Structures, Montréal: 153–159

Makkonen L (2006) Plotting Positions in Extreme Value Analysis. Journal of Applied Meteorology and Climatology, vol 45: 334–340

McComber P, Latour A, Druez J, Laflamme J (1996) The Icing Rate Meter, an Instrument to Evaluate Transmission Line Icing. In: Proc of the 7th International Workshop on Atmospheric Icing of Structures, Chicoutimi: 159–168

Popolansky F, Kruzik J, Lehky P, Hrabanek J, Lago J (1998) Ice Monitoring at Stand Studnice: Tuned Vibration Control Overhead Line Conductors. CIGRÉ Paper 22–105, Session-1998, Paris

Savadjiev K, Farzaneh M (2005) Characterization of Icing Events Based on Statistical Analysis of Field Data. In: Proc of the 11th International Workshop on Atmospheric Icing of Structures, Montréal: 131–136

Savadjiev K, Farzaneh M (2004) Modeling of Icing and Ice Shedding on Overhead Power Lines Based on Statistical Analysis of Meteorological Data. IEEE Transactions on Power Delivery, vol 19, no 2: 715–721

Savadjiev K, Farzaneh M (2003) Probabilistic Model of Combined Wind and Ice Loads on Overhead Power Line Conductors. Canadian Journal of Civil Engineering, 30: 704–710

Savadjiev K, Farzaneh M (2001) Study of Icing Rate and Related Meteorological Parameter Distributions during Atmospheric Icing Events. In: Proc of the 11th International Offshore and Polar Engineering Conference, vol I, Stavanger: 665–670

Savadjiev K, Farzaneh M (1999) Analysis and Interpretation of Icing Rate Meter and Load Cell Measurements on the Mt. Bélair Icing Site. In: Proc of the 9th International Offshore and Polar Engineering Conference, vol II, Brest: 607–611

Savadjiev K, Farzaneh M, Druez J (1998) Statistical Analysis of Two Probabilistic Models of Ice Accretion on Overhead Line Conductors. In: Proc of the 8th International Offshore and Polar Engineering Conference (ISOPE), Montrél: 530–536

SN-40.1 (1993) Critères de conception des lignes de transport et répartition d'Hydro-Québec. Service Études et Normalisation, Hydro-Québec, Montréal

Chapter 3
Numerical Modelling of Icing on Power Network Equipment

Lasse Makkonen and Edward P. Lozowski

3.1 Introduction

To the reader of this book, it is hardly necessary to offer justification for considering icing as a problem on power network equipment. Suffice it to say that, responding to the disastrous Ice Storm of 1998 in Canada (Abley 1998; Anon 1999), in which 1,300 high voltage transmission line towers and 35,000 distribution line structures were destroyed, should offer ample motivation to inquiring minds, who seek to improve our understanding of ice storms, for the benefit of the citizens of all countries subjected to icing storm episodes.

Nevertheless, the reader might wonder why an understanding of icing should be sought via numerical modelling, rather than through observations and measurements alone. The Ice Storm of 1998 illustrates the limitations of relying on direct icing measurements for determining ice loads for structural design. The Ice Storm caused unexpected damage in the only region of the world where regular icing measurements have been made for a long time using a dense observation network (Laflamme and Periard 1996).

The limited value of icing measurements arises from the rarity of icing events and their complex dependence on combinations of atmospheric and geographical factors. Consequently, icing events occur sporadically and they are highly variable in space. They may not follow a probability distribution that is readily derivable from measured icing data. Moreover, such data do not exist at all in many areas of the world. Furthermore, for tall structures, ice data is required high above the ground, and such data cannot usually be collected until the structure has been erected. Then it may be too late, as illustrated by frequent power line failures and the 140 ice-induced distribution tower collapses during the last 40 years in the United States alone (Mulherin 1998).

These deficiencies in icing measurements illustrate the importance of modelling ice accretion. The major advantages of modelling are twofold. First, climate and

L. Makkonen
Technical Research Centre of Finland, 02044 VTT, Finland
e-mail: lasse.makkonen@vtt.fi

M. Farzaneh (ed.), *Atmospheric Icing of Power Networks*,

weather data, which are much more extensive in time and space than icing data, can be utilized as model input in order to generate synthetic icing data with much improved spatial and temporal resolution. Second, sound theoretical model simulations can be extended beyond the range of our limited empirical verifications. The latter advantage is particularly important in structural design, because the designer is typically interested in extreme events that we may not have experienced yet.

Theoretical icing modelling is also required for developing anti-icing and de-icing systems. Such systems are necessary to protect facilities, whose safe operation depends critically on eliminating even small amounts of accreted ice, due to its effect on aerodynamic behaviour. Aeroplanes, helicopters and wind turbines are typical examples. Aircraft icing modelling has a long tradition (Gent et al. 2000), while increasing wind energy production in cold, hilly regions of the world has recently created a demand for anti-icing of wind turbines (Peltola et al. 2004).

This chapter is not intended to be a "cookbook" on how to create an icing model. Rather, we document the progress of ice accretion modelling for power network equipment, starting with empirical approaches and ending with modern, 3D simulations that are based on emulating the natural physical processes of icing. We also reflect on the need for future model development and testing (for further details, see Lozowski and Makkonen 2005). In addition, we provide a description of the process of icing modelling, mainly from a physical point of view. Accordingly, purely statistical models are not considered and the discussion does not go deeply into the mathematical and numerical methods used in icing models. Such details can be found in the individual papers we reference and in Dranevic (1971), Gartzman (1987) and Poots (1996).

Fig. 3.1 Rime on a 22 kV overhead line in Voss, Norway, 18 April 1961. This event is the highest ice load in the world recorded on power lines, 305 kg/m on each span (Photo: O. Wist. Reproduced by permission of S. M. Fikke)

Fig. 3.2 Ice on aerial cables after a freezing rain event in Slovenia (Photo: IBE, Ljubljana)

Atmospheric icing occurs in a variety of forms as a result of the interplay of numerous physical processes. Meteorological variables play a key role in determining the amount and type of ice accretion, especially air temperature, wind speed and hydrometeor phase, morphology, mass flux and size distribution. A reader interested in the subject, but having little experience with it, may find it useful to have a visual image of the ice deposits whose growth this paper tries to describe theoretically. In order to help the reader and provide food for her imagination, photographs of the basic forms of atmospheric icing are presented in Figs. 3.1 to 3.4.

3.2 The Fundamental Equation of Icing

The sources of natural ice that forms on power network equipment include supercooled cloud and fog droplets, freezing rain and drizzle drops, wet snow and water vapour. It can be shown (Makkonen 1984a, pp. 26–27; Makkonen 2007) that mass accumulation due to the deposition of water vapour in the form of hoarfrost is

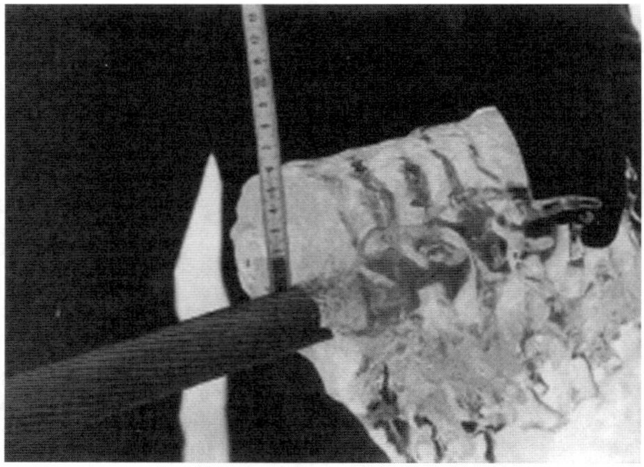

Fig. 3.3 Glaze ice sample taken after the same freezing rain event shown in Fig 3.2. The icicles have been partly broken (Photo: IBE, Ljubljana)

usually negligible, compared to typical mass accumulation rates due to the impingement of liquid water droplets and snow particles.

Significant ice loads can form due to particles in the air colliding with some power network component. These particles can either be liquid, solid or a mixture of water and ice. The maximum rate of icing (expressed as a mass flux density or mass per unit time per unit cross-sectional area of the object) is limited by the corresponding mass flux density of the impinging particles. The mass flux density of the impinging particles, F, may be expressed as the product of the aerial mass

Fig. 3.4 Wet snow accretion on a 300 kV power line in Dale-Fana, Norway (Photo: Satnett)

concentration of the particles, w, and the effective velocity [1], v, of the particles with respect to the object, i.e. $F = wv$. The rate of icing, expressed as the rate of change of the ice mass (M) accumulated on the object, is then obtained from the equation

$$\frac{dM}{dt} = \alpha_1 \alpha_2 \alpha_3 w v A \quad (3.1)$$

where A is the effective cross-sectional area of the object[2]. The factors α_1, α_2 and α_3 are less than unity and represent processes that may reduce *dM/dt* from its maximum potential value. They are:

α_1 the collision efficiency,
α_2 the coalescence efficiency,
α_3 the accretion efficiency.

Factor α_1 is the mean collision efficiency of the particles, i.e. the ratio of the impinging mass flux density to the aerial mass flux density. The collision efficiency may be considerably less than unity, because small particles tend to follow the air streamlines and may be deflected around the object, as shown in Fig. 3.5.

Factor α_2 is the mean coalescence efficiency of those particles that hit the object, i.e. the ratio of the mass flux density of the particles that stick to the object, to the impinging mass flux density. The coalescence efficiency is less than unity when

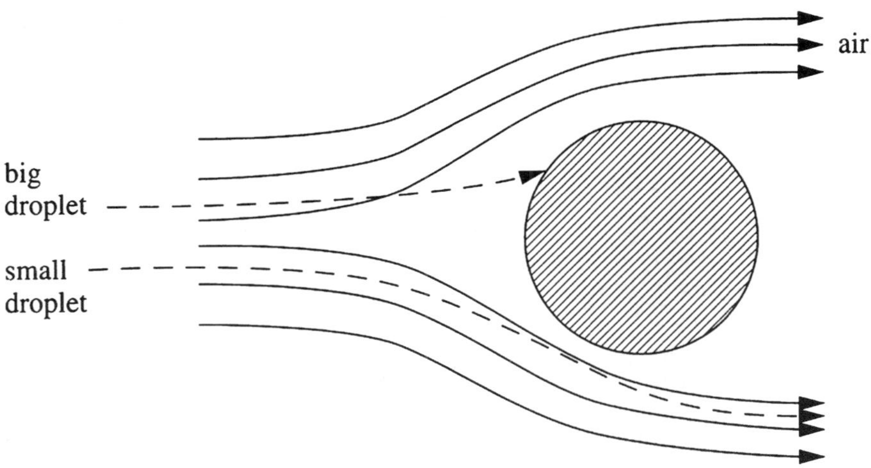

Fig. 3.5 Air streamlines and droplet trajectories around a cylinder

[1] Because particles of different sizes and phases will have differing aerial velocities, the effective velocity is essentially a mass-weighted mean velocity of all of the impinging particles.
[2] Because particles of differing sizes and phases will impinge from different directions, the effective cross-sectional area is essentially the mass-weighted average cross-sectional area seen by each particle.

particles bounce from the surface. For our purposes, particles are considered to stick when they are permanently collected, or, equivalently, when their residence time on the object is sufficient to affect the icing rate due to heat exchange with the object or airstream.

Factor α_3 is the accretion efficiency, i.e. the ratio of the ice mass flux density to the mass flux density of the particles that stick to the surface. The accretion efficiency α_3 is less than unity when the heat flux from the accretion is too small to freeze all the liquid particles that stick to the accretion. When this happens, part of the sticking mass may be lost from the surface[3] by shedding, under the influence of aerodynamic and gravitational forces.

When there is no permanent surface liquid and no run-off ($\alpha_3 = 1$), the process is called "dry growth". This situation is schematically shown in Fig. 3.6. The ice resulting from dry growth is customarily called "rime". When the situation shown in Fig. 3.7 develops ($\alpha_3 < 1$), there is a liquid layer on the surface of the accretion[4] and freezing takes place beneath this layer. This is called "wet growth". The ice resulting from this process is customarily called "glaze". The terms "collection efficiency" for the product $\alpha_1\alpha_2$ and "sticking efficiency" for α_3 are sometimes used in the literature (especially by cloud physicists).

One should note that, although we speak of "icing" and "icing rate", the accretion that forms may be a mixture of ice and liquid water (Macklin 1961). In fact, when a liquid film forms at the accretion surface (Fig. 3.7), the growing ice always initially entraps a considerable amount of liquid water (Knight 1968; Makkonen 1987). This is called "spongy ice". Accretion of wet snow and the growth of icicles (Makkonen 1988) also result in deposits that include liquid water. Liquid water

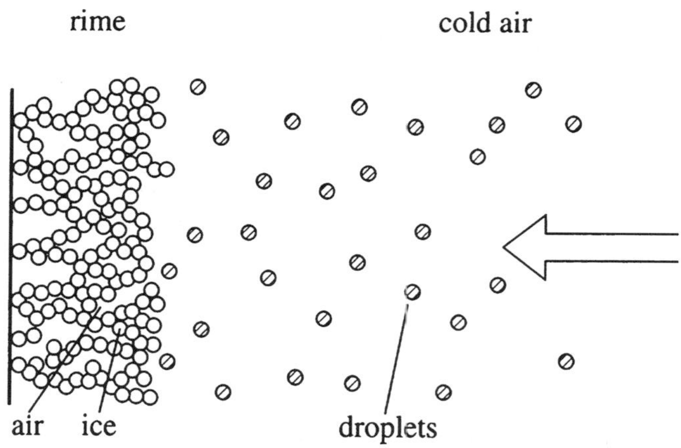

Fig. 3.6 Growth of rime ice (dry growth)

[3] Recently, Lozowski et al. (2005) have confirmed that substantial unfrozen liquid may reside in the ice accretion, trapped within a porous matrix of ice known as "spongy ice."

[4] Depending on the fluid dynamics of this layer, it may or may not be continuous. A non-continuous surface liquid mass may consist of an ensemble of sessile droplets or a series of rivulets.

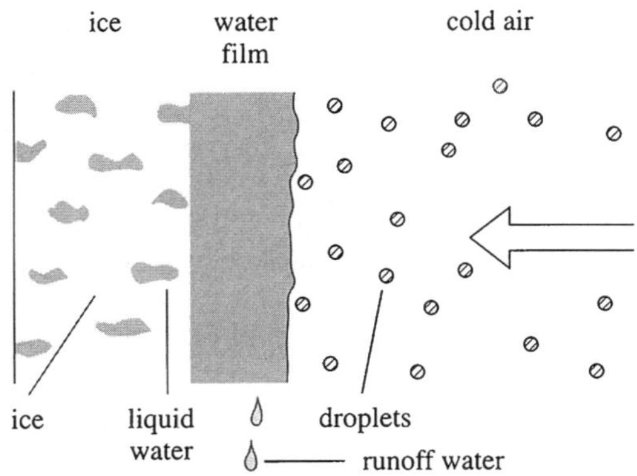

Fig. 3.7 Growth of glaze ice (wet growth)

within ice accretions is, however, seldom detected, because the deposits often completely freeze soon after the icing storm is over.

In the estimation of ice loads, one should note that a very important factor is the duration of the icing event. Thus, in addition to the parameters that determine the rate of icing, the key element of successful icing modelling is a good understanding of the sequence and combinations of weather elements during the icing event. This is not a trivial problem, but it can be roughly summarized as follows:

- For freezing precipitation, one must account for situations where the wet bulb temperature $< 0\,^{\circ}\mathrm{C}$ and there is liquid precipitation.
- For rime, one must account for situations where the air temperature $< 0\,^{\circ}\mathrm{C}$ and there is fog, or the location of interest is higher than the cloud base.
- For wet snow, one must account for situations with heavy snow fall, when the wet bulb temperature $> 0\,^{\circ}\mathrm{C}$.

The simulation of ice accretion, for practical purposes, requires careful consideration of all these criteria (Sundin and Makkonen 1998).

3.3 Computing the Rate of Icing

3.3.1 Implementing Icing Physics in the Fundamental Equation

Equation 3.1 points out the basic problems of estimating icing rates on objects. The three factors, α_1, α_2 and α_3, all lying between 0 and 1, must be determined. In addition, the mass concentration of particles in the air, w, the effective particle velocity, v, and the effective cross-sectional area of the object, A, must also be determined. Determination of the atmospheric parameters for a location of interest, or forecasting them, is a problem that will not be discussed here. It should be noted, however, that the mass concentration, w, is not a routinely measured parameter, and that its estimation or forecasting is a difficult problem in its own right. Moreover, the effective particle velocity, v, depends on the wind speed and the often unknown terminal velocities of the various types of water/ice particles that comprise the precipitation or airborne aerosol.

The complexity of the icing process means that there is little hope of making successful predictions of the icing rate using simple empirical methods. Nevertheless, a number of attempts to do so have been made. To a present-day icing modeller, some of these may appear primitive. However, one must acknowledge that there was a need for extreme design ice load estimates long before our present understanding of icing physics had been developed, not to mention modern computing capabilities. Moreover, due to the paucity of the necessary input data for icing models, much empiricism must still be included in the modelling process. This will become evident as the chapter progresses.

The classical, empirical approach to estimate the rate of rime icing ($\alpha_2 = \alpha_3 = 1$) is to apply Eq. 3.1 with α_1wA = constant. Under this assumption, the icing rate depends linearly on the wind speed only. The slope constant for this equation has been determined empirically by Makkonen (1984a), based on data by Rink (1938), Waibel (1956), Ahti and Makkonen (1982), Tammelin and Säntti (1996) and Dobesch et al. (2005). As one might expect, the slopes vary with location; hence, there tends to be a low correlation coefficient between model and observations. A non-linear wind speed dependence has also been proposed (e.g. Diem 1956), but these relations too have little predictive value, for reasons that will become evident when we consider theoretical means to determine the factors α_1, α_2, α_3 and A. Nevertheless, since errors in the icing rate tend to average out over the long-term, provided a predictive method has no bias, empirical approaches may have some validity in predicting cumulative ice loads, as shown by Zavarina et al. (1976) and Sundin and Makkonen (1998).

Attempts to empirically estimate glaze icing due to freezing rain have quite naturally been made by using precipitation amount as the predictor (e.g. Lenhard 1955). Subsequent studies have shown, however, that the correlation between precipitation amount and ice load is very low (McKay and Thompson 1969; Snitkovskii 1977), suggesting that modelling based on icing physics is clearly required.

As for wet snow, empirical methods are still "state-of-the-art", and they form the basis for modelling, as discussed in Section 3.3.3.

3.3.2 Collision of Atmospheric Particles

When a droplet moves with the air stream toward an icing object, its trajectory is determined by aerodynamic drag and inertia [5]. When inertia is small (small particles), drag dominates and the droplets closely follow the airflow streamlines

---

[5] Other forces may also play a role, particularly for the high speed collisions typical of wind turbine and aircraft icing (Landau and Lifshitz 1959).

(Fig. 3.5). Since air must go around the object, the droplets also tend to do so. The impinging flux density in this case can be much smaller than the aerial flux density of the spray. For large droplets, on the other hand, inertia dominates and the droplets tend to hit the object without substantial deflection (Fig. 3.5).

The relative magnitude of the inertia and drag of aerosol particles depends on particle size and shape, airstream velocity and the size and shape of the icing object. When these are known, the collision efficiency, α_1, in principle, can be theoretically determined by numerically solving for the airflow around the object and for the particle motion in the air flow (Lock 1990; Bourgault et al. 2000). This approach was pioneered by Langmuir and Blodgett (1946) and Mazin (1957). The trajectories must, in principle, be determined for each particle. However, it is often possible to bin particles based on their phase, size and shape and hence solve only for a smaller number of particles, representative of each bin, in order to determine the overall collision efficiency α_1. Such computations can be exceedingly complicated and time consuming, but there are several ways to simplify them, particularly for simple object and particle geometries.

Firstly, the collision efficiency can be parameterized using two dimensionless parameters, the Stokes number, K and the Langmuir number, φ, where

$$K = \frac{\rho_w\, d^2 v}{9\mu D} \tag{3.2}$$

and

$$\varphi = \frac{Re^2}{K} \tag{3.3}$$

with the droplet Reynolds number, Re, based on the free stream velocity, v

$$Re = \frac{\rho_a v d}{\mu} \tag{3.4}$$

where d is droplet diameter, D cylinder diameter, ρ_w water density, μ absolute viscosity of air, and ρ_a air density.

When the icing object is cylindrical, there exists an analytical solution for the potential air flow[6] around it, based on which numerical trajectory computations can be made. Finstad et al. (1988b) developed the following empirical fit to their numerical trajectory data, for spherical water droplets

$$\alpha_1 = A - 0.028 - C(B - 0.0454) \tag{3.5}$$

[6] The reader should be aware that flow separation in the wake of the cylinder, at high cylinder Reynolds number, can distort the upstream potential flow.

where

$$A = 1.066K^{-0.00616}exp(-1.103K^{-0.688})$$
$$B = 3.641K^{-0.498}exp(-1.497K^{-0.694})$$
$$C = 0.00637(\varphi - 100)^{0.381} \quad (3.6)$$

Finstad et al. (1988c) have also shown, both theoretically and empirically, that with good accuracy, a single parameter, the median volume diameter (MVD), can be used in the calculations (as d in Eqs. 3.2 and 3.4), without having to calculate α_1 separately for each droplet size category[7].

The collision efficiency, α_1, depends strongly on the particle size, and, for sufficiently large MVD, such as in freezing rain or wet snow, one can estimate that $\alpha_1 = 1$ in most practical applications, unless the structure is extremely big. Therefore, α_1 usually needs to be computed only when icing is caused by cloud droplets. It is, however, quite feasible to calculate the collision efficiency of wet snow particles by snowflake trajectory computations (Skelton and Poots 1991).

3.3.3 Sticking

When a supercooled water droplet hits a dry ice surface, it rapidly freezes and, typically, does not bounce (Fig. 3.6). When a liquid layer sits on the surface, the droplet tends to spread on the surface and again there is, typically, no bouncing (Fig. 3.7). Nevertheless, small droplets or ice fragments leaving the surface may be created in these processes (List 1977; Choularton et al. 1980; Dong and Hallett 1989). Their cumulative volume is, however, so small that their effect on icing is usually insignificant. Therefore, liquid water droplets can generally be considered not to bounce, i.e. for water droplets, $\alpha_2 \approx 1$.

Ice crystals, however, bounce very effectively (Wakahama et al. 1977). For completely solid precipitation particles, e.g. dry snow, the coalescence efficiency, α_2, is essentially zero. However, at wind speeds less than about 2 ms^{-1}, dry snow may settle on power line cables (Sakamoto 2000), whereafter a cylindrical sleeve may form by sliding and creep of the snow. Typically, a snow load resulting from dry snow will not be heavy, in comparison with the loads from wet snow discussed below. This is a consequence of the following factors. First, the flux of snow to the cable is relatively small at the low wind speeds necessary for dry snow to stick. Since the terminal velocity of snowflakes is small, the flux is determined largely by the wind speed, unless the wind speed is significantly less than about 1 ms^{-1}. Secondly, the density of dry snow accretions is typically less than 0.1 gcm^{-3}. This

[7] We suspect that, for freezing rain and drizzle, this should be the median mass flux diameter rather than the median volume diameter. These two parameters will be virtually identical for cloud and fog droplets.

low density is associated with a small cohesion strength, which causes dry snow deposits to drop off easily. Consequently, dangerous combinations of high snow load and high wind speed do not occur with dry snow.

When there is liquid on the surface of the snow particles, they stick more effectively, so that at small impact speeds and favourable temperature and humidity conditions, α_2 is close to unity. This is the situation when wet snow accretes on cables. Presently there is no detailed theory for the coalescence efficiency of wet snow. The available approximate methods for estimating α_2 are empirical equations based on laboratory simulations and some field observations (Sakamoto and Miura 1993; Poots 1996; Sakamoto 2000). The best approximation for α_2, for cylindrical targets, is probably (Admirat et al. 1988)

$$\alpha_2 = \frac{1}{v} \tag{3.7}$$

where v is wind speed in ms^{-1}. When $v < 1\ ms^{-1}$, $\alpha_2 = 1$.

Air temperature, humidity and precipitation intensity also affect α_2 for wet snow, as shown by several experimental studies of this problem (Poots 1996; Sakamoto 2000). It is important to note that $\alpha_2 > 0$ only when the snow particle surface is wet, so that snow does not accrete effectively when the wet-bulb temperature is below 0 °C (Makkonen 1989). This criterion is very important because it facilitates a determination of the duration of wet snow events using climate and weather data.

3.3.4 Heat Transfer

Here we consider icing arising from the impingement of supercooled water drops. Icing arising from other mechanisms may give rise to heat transfer terms that can be formulated in a way analogous to those presented here, mutatis mutandis. In dry icing (Fig. 3.6), all the impinging water droplets freeze and the accretion efficiency, $\alpha_3 = 1$. In wet icing (Fig. 3.7), the freezing rate is controlled by the rate at which the latent heat released in the freezing process can be transferred away from the freezing liquid. The portion of the impinging water that cannot be frozen or incorporated within the ice matrix is removed by gravity and wind stress. The heat transfer problem for icing is typically formulated as a heat balance for the "icing surface", in which the latent heat of freezing is equated to the sum of all other heat transfer terms. Simplified formulations of the heat balance were used in early glaze ice models (Kuroiwa 1965), but we will consider a more comprehensive heat balance.

The heat balance equation for the icing surface can be written for wet icing as

$$Q_f = Q_c + Q_e + Q_l + Q_s - Q_v \tag{3.8}$$

where

Q_f = latent heat released during freezing at the freezing temperature (0 °C)
Q_v = frictional and compressive heating by the airstream

Q_c = loss of sensible heat to the airstream
Q_e = heat loss due to evaporation
Q_l = heat loss in warming the impinging supercooled water to the freezing temperature
Q_s = heat loss or gain due to net radiation (both long- and shortwave)

The terms of the heat balance equation can be parameterized using meteorological and structural variables.

The heat released during freezing is transported from the ice-water interface, through unfrozen liquid water on the ice surface, into the airstream. Consequently, there is a negative temperature gradient ahead of the growing ice, giving rise to supercooling, which results in dendritic ice growth and some liquid entrapment within the ice matrix (spongy ice). Since the unfrozen water can be entrapped without releasing any latent heat, the term Q_f in Eq. 3.8 is

$$Q_f = (1 - \lambda)\alpha_3 F L_f \tag{3.9}$$

where λ is the liquid fraction of the accretion, F is the mass flux density of water to the surface ($F = \alpha_1 \alpha_2 w v$) and L_f is the specific latent heat of freezing at 0° C. Attempts to determine the liquid fraction, λ, have been made both theoretically (Makkonen 1987; Makkonen 1990; Blackmore and Lozowski 1996; Blackmore et al. 2002; Makkonen 2008) and experimentally (Lesins et al. 1980; Gates et al. 1986; Lock and Foster 1990; Blackmore et al. 2002; Lozowski 2005). These studies suggest that λ can be rather insensitive to the growth conditions, and that a value of λ around 0.3 is a reasonable first approximation for power network icing under wet conditions.

Another important consequence of spongy ice is that heat conduction into the ice or substrate does not need to be considered in the heat balance, since a mixture of ice and water is at a uniform temperature of 0 °C and, therefore, there is no heat conduction through it[8].

The kinetic heating by the airstream, Q_v, is a relatively small term (except at aircraft speeds), but, since it is easily parameterized by

$$Q_v = \frac{h r v^2}{2C_p} \tag{3.10}$$

it is usually included in the heat balance. Here h is the convective heat transfer coefficient, r is the recovery factor ($r = 0.79$ for a cylinder), v is the wind speed and C_p is the isobaric specific heat capacity of air. Kinetic heating by the impinging droplets is insignificant and is ignored. The convective heat transfer is

[8] This said, there can be heat conduction out of a spongy ice layer if the substrate is sufficiently cold or if there is heat conduction through the substrate. This will lead to freezing of some of the entrapped liquid in the spongy ice.

$$Q_c = h(t_s - t_a) \tag{3.11}$$

where t_s is the temperature of the icing surface, assumed to be the equilibrium freezing temperature ($t_s = 0\,^\circ\mathrm{C}$ for pure water) and t_a is the air temperature.

The evaporative heat transfer is parameterized as

$$Q_e = \frac{h\varepsilon L_e}{pC_p}(e_s - e_a) \tag{3.12}$$

where ε is the ratio of the molecular weights of water vapour and dry air ($\varepsilon = 0.622$), L_e is the specific latent heat of vaporization at $0\,^\circ\mathrm{C}$, e_s is the saturation water vapour pressure at $0\,^\circ\mathrm{C}$, e_a is the ambient vapour pressure in the airstream and p is the static air pressure. Here, e_s is a constant (6.17 hPa) and e_a is a function of the temperature and relative humidity of the airstream.

The term Q_l arises from the temperature difference between the impinging droplets and the icing surface

$$Q_l = FC_w(t_s - t_d) \tag{3.13}$$

where C_w is the specific heat capacity of water and t_d is the temperature of the droplets at impact.

For cloud and fog droplets, $t_d = t_a$ is likely a very good assumption. This assumption is usually also made for supercooled rain and drizzle drops, although they may not have fully adjusted to the air temperature in the surface layer on impingement.

The heat loss due to long wave radiation may be linearized as (Makkonen 1981)

$$Q_s = \sigma a(t_s - t_a) \tag{3.14}$$

where σ is the Stefan-Boltzmann constant (5.67×10^{-8} Wm^{-2} K^{-4}) and a is a radiation linearization constant (8.1×10^7 K^3). This equation takes into account only long-wave radiation and assumes emissivities of unity for both the icing surface and the airstream. Shortwave radiation from the sun is neglected, because atmospheric icing tends to occur in cloudy weather with much diminished (though possibly not negligible) solar irradiances.

Using the parameterizations of Eqs. 3.9, 3.10, 3.11, 3.12, 3.13, 3.14 in the heat balance, Eq. 3.8, and solving for the accretion fraction results in

$$\alpha_3 = \frac{1}{F(1-\lambda)L_f}\left[(h+\sigma a)(t_s - t_a) + \frac{h\varepsilon L_e}{C_p p}(e_s - e_a) - \frac{hrv^2}{2c_p} + FC_w(t_s - t_d)\right] \tag{3.15}$$

So far we have said nothing about determining the convective heat transfer coefficient, h, in Eqs. 3.10, 3.11, 3.12 and 3.15. There are standard methods to estimate both local and overall h for simple, smooth objects of various sizes and shapes

(Schlichting 1979). In recent years, Computational Fluid Dynamics (CFD) has been used very successfully by aircraft icing modellers, to estimate the local heat transfer coefficient over aircraft components and indeed entire airframes. In most wet growth icing models, starting from the very first ones (Ludlam 1951; Imai 1953; Messinger 1953), one has assumed that the heat transfer coefficient of an uncontaminated cylinder or airfoil represents the icing heat transfer well enough. However, even for simple shapes, roughness of the ice surface complicates the problem and can enhance the local heat transfer considerably. The effect of surface roughness on h has been studied theoretically and computationally (Makkonen 1985; Szilder et al. 2005), and these methods and results can be incorporated into an icing model. However, the factors that determine and control the roughness of an accreting surface are presently poorly understood, so that empirical estimates of the roughness scale parameters are still necessary (Makkonen et al. 2001).

With an estimate of h, Eq. 3.15 can be used to determine the accretion efficiency α_3, and thereby the rate of icing in Eq. 3.1. It should be noted that Eq. 3.15 has been formulated as a means to determine the global accretion efficiency of the entire ice accretion, and hence it is written in terms of the mass flux density of the impinging particles, F. Nevertheless, it is also valid locally on an icing surface, provided that F represents both the directly impinging local mass flux and the runback mass flux from other parts of the surface. With surface liquid flow, the mean temperature of the total water flux will differ from the temperature of the droplets in the air, and this must be taken into account, typically by formulating and solving the dynamics and thermodynamics of the surface liquid (Myers et al. 2002). In order to predict not only the overall mass of the accretion, but also its shape and spatial distribution, aspects of formulating the local heat balance have been included in some icing models (Lozowski et al. 1983; Lozowski et al. 1987; Szilder et al. 1987; Makkonen et al. 2001).

We noted earlier that a criterion for wet snow accretion is that the wet bulb temperature should exceed 0 °C, while heavy snow is falling. Under these circumstances, the direction of the heat flux is from the air to the snow deposit. Thus, during wet snow accretion, melting of the snow may occur and this may reduce the accretion rate. The thermal balance required to estimate the accretion efficiency, α_3, in such a case, may be determined in an analogous way to our derivation of Eq. 3.15 (Grenier et al. 1986; Admirat and Sakamoto 1988; Sakamoto 2000). The problem of determining α_3 for wet snow is, however, very complex, because the excess melt water may, instead of dripping, be drawn into the snow matrix by capillary forces. Moreover, the extra water in the snow may destroy the network of interconnected ice grains that holds together the snow structure (Colbeck and Ackley 1983). Consequently, it is likely that the upper temperature limit of wet snow accretions is set by shedding, arising from the collapse of the integrity of the snow accretion, rather than by overall melting.

One method to de-ice electrical conductors is to heat them using a higher than usual current. The model described above and in Eq. 3.8 then needs to be modified by adding a heating term Q_h to the right-hand-side. Modelling of the time required to de-ice the conductors by melting is then possible, in both icing and post-icing conditions (Personne and Gayet et al. 1988; Huneault et al. 2005). In the latter case, solar radiation and conduction of heat within the accretion must be considered.

3.3.5 Icicle Growth

We have already alluded to the necessity of dealing with surface fluxes of unfrozen water in icing modelling. The traditional approach has been to ignore the dynamics of the surface flow and apply mass conservation only (i.e. the continuity equation). This approach typically suffices for 2D models, but it fails in 3D. Moreover, even for 2D models, one needs to solve for the flow dynamics in order to determine the flow thickness, which is a critical factor in determining the surface temperature, t_s. Ultimately, unfrozen surface liquid that does not freeze at some location must be shed, either as a result of gravity or wind stress. Gravitational shedding can lead to icicle formation, which is a particularly interesting and practical consideration for icing modelling.

Sometimes icicles are formed during freezing precipitation, because air temperatures are typically near freezing (see Figs 3.2 and 3.3). Until recently, icicles were ignored in icing models, or it was considered that icicles grow only from the freezing of run-off water from the rest of the accretion. For example, in Goodwin et al. (1983), icicles are neglected, and only the overall heat balance of the object is considered in determining the accretion efficiency, or the accretion efficiency is assumed to be unity (Goodwin et al. 1983). However, a recent numerical simulation of glaze icing in freezing rain with icicle growth (Makkonen 1998) has, somewhat surprisingly, shown that when the air temperature is high enough for icicle growth, the total load may be much higher than at any other temperature, because of the capture of airborne particles by the icicles themselves. Thus, while icicle growth in freezing rain is rarely observed, it is still a relevant factor for extreme icing events. Consequently, we discuss icicle modelling below.

When there is a source of water at the root (top) of the icicle, a liquid film forms on the icicle surface and flows towards the tip due to gravity or wind drag. Water spreads effectively on an icicle surface, so that a liquid water film tends to cover the entire icicle surface, unless the flux of water is extremely small – of the order of 10 μl/s or less. The thickness of the liquid film on the icicle surface during its growth is typically 40–100 μm (Maeno and Takahashi 1984).

When water flows down the icicle surface, a part of it freezes but, if the water supply is sufficient, a pendant drop forms at the tip of the icicle. The pendant drop grows until it reaches a certain size, then falls, whereafter another drop starts to grow. Measurements under calm conditions show that the diameter of the pendant drop and the diameter of the tip of the icicle lie between 4.8–5.0 mm, regardless of growth conditions (Maeno and Takahashi 1984).

When an icicle grows, the latent heat of fusion released in the freezing of ice beneath the water film must be removed from the ice/water interface. The rate of heat loss from the surface to the environment, therefore, controls the growth rate of the ice. This situation is analogous to "wet growth". The heat loss from the surface to the air is by thermal convection and evaporation, since icicles tend to grow in light wind conditions. Outgoing radiation is small and heat conduction to the interior of the icicle is negligible. If the pendant drop supercools significantly, it may release heat to the tip of the icicle.

Makkonen (1998) has shown that the observed vertical growth rate of an icicle is much higher than its horizontal growth rate, because the vertical dimension of the tip grows fast, but not the mass of ice at the tip. In other words, the tip of the icicle grows vertically as a thin shell of ice enclosing unfrozen water. The growing tube of ice within the pendant drop at the icicle tip has a wall thickness of less than 100 μm, depending on the growth rate (Maeno et al. 1994). As a result of this growth mechanism, the surface area from which the heat loss leading to vertical growth takes place (the hemispherical surface of the pendant drop) is much larger than the surface area of the ice that is growing vertically and releasing latent heat of fusion (the base of the thin hollow ice shell). The situation is shown schematically in Fig. 3.8.

The liquid water in the interior of an icicle cannot be frozen by processes that transfer heat downwards or horizontally, because there can be no temperature gradient in these directions, since the temperature of the walls and tip of an icicle is at

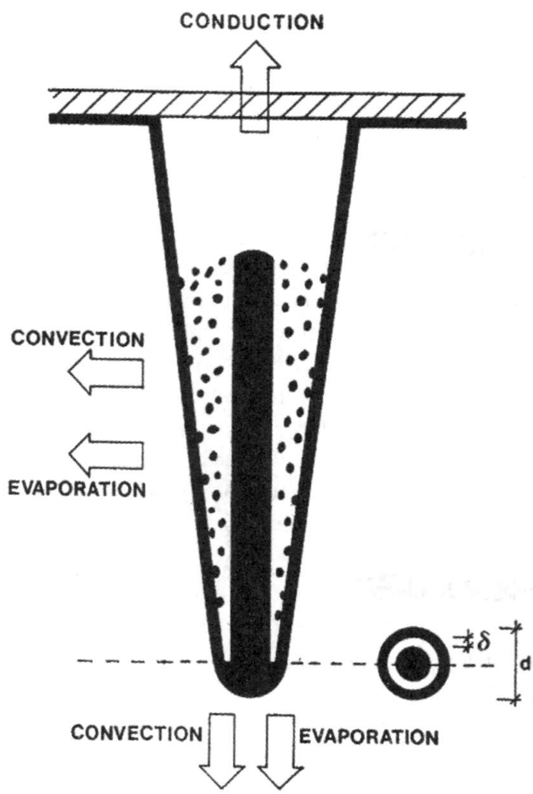

Fig. 3.8 Schematic cross-section of a growing icicle (Makkonen 1988). *Dark* is liquid water and *white* is ice. Dimensions are exaggerated (Reproduced from the Journal of Glaciology with permission of the International Glaciological Society)

the freezing point. Consequently, the only mechanism by which the entrapped liquid water can be frozen is heat conduction upwards through the root of the icicle. By this process, the liquid water inside the icicle freezes slowly from the root downwards (Fig. 3.8). The conduction through the root of the icicle does not affect the icicle growth rate, as it does not contribute to the heat balance of the icicle surface. Conduction determines only how fast the interior of the icicle changes from liquid to solid. Once the thin cover of ice at the tip of an icicle has grown vertically, it also starts to grow horizontally outwards, due to heat loss to the air at the icicle walls. The icicle then grows in width. As the ice on the walls grows beneath a supercooled water film, the ice/water interface assumes a dendritic sub-structure and entraps some liquid water into a spongy ice matrix.

A comprehensive numerical model of icicle growth, based on the physical description above, was developed by Makkonen (1988) and verified by Maeno et al. (1994). The model simulations point out that the growth rate of an icicle, under constant conditions, is strongly time-dependent. The elongation rate increases with time under fixed atmospheric conditions and water-supply rate. This is mainly due to the decreasing drip rate, as more of the supply water is frozen on the walls as the icicle grows. The growth rate in width slowly decreases with time, mainly because the heat-transfer coefficient decreases with increasing icicle diameter. Under fixed conditions, the growth rate of icicle mass considerably increases with time, until the icicle grows so large that there is no drip and no growth in length. By that time, all the supply water is collected by the icicle, and the mass-growth rate is constant.

The icicle model predicts that an icicle's length growth rate should decrease with increasing water-supply rate, while the icicle's radial growth rate is only slightly affected by it. The net result is that the mass of an icicle decreases with increasing water-supply rate, provided that the water supply is always sufficient for the icicle to elongate. This result is contrary to what one might intuitively expect, but it is in accordance with data from laboratory tests (Maeno et al. 1994). The explanation of decreasing elongation rate with increasing water supply rate is the warming caused by drip water, since the pendant drop leaves the tip more supercooled than it arrives.

The model suggests that there is no upper limit for the size of an icicle, provided conditions for its continuous growth prevail. Under natural conditions there are, however, several factors that limit icicle size. If the water-supply rate is high, the icicle initially grows slowly and is unlikely to grow large. On the other hand, if it is low, then the icicle soon ceases to elongate due to no water flow to the icicle tip. Really big icicles can, therefore, form only under conditions in which the water supply rate is at first small and then increases. Another limiting factor is that very low air temperatures, favouring rapid icicle growth, seldom occur during freezing rain (Stallabrass 1983). They also do not promote melting of already formed ice or snow accretions above, that might produce the necessary melt water flux for icicles to grow.

This icicle model has been incorporated into a freezing rain model for a cable (Makkonen 1998), by taking into account the fact that the icicles themselves also collect water droplets directly. This feedback effect is very significant at high wind speeds, because it provides the above-mentioned continuous increase of the water supply.

The spacing of icicles, i.e. the number of icicles per unit length of a conductor cable, is an important factor in determining ice load (see Fig. 3.2). Not only is the total load determined by the number of icicles, but also their size strongly depends on the spacing, which controls how much run-off water enters the root of each icicle. The icicle spacing problem was analytically solved by Makkonen and Fujii (1993), who also verified the solutions using icing wind tunnel experiments. Their results, confirmed by a more recent analysis by De Bruyn (1997), show that the spacing of icicles is slightly more than 2 cm, regardless of the growth conditions. It is, therefore, not uncommon for large icicles to grow into contact with each other. They may then form a continuous sheet of ice, as seen in Fig. 3.3.

While the model of Makkonen (1988) is a numerical icicle simulation algorithm, based on the fundamental physics, other types of icicle models have also been developed. Szilder and Lozowski (1994) suggest a simple analytical model and Szilder and Lozowski (1995b) propose a three-dimensional, random walk model. In the latter model, water flow along the icicle surface is divided into fluid elements, which follow a stochastic path towards the icicle tip. During its motion, an element may freeze at lattice positions, according to a specified freezing probability. Some physics (e.g. following the heat balance approach discussed in Section 3.3) is used to determine the freezing probabilities. The model of Szilder and Lozowski (1995b) is particularly interesting because it is not limited to any specific object geometry (see also Szilder 1995a).

3.4 Numerical Modelling

3.4.1 Physical Models

Solving for the icing rate analytically from Eqs. 3.1 and 3.15 is not practical, because non-linear equations for the dependence of the specific heats and the saturation water vapour pressure on temperature, as well as a complex procedure for determining h, are involved. Numerical methods must also be used because icing is a time-dependent process, so that the object dimension (A in Eq. 3.1) changes during icing, and this change affects the collision efficiency α_1 and the accretion efficiency α_3 (via the heat transfer coefficient h). Furthermore, a change in the ice accretion size and shape may cause the icing process to transform from dry growth to wet growth, even under constant atmospheric conditions (Makkonen 1984b). All this makes the process of icing rather complicated and challenging to model.

Modern personal computers provide ample means to readily obtain elegant results from complex icing models (e.g. Finstad et al. 1988a; Fu 2004). In order to simplify the modelling, the problem of the changing accretion shape may be avoided by assuming that the ice deposit maintains its cylindrical geometry. Based on field data, this is a particularly reasonable assumption for rime icing and wet snow accretions on power line cables (Dranevic 1971).

Time-dependent numerical models of icing also require modelling of the density of the accreted ice, since the ice volume varies inversely with its density. Rime icing can be simulated numerically using ballistic models (Personne and Duroure 1987; Gates et al. 1987; Szilder 1993). Such stochastic models of ice accretion microstructure have been expanded to include rotating objects (Personne et al. 1990) and three-dimensional modelling (Porcu et al. 1995; Szilder and Lozowski 1996).

The ballistic modelling approach, where the impinging droplets are supposed to remain spherical at impact, cannot account for the non-spherical shape of natural frozen droplets at high ice surface temperatures and impact speeds (Macklin and Payne 1968; Macklin and Payne 1969). The droplet spreading process (Obreiter 1987) is difficult to model, and, therefore, empirical formulations for rime density are mostly used in rime icing modeling. Such equations have been proposed by Macklin (1962), Jones (1990) and Levi et al. (1991). For example, the following best-fit equation, based on wind tunnel experiments by Makkonen and Stallabrass (1984), is adequate for the density ρ of rime ice (g cm^{-3}) in dry growth, on a slowly rotating cylinder

$$\rho = 0.378 + 0.425 log R - 0.0823 (log R)^2 \tag{3.16}$$

Here, R is the Macklin (1962) parameter, defined by

$$R = -\frac{v_0 d_m}{2t_s} \tag{3.17}$$

where v_o is the droplet impact speed for the median volume droplet of diameter d_m, and t_s is the surface temperature of the accretion in degrees Celsius. Equations to calculate the droplet impact velocity, v_o, can be found in Finstad et al. (1988a). The surface temperature, t_s, must be solved numerically from the heat balance equation. This can be done by using Eq. 3.8 and solving Eq. 3.15 for t_s, setting $\alpha_3 = 1$. However, because the accretion rate is often too small to release enough latent heat of freezing to significantly heat the icing surface, t_s can, in many cases, be approximated by the air temperature, t_a. For an accurate simulation of the shape of the accretion, rime density may also be calculated on the icing surface (Bain and Gayet 1983; Finstad and Makkonen 1996; Szilder and Lozowski 1996).

For glaze ice (wet growth) and icicles, density variations are small and a value of 0.9 g cm^{-3} can be assumed. For wet snow, quantitative estimation of the density is uncertain at present. Some empirical results show that the density of wet snow depends mainly on the air temperature (Sakamoto and Ishihara 1984), while others suggest that it depends on wind speed (Sakamoto 1993). Apparently, local climatic conditions play a significant role in these relationships (Admirat 1988), perhaps due to different typical precipitation intensities during accretion. Sakamoto (2000) has made an effort to combine all these variables into an empirical equation for the density of wet snow. In practical modelling, one usually has to assume a constant empirical value of the wet snow density, e.g. 0.4 g cm^{-3} (Koshenko and

Bashirova 1979), although the scatter is large in natural conditions (Eliasson and Thorsteins 1996).

When the above estimates of the density of accretions are included, a numerical model can be developed to simulate time-dependent icing of an object. Various physical phenomena can be included in the model as sub-routines and run selectively, according to the input data and the state of the simulated process. Calculation progresses in a step-wise manner. Since accreted ice amounts depend on many atmospheric parameters, it is possible, under certain conditions, to perform icing modelling in reverse. This then allows one to estimate these parameters from icing measurements. For example, reverse modelling from icing data, initially proposed by the pioneers of icing modelling (Langmuir and Blodgett 1946), and subsequently developed by Rogers et al. (1983), Howe (1991) and Makkonen (1992), provides more accurate estimates of the median volume droplet diameter and liquid water content of supercooled clouds than any other measurement method.

A real structure, such as a lattice tower, usually consists of small structural members of different sizes. Modelling icing on such a complex structure may be done by breaking the structure into an ensemble of smaller elements, calculating the ice load separately for each of them, and finally summing up the results to get the total ice load. One may then need to take into account situations where the elements shadow each other, where there is run-off from one element to another, or where the simulated ice on them grows together, forming a single, big element.

So far, we have discussed the simulation of the total ice load on an object. Numerical modelling makes it possible to simulate not only the overall collision, accretion and sticking efficiency of objects, but also their local values along the exposed surfaces. Thus accretion shape can be simulated. This is of less importance for general ice load estimates, but it is very useful in those applications where the aerodynamic forces, as modified by the accreted ice, are significant. These include aircraft wings and wind turbine blades, as well as aerial cables, due to the possibility of aerodynamically induced galloping.

Modelling of the local collision efficiency requires simulation of the air flow over ice accretion shapes and the simulation of droplet trajectories in detail. Several such models (Poots and Rodgers 1976; Ackley and Templeton 1979; McComber and Touzot 1981; Scott et al. 1987; Jones and Egelhofer 1991; Shin et al. 1994; Finstad and Makkonen 1996; Makkonen et al. 2001) have been developed. Most of them are designed specifically for airfoils (Gent et al.2000).

The coalescence efficiency, relevant for wet snow accretion modelling, is poorly known in general, and hence its local values have often been based on ad-hoc assumptions, such as a cosine law (Poots and Skelton 1995). Rather detailed numerical models of wet snow accretion on power lines have been developed based on such assumptions (Skelton and Poots 1991; Poots 1996; Poots 1998), but, because these models are based on very simplistic assumptions about the dependence of α_2 on atmospheric conditions, the use of such detailed modelling is of limited value in practice. Sakamoto and Miura (1993) and Sakamoto (2000) have developed a comprehensive empirical model for wet snow accretion, which seems to have some prediction skill (Kitashima et al. 2005). Overall, however, problems related to determining the coalescence efficiency of wet snow, hamper the estimation of snow loads on cables significantly.

The snow mass concentration in the air, w, (see Eq. 3.1) must, of course, be estimated for snow accretion modelling. It depends on the fall speed of the snowflakes, for a given (measured) precipitation rate. This dependence produces an additional complexity in the modelling, because the fall speed depends on the shape, size and wetness of the snow particles. One way to circumvent this problem is to try to correlate the snow content with visibility and to use the latter as the predictor for wet snow accretion. Based on this idea, Makkonen (1989) proposed a simple equation for the wet snow accretion rate R_w (kg m^{-2} s^{-1})

$$R_w = 2.1\lambda^{-1.29} \tag{3.18}$$

where λ is the visibility in snow, in metres.

The local accretion efficiency of wet growth icing can be simulated using empirical, boundary-layer or full Navier-Stokes models for local heat transfer (e.g. Lozowski et al. 1983; MacArthur 1983; Makkonen 1985; Szilder et al. 2005). A major complexity arises when there is runback water on the surface. Numerical methods to deal with surface liquid fluxes have been developed recently, particularly for airfoils (Lozowski et al. 1999; Gent et al. 2000; Myers et al. 2002; Vargas 2007).

Various novel methods, such as stochastic modelling, have also been developed to predict the shape of ice accretions (e.g. Lozowski et al. 1983; Szilder et al. 1987; Szilder and Lozowski 1995a), but they will be of limited practical use in power network icing, until the factors α_1, α_2 and α_3 in Eq. 3.1 can be predicted locally for complex shapes. Nevertheless, these models, discussed below, are a significant step forward, because the shape of the accretion is very important in determining the wind drag and lift on iced structures.

3.4.2 Enhanced Physical Models

We discuss here enhanced physical models that were originally developed in the 1980s for the modelling of aircraft icing (Morency et al. 1999; Bourgault et al. 2000). It is only very recently that such models have been developed specifically, and independently, for modelling icing on cables (Fu 2004; Fu et al. 2005). We will here describe the generalities of enhanced physical models, using aircraft icing models as references, and pointing out some similarities and differences with Fu (2004) along the way.

While all of these models are fundamentally similar to the models by Lozowski et al. (1983) and Makkonen (1984b), inasmuch as they endeavour to represent physical processes explicitly, their physical verisimilitude and computational procedures have generally been substantially enhanced. A number of the models are three-dimensional and time-dependent, in the sense of accounting for the feedback between the growing ice accretion and the multi-phase flow around it. Fu (2004) is time-dependent but two-dimensional. Aliaga et al. (2007) is three-dimensional and removes the quasi-steady approximation by coupling the air and droplet flow with the icing model in an unsteady manner. Typically, these models use modern CFD methods to solve for the airflow, and either Lagrangian or (more efficiently) Eulerian techniques to solve for the local collision efficiency. Fu (2004) computes a potential airflow, with separation, using the boundary element method, and couples it with an integral boundary layer model, to determine the heat transfer distribution over the ice accretion. Some models, however, solve the full Navier-Stokes equations for both the flow and the heat transfer (Szilder et al. 2005). A few icing models also solve the thin film equations for unfrozen surface liquid, typically making various assumptions, e.g. lubrication theory (Wang and Rothmayer 2005) in order to do so (Fu et al. 2005, Myers et al. 2002; Feo and Tsao 2007). Many models, however, still treat the runback of unfrozen liquid using a control volume approach, which works satisfactorily in two dimensions but encounters difficulties in three dimensions. Without an explicit thin film, however, the model accretion surface temperature, under glaze icing conditions, can be in error by several degrees. One of the consequences of surface film flow is a rough ice surface. Attempts to simulate the roughness of ice have also been made in some recent models (Arnold et al. 1997; Makkonen et al. 2001).

Early physical models consisted of only a few hundred lines of code, typically written in BASIC. Codes for enhanced physical models are typically much lengthier, and this leads to additional challenges beyond those of representing the physics. Code validation, code maintenance, code updating, code interfacing and visualization of the results are among them. Fu (2004) has addressed some of these issues by using object oriented programming in C++.

3.4.3 Morphogenetic Models

Morphogenetic models are a very different class of icing model that adopt a unique approach to representing ice formation physics. They were originally conceived by Szilder (1992) as a way to estimate ice properties, such as density, and to deal with complex, discrete ice accretion structures such as rime feathers and icicles. So far, at least, the explicit simulation of 3D rime feathers and icicles has defied even the most advanced enhanced physical models. Yet they occur quite naturally in morphogenetic models.

The roots of morphogenetic modelling can be found in other fields of endeavour, including cellular automata and discrete particle methods in fluid dynamics, for example in lattice gas hydrodynamics – LGH (Rivet and Boon 2001). Analogous discrete particle models have been introduced into thin film nano-engineering (Smy et al. 2001), and there are some analogies with algorithmic botany (Prusinkiewicz 2000). Unlike LGH, however, morphogenetic modelling has not yet been shown to be fully equivalent to solving the classical physico-mathematical equations that govern icing, along with their appropriate initial and boundary conditions. As a result, many morphogenetic modellers of icing prefer to use the term *emulation* rather than

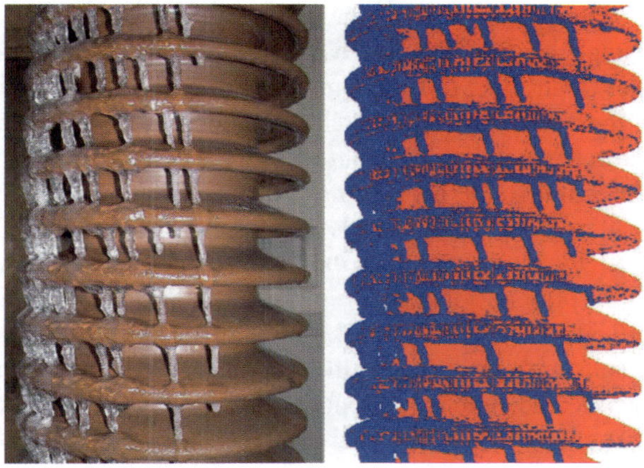

Fig. 3.9 Example of morphogenetic model prediction (*right*). More than 1,000,000 particles were used. Ice accretion produced in the CIGELE Freezing Precipitation Simulation Laboratory (*left*) (Reproduced by permission of Wladyslaw Rudzinski, University of Alberta)

simulation, to describe the results of morphogenetic modelling. And the results can be quite impressive (see Rudzinski et al. (2005), Fig. 3.9, and Lebatto et al. (2005), for example). Nevertheless, morphogenetic modelling is still in its infancy, and in the future, rigorous theoretical underpinnings will need to be developed, to establish the physical and mathematical equivalence of morphogenetic emulations and natural icing. For now, however, we content ourselves with describing their similarities. Morphogenetic models have been produced in two-dimensional and three-dimensional versions. We describe them below, as if they were three-dimensional.

The essence of morphogenetic modelling is that an ice accretion is built up using discrete particles, one at a time. Depending upon their size, these discrete particles can be thought of either as individual droplets or as ensembles of droplets that behave in unison. A model can be lattice free, but it is more typical to construct the model on a three-dimensional, rectangular lattice with cubic cells. The cells may be empty or occupied by substrate or liquid/solid particles (henceforth simply particles). Each cell holds a single particle. Boundary conditions for the problem are established by first filling appropriate cells with substrate particles and then specifying an algorithm that prevents liquid/solid particles from moving away from the substrate or into it, unless specific requirements are fulfilled that allow them to *drip* or to seek an internal *cradle location*. Initial conditions are established by specifying the impact location of a particle. This is done, as in physical models, by solving the Lagrangian trajectory equation for the particle, or by using a parameterization or Eulerian shortcut to determine the local collision efficiency distribution.

Once a particle has impacted, it begins a solitary random walk. Instead of solving a Lagrangian equation for particle motion on the surface, a particle's behaviour is determined stochastically, under the influence of certain behavioural tendencies. Since the acceleration of surface liquid flow is typically small, its fluid dynamics consists of quasi-equilibrium behaviour, involving a balance of gravitational, viscous, surface tension and wind stress forces. In addition, heat transfer can lead to freezing.

The combination of these processes is emulated in morphogenetic models by using a Monte Carlo approach, which gives rise to a "random walk" for each particle. Controlling this random walk are probabilities of motion that are related to the force balance and a probability of freezing related to heat transfer. Very importantly, particles may leave the surface by dripping, provided that they satisfy conditions that emulate the pendant drop formation that gives rise to icicles. This algorithm leads to the growth of icicles in morphogenetic models. For now, however, shedding from locations other than the tip of an icicle is not allowed.

At present, some of the microscopic model parameters are estimated using theoretical considerations, while others are deduced empirically by comparing the model results with experiments. In addition to simulating the detailed morphology of complex ice accretions, morphogenetic models have an intrinsic stochastic variability that emulates natural variability. Having been developed for simulating power network equipment icing, morphogenetic models are now being adapted for simulating aircraft icing (Szilder and Lozowski 2002).

3.5 Conclusions

The twentieth century witnessed two stages in the development of models to estimate the accumulation of ice on power network equipment. The first stage consisted of analytical models, expressed as simple equations, which could be readily solved by hand or with a calculator. In order to achieve simplicity, these models relied on strong assumptions or constraints about the nature of the icing process. Some of them also incorporated experimentally based empiricisms. For all of them, the goal was to predict bulk ice accretion properties such as ice load. They did not consider details of the icing process.

The second stage, consisting of simple numerical models, began around 1980. New models were developed which endeavoured to account for the physical details of the icing process. These models required personal computers for their implementation. Nevertheless, numerous assumptions and simplifications were still made in order to keep the computational problem tractable.

With the advent of ever increasing computational power, the twenty-first century has seen the development of yet another generation of models which we may call "supercomputer models". Some of these numerical models are by far the most complete in terms of accounting for the physical processes of icing. Others adopt a radically different approach to icing simulation. Nevertheless, the full potential of these third-stage models can only be fully realized and utilized, when quality controlled field and laboratory data are available to verify them. Sadly, it is the general unavailability of field data specifying both environmental conditions and icing characteristics that has hampered widespread and thorough verification of icing models (Dobesch et al. 2005; Fikke et al. 2007).

Many fundamental aspects of the state-of-the-art theory of ice accretion on structures have indeed been successfully verified (Macklin and Payne 1968;

Makkonen and Stallabrass 1984; Gates et al. 1986; Makkonen and Stallabrass 1987; Maeno et al. 1994; Shin et al. 1994; Lu et al. 1998; Makkonen 1998; Vargas 2007). However, there remain several poorly understood areas that require more verification and development. These are outlined below.

3.5.1 Collision Efficiency

A major uncertainty is involved when the collision efficiency is very small ($\alpha_1 < 0.1$). In such cases, the theory tends to predict too small values of α_1 (Lehtonen et al. 1986; Personne and Gayet 1988; Yano 1988). It has been suggested that this may occur because of roughness elements on the surface, which act as individual collectors (Personne et al. 1988), or because of condensation of water vapour (Makkonen 1992), and because of omission of the history term in the equations of droplet motion (Oleskiw and Lozowski 1983). However, when α_1 is small, the icing rate is also small, so that this problem does not generally hamper the estimation of extreme ice loads for structural design. On the other hand, when the size of the structure and the icing duration are very large, the growth rate and total ice load may be substantial, even at low α_1. Estimates of icing for very large objects, particularly at low wind speeds, should, therefore, be made with caution. There is not much hope of improving the estimation methods in this respect, because small values of α_1 are so sensitive to changes in the droplet size, that its accurate determination is impossible, due to errors in measuring the median volume diameter of cloud droplets. Thus, empirical methods may still need to be used in the future, when estimating rime icing on large structures.

Another relatively poorly known aspect of ice accretion theory is the effect of droplet trajectory angle on collision efficiency. This uncertainty is related to difficulties in modelling the air flow and droplet trajectories around objects with very complex shape (insulator strings for example). There are also some unsolved problems that are of great practical importance that should yield to modelling with modern methods. These include the effect of wind direction relative to power cables. So far, only very approximate empirical equations are available for estimating this important effect (Nikiforov 1983).

3.5.2 Coalescence Efficiency

Estimation of the coalescence efficiency, α_2, of wet snow is presently quite inaccurate, so that Eq. 3.7 should be seen only as a first approximation, until more sophisticated methods have been developed. Such developments could be based on detailed heat balance considerations of falling snowflakes following Matsuo and Sasyo (1981). Until then, the theoretical models of wet snow accretion (Skelton and Poots 1991; Poots 1996), which are otherwise quite sophisticated, but ignore the crucial coalescence efficiency problem, cannot be effectively used. For the present, simple empirical models (Admirat and Sakamoto 1988; Makkonen 1989; Sakamoto and Miura 1993), that at least try to estimate α_2 are to be preferred.

3.5.3 *Surface Roughness*

Theoretical calculation of glaze formation is relatively reliable, provided that the model has the correct input. The weakest part of glaze ice modeling is presently the lack of data on ice surface roughness for heat transfer calculations. Further studies should be made on the kinetics of the icing surface, along the lines of Olsen and Walker (1986) and Hansman and Turnock (1988). When icicles may contribute to the ice load, a model of icicle growth needs to be incorporated into the modelling framework.

3.5.4 *Legacy Models*

It is important to point out that many current glaze icing models for power lines are conceptually suspicious or include errors of various kinds (Makkonen 1998). Unfortunately, some of these models are still in widespread use, for example, as parts of national building codes. The recommended simple model to be used for estimating glaze ice loads on overhead lines without icicle growth is that of Goodwin et al. (1983), as shown by Makkonen (1998) and Lu et al. (2000).

3.5.5 *Ice Shedding and Melting*

Some specific properties of icing objects may hamper the modelling of ice loads on them. For example, while the diameter (Makkonen 1986) and the torsional stiffness (McComber 1984; Finstad et al. 1988a; Poots 1996) of a cable can readily be taken into account in the modelling of icing, their effects on ice shedding cannot. Ice shedding mechanisms related to cable twisting are inadequately understood, considering that they appear to be the primary cause of different ice loads on cables with different torsional properties (Holodov and Popov 1976; Govoni and Ackley 1983). Studying this problem in the future will be a challenging task, since there is strong feedback between the ice accretion growth and twisting of the cable. This is not only because of the aerodynamics of the accretion, but also because the asymmetrical ice load itself initiates torsional motion, due to increased cable tension (McComber et al. 1995). Further complications regarding icing of power line cables include the heating effect of an energized cable and the effect of the electric field on droplet trajectories and ice properties (Phan and Laforte 1981; Farzaneh 1999).

3.5.6 *Complex Structures*

When modelling icing of complex structures, such as lattice transmission line towers, some components of the structure may be sheltered from ice accretion by other components. Also, different parts of the structure may completely freeze together, whereafter they should be modelled as a single object. Such aspects must be considered individually for each structure. Due to the difficulties in doing this numerically, attempts have been made to model the icing of complex structures physically, using small-scale icing experiments (Makkonen and Oleskiw 1997).

3.5.7 Environmental Data

As to the use of theoretical icing models in predicting design ice loads on structures, the major problem is the input requirement. The median volume droplet size and liquid water content, which are not routinely measured, are less significant when considering atmospheric glaze icing (Makkonen 1981), but they critically affect rime icing. Obtaining correct wind velocity data is often uncertain, because of the adverse effects of icing on anemometers (Makkonen et al. 2000; Fikke et al. 2007). Furthermore, rather accurate values of the air temperature and cloud base height are required, in order to correctly detect the start and end of natural icing episodes (Glukhov 1989; Sundin and Makkonen 1998; Harstveit 2002). Extrapolation of these and other required input parameters to the often remote sites of interest is extremely difficult. The future usefulness of theoretical modelling of icing essentially depends on progress in this area. High resolution numerical weather prediction models may be helpful in this regard.

3.5.8 Icing Life Cycle

Due to the limitations of theoretical modelling, much emphasis in practical design has to be given to empirical models of icing. Considerable progress in this area has taken place recently, but the success of modelling of design loads, with low probability and long return periods, depends on being able to model the disappearance of ice as well. Field studies (Govoni and Ackley 1983; Lehtonen et al. 1986; Sundin and Makkonen 1998) have shown that both melting and shedding, caused by wind and structural dynamics, are important in determining cumulative ice loads. Detailed modelling of these processes (McComber 1990; Scavuzzo et al. 1994) has been limited so far.

3.5.9 Shedding and Wind Drag

In many cases, the relevant design load of a structure depends on the combination of extreme ice and extreme wind. It is not possible to estimate the probability of combined ice and wind loads over long periods, without data on the recurrence of ice. Thus, future safe design of towers, for example, clearly depends on better modelling of ice melting and shedding events. Finally, it is noteworthy that the shedding events themselves may, in some cases, be more damaging to the structure than the static ice load (Jamaleddine et al. 1996; Fekr et al. 1998) and that falling ice is always a hazard to objects and people below (ISO 2001). Another key aspect of modelling the combined ice and wind load is the drag coefficient of an iced object. This aspect should be studied by further experiments (Golikova 1976) and by developing existing theoretical ideas (McComber and Bouchard 1986).

3.5.10 Icing Forecasting

The main practical significance of the research reviewed here is in estimating extreme ice loads for structural design, thereby promoting safe and economical engineering. Another ambitious application, already studied preliminarily (Fuchs et al. 1998; Vassbo et al. 1998; Shao 1999), is to make icing forecasts using gridded, mesoscale numerical weather prediction model output as environmental input to an icing model. Such forecasts will require specific high resolution meteorological models that are able to predict, inter alia, the liquid water content and occurrence of freezing rain. These aspects of icing forecasting are also being intensively studied (Zerr 1997; Reissner et al. 1998; Nygaard et al. 2007).

3.5.11 Ice Control Methods

Icing modelling and forecasting may be applied to optimize anti-icing and de-icing equipment and strategies. Such ice prevention modelling efforts have been made for power line cables by Grenier et al. (1986) and Huneault et al. (2005), and for wind turbines by Marjaniemi and Peltola (1996). These studies, among others, point out the necessity for a deep understanding of icing processes. Uncontrolled heating of a surface, for example, may allow dry ice particles to stick and transform the icing process from dry to wet, causing icicle growth. It may also unfavourably alter the aerodynamics of ice deposits. De-icing at the wrong moment may cause ice to fall and cause danger. It may also initiate a disastrous dynamic response of the structure. Overall, poorly designed or uncontrolled ice prevention may not only increase the ice loads, but also initiate icing events that would not otherwise take place, thereby directly contributing to structural damage.

References

Abley M (1998) The Ice Storm. McLelland & Stewart, Toronto

Ackley SF, Templeton MK (1979) Computer modeling of atmospheric ice accretion. U.S. Army Cold Regions Research & Engineering Laboratory, CRREL Report 79–4

Admirat P, Sakamoto Y (1988) Wet snow on overhead lines: State-of-art. In: Proc 4th International Workshop on Atmospheric Icing of Structures: 8–13

Admirat P, Maccagnan M, de Goncourt B (1988) Influence of joule effect and of climatic conditions on liquid water content of snow on conductors. In: Proc 4th International Workshop on Atmospheric Icing of Structures: 367–371

Ahti K, Makkonen L (1982) Observations on rime formation in relation to routinely measured meteorological parameters. Geophysica 19: 75–85

Aliaga CN, Aube MS, Baruzzi GS, Habashi WG, Nadarajah S (2007) A third-generation in-flight icing code: FENSAP-ICE-Unsteady. SAE 2007 Aircraft & Engine Icing International Conference, Seville, Paper 2007-01-3339

Anon (1999) Facing the Unforseeable. Lessons from the Ice Storm of '98. Les publications du Quebec, ISBN 2-551-18089-9

Arnold K, Tetzlaff G, Raabe A (1997) Modeling of ice accretion on a non-rotating cylinder. Meteorol. Zeitschrift 6: 120–129

Bain M, Gayet JF (1983) Contribution to the modeling of the ice accretion process – Ice density variation with the impacted surface angle. Ann Glaciol 4: 19–23

Blackmore RZ, Lozowski EP (1996) A freshwater spongy spray icing model with surficial structure. In: Proc 7th International Workshop on Atmospheric Icing of Structures: 33–43

Blackmore RZ, Makkonen L, Lozowski EP (2002) A new model of spongy icing from first principles. J Geophys Res 107: D21, AAC9-1-AAC9-15

Bourgault Y, Boutanios Z, Habashi WG (2000) An Eulerian approach to 3-D droplet impingement simulation using FENSAP-ICE. Part I: Model, algorithm and validation. Journal of Aircraft 37: 95–103

Bourgault Y, Beaugendre H, Habashi WG (2000) Development of a shallow water icing model in FENSAP-ICE. Journal of Aircraft 37: 640–646

Choularton TW, Griggs DJ, Humood BY, Latham J (1980) Laboratory studies of riming, and its relation to ice splinter production. Quart J Roy Meteor Soc 106: 367–374

Colbeck SC, Ackley SF (1983) Mechanism for ice bonding in wet snow accretions on power lines. In: Proc First International Workshop on Atmospheric Icing of Structures: 25–30

De Bruyn JR (1997) On the formation of periodic arrays of icicles. Cold Regions Sci Technol 25: 225–229

Diem M (1956) Ice loads on high voltage lines in the mountains. Arch Mat Geoph Biokl, Ser B 7: 84–85 (in German)

Dobesch H, Nikolov D, Makkonen L (2005) Physical processes, modelling and measuring of icing effects in Europe. Österreichische Beiträge zu Meteorologie und Geophysik 34, 71 p (ISSN 1016–6254)

Dong YY, Hallett J (1989) Droplet accretion during rime growth and the formation of secondary ice crystals. Quart J Roy Meteor Soc 115: 127–142

Dranevic EP (1971) Glaze and Rime. Gidrometeorologitsheskoe Izd. Leningrad (in Russian)

Eliasson AJ, Thorsteins E (1996) Wet snow icing combined with strong wind. In: Proc 7th International Workshop on Atmospheric Icing of Structures: 131–136

Farzaneh M (1999) Ice accretions on high voltage conductors and insulators and related phenomena. Phil Trans R Soc Lond A 358: 2971–3005

Fekr MR, McClure G, Hartmann D (1998) Investigation of transmission line failure due to ice shedding effects using dynamic analysis. In: Proc 8th International Workshop on Atmospheric Icing of Structures: 11–16

Feo A, Tsao JC (2007) The water film Weber number in glaze icing scaling. SAE 2007 Aircraft & Engine Icing International Conference, Seville, Paper 2007-01-3295, 12 p

Fikke S, Ronsten G, Heimo A, Kunz S, Ostrozlik M, Persson PE, Sabata J, Waering B, Wichura B, Chum J, Laakso T, Säntti K, Makkonen L (2007) COST 727: Atmospheric icing on structures, Measurement and data collection on icing: State of the art. Publication of MeteoSwiss, No. 75, 110 p (ISSN: 1422-1381)

Finstad KJ, Makkonen L (1996) Improved numerical model for wind turbine icing. In: Proc 7th International Workshop on Atmospheric Icing of Structures: 373–378

Finstad KJ, Fikke SM, Ervik M (1988a) A comprehensive deterministic model for transmission line icing applied to laboratory and field observations. In: Proc 4th International Workshop on Atmospheric Icing of Structures: 227–231

Finstad KJ, Lozowski EP, Gates EM (1988b) A computational investigation of water droplet trajectories. J Atmos Oceanic Technol 5: 160–170

Finstad KJ, Lozowski EP, Makkonen L (1988c) On the median volume diameter approximation for droplet collision efficiency. J Atmos Sci 45: 4008–4012

Fu P (2004) Modelling and simulation of the ice accretion process on fixed or rotational cylindrical objects by boundary layer element method. Ph.D. dissertation, Université du Québec à Chicoutimi

Fu P, Farzaneh M, Bouchard G (2004) Simulation of ice accumulation on transmission line cables based on time-dependent airflow and water droplet trajectory calculations. In: Proc 23rd International Conference on Offshore Mechanics and Arctic Engineering

Fu P, Farzaneh M, Bouchard G (2005) Modelling a water flow on an icing surface. In: Proc 11th International Workshop on Atmospheric Icing of Structures

Fuchs W, Lutkebohmert M, Schickel K-P (1998) Forecast of icing of structures by use of boundary layer model products. In: Proc 8th International Workshop on Atmospheric Icing of Structures: 179–183

Gartzman LB (1987) Probabilities of ice, wind and temperature loads on electrical transmission lines. Gidrometeoizdat, Leningrad (in Russian)

Gates EM, Narten R, Lozowski EP, Makkonen L (1986) Marine icing and spongy ice. In: Proc 8th IAHR Symposium on Ice 2: 153–163

Gates EM, Liu A, Lozowski EP (1987) A stochastic model of atmospheric icing. J Glaciol 34: 26–30

Gent RW, Dart NP, Cansdale JT (2000) Aircraft icing. Phil Trans R Soc Lond A 358:2873–2911

Glukhov VG (1989) Tall building icing on plain territory of the USSR. Gidrometeoizdat, Leningrad (in Russian)

Golikova TN (1976) Aerodymamic properties of ice covered cables. Trudy GGO 379:20–29 (in Russian)

Goodwin EJ III, Mozer JD, Di Gioia AM Jr, Power BA (1983) Predicting ice and snow loads for transmission lines. In: Proc First International Workshop on Atmospheric Icing of Structures: 267–273

Govoni JW, Ackley SF (1983) Field measurements on combined icing and wind loads on wires. In: Proc First International Workshop on Atmospheric Icing of Structures: 205–213

Grenier JC, Admirat P, Maccagnan M (1986) Theoretical study of the heat balance during the growth of wet snow sleeves on electrical conductors. In: Proc 3rd International Workshop on Atmospheric Icing of Structures: 125–130

Hansman RJ, Turnock SR (1988) Investigation of microphysical factors which influence surface roughness during glaze ice accretion. In: Proc 4th International Workshop on Atmospheric Icing of Structures: 139–146

Harstveit K (2002) Using routine meteorological data from airfields to produce a map of icing risk zones in Norway. In: Proc 10th International Workshop on Atmospheric Icing of Structures

Holodov VU, Popov NI (1976) The effect of structural parameters of electric overhead lines on ice loads on cables. Trudy GGO 379: 30–36 (in Russian)

Howe JB (1991) Rotating multicylinder method for the measurement of cloud liquid-water content and droplet size. U.S. Army Cold Regions Research & Engineering Laboratory, CRREL Report 91–2

Huneault M, Langheit C, Caron J (2005) Combined models for glaze ice accretion and de-icing of current-carrying electrical conductors. IEEE Transactions on Power Delivery 20: 1611–1616

Imai I (1953) Studies on ice accretion. Res Snow Ice 1: 35–44 (in Japanese)

ISO (2001) Atmospheric icing of structures. International Standard, International Standards Organization, ISO 12494: 2001(E)

Jamaleddine A, Beauchemin R, Rousselet J, McClure G (1996) Weight-dropping simulation of ice-shedding effects on an overhead transmission line model. In: Proc 7th International Workshop on Atmospheric Icing of Structures: 44–48

Jones KF (1990) The density of natural ice accretions related to non-dimensional icing parameters. Quart J Roy Meteor Soc 119: 599–604

Jones KF, Egelhofer KZ (1991) Computer model of atmospheric ice accretion on transmission lines. U.S. Army Cold Regions Research & Engineering Laboratory, CRREL Report 91–3

Kitashima T, Miura A, Sakamoto Y (2005) A new attempt to estimate wet snow accretion on overhead lines. In: Proc 11th International Workshop on Atmospheric Icing of Structures

Knight CA (1968) On the mechanism of spongy hailstone growth. J Atmos Sci 25: 440–444
Koshenko AM, Bashirova L (1979) Recommendations on forecasting the precipitation and deposition (sticking) of wet snow. Trudy UkrNIGMI 176: 96–102 (in Russian)
Kuroiwa D (1965) Icing and snow accretion on electric wires. U.S. Army Cold Regions Research & Engineering Laboratory, CRREL Research Report 123
Laflamme JN, Periard G (1996) The climate of freezing rain over the province of Quebec in Canada: A preliminary analysis. Atmos Res 46: 99–111
Landau L, Lifshitz EM (1959) Fluid Mechanics. Pergamon Press, London
Langmuir I, Blodgett KB (1946) A mathematical investigation of water droplet trajectories. Pergamon Press, Collected Works of Irving Langmuir, 10: 335–393
Lebatto E, Farzaneh M, Lozowski EP (2005) Three-dimensional morphogenetic model of ice accretion on a non-rotating cylinder. In: Proc 11th International Workshop on Atmospheric Icing of Structures
Lehtonen P, Ahti K, Makkonen L (1986) The growth and disappearance of ice loads on a tall mast. In: Proc 3rd International Workshop on Atmospheric Icing of Structures: 363–368
Lenhard RW (1955) An indirect method for estimating the weight of glaze on wires. Bull Am Meteor Soc 36: 1–5
Lesins GB, List R, Joe PI (1980) Ice accretions, Part I: Testing of new atmospheric icing concepts. J Rech Atmos 14: 347–356
Levi L, Nasello OB, Prodi F (1991) Morphology and density of ice accreted on cylindrical collectors at low values of impaction parameter. I: Fixed deposits. Quart J Roy Meteor Soc 117: 761–782
List R (1977) Ice accretion on structures. J Glaciol 19: 451–465
Lock GSH (1990) The growth and decay of ice. Cambridge Univ Press, Cambridge
Lock GSH, Foster IB (1990) Experiments on the growth of spongy ice near a stagnant point. J Glaciol 36: 143–150
Lozowski EP, Stallabrass JR, Hearty PF (1983) The icing of an unheated, nonrotating cylinder. Part I: A simulation model. J Climate Appl Meteor 22: 2053–2062
Lozowski EP, Gates EM, Makkonen L (1987) Recent progress in the incorporation of convective heat transfer into cylindrical ice accretion models. In: Proc International Symposium on Cold Regions Heat Transfer: 17–24
Lozowski EP, Szilder K, Makkonen L (1999) Computer simulation of marine ice accretion. Phil Trans R Soc Lond A 358: 2811–2846
Lozowski EP, Oleskiw M, Blackmore RZ, Karev A, Kolar L, Farzaneh M (2005) Spongy icing revisited: measurements of ice accretion liquid fraction in two icing wind tunnels. 43rd AIAA Aerospace Sciences Meeting and Exhibit, AIAA-2005-658
Lozowski EP, Makkonen L (2005) Fifty years of progress in modelling the accumulation of atmospheric ice on power network equipment. In: Proc 11th International Workshop on Atmospheric Icing of Structures
Lu ML, Popplewell N, Shah AH, Barrett W, Au A (1998) Mass of ice accretion from freezing rain simulations. In: Proc 8th International Workshop on Atmospheric Icing of Structures: 89–94
Lu ML, Popplwell N, Shah AH (2000) Freezing rain simulations for fixed, unheated conductor samples. J Appl Meteor
Ludlam FH (1951) The heat economy of a rimed cylinder. Quart J Roy Meteor Soc 77: 663–666
MacArthur CD (1983) Numerical simulation of airfoil ice. In: Proc 21st Aerospace Sciences Meeting, AIAA-83-0112
Macklin WC (1961) Accretion in mixed clouds. Quart J Roy Meteor Soc 87: 413–424
Macklin WC (1962) The density and structure of ice formed by accretion. Quart J Roy Meteor Soc 88: 30–50
Macklin WC, Bailey IH (1968) The collection efficiency of hailstones. Quart J Roy Meteor Soc 94: 393–396
Macklin WC, Payne GS (1968) Some aspects of the accretion process. Quart J Roy Meteor Soc 94: 167–175

Macklin WC, Payne GS (1969) The spreading of accreted drops. Quart J Roy Meteor Soc 95: 724–731.
Maeno N, Takahashi T (1984) Studies on icicles. In: General aspects of the structure and growth of an icicle. Low Temperature Sci, Ser A 43: 125–138 (in Japanese)
Maeno N, Makkonen L, Nishimura K, Kosugi K, Takahashi T (1994) Growth rate of icicles. J Glaciol 40: 319–326
Makkonen L (1981) Estimating intensity of atmospheric ice accretion on stationary structures. J Appl Meteor 20: 595–600
Makkonen L (1984a) Atmospheric icing on sea structures. U.S. Army Cold Regions Research & Engineering Laboratory, CRREL Monograph 84–2
Makkonen L (1984b) Modeling of ice accretion on wires. J Climate Appl Meteor 23: 929–939
Makkonen L (1985) Heat transfer and icing of a rough cylinder. Cold Regions Sci Technol 10: 105–116
Makkonen L (1986) The effect of conductor diameter on ice load as determined by a numerical icing model. In: Proc 3rd International Workshop on Atmospheric Icing of Structures: 81–89
Makkonen L (1987) Salinity and growth rate of ice formed by sea spray. Cold Regions Sci Technol 14: 163–171
Makkonen L (1988) A model of icicle growth. J Glaciol 34: 64–70
Makkonen L (1989) Estimation of wet snow accretion on structures. Cold Regions Sci Technol 17: 83–88
Makkonen L (1990) The origin of spongy ice. In: Proc 10th IAHR Symposium on Ice 2: 1022–1030
Makkonen L (1992) Analysis of rotating multicylinder data in measuring cloud droplet size and liquid water content. J Atmos Oceanic Technol 9: 258–263
Makkonen L (1998) Modeling power line icing in freezing precipitation. Atmos Res 46: 131–142
Makkonen L (2007) A model of hoarfrost formation. In: Proc 12th International Workshop on Atmospheric Icing of Structures, Yokohama, CD-ROM
Makkonen L (2008) Self-regulating solidification. J Mater Res (submitted, 2008)
Makkonen L, Fujii Y (1993) Spacing of icicles. Cold Regions Sci Technol 21: 317–322
Makkonen L, Lehtonen P, Helle L (2000) Anemometry in icing conditions. J Atmos Oceanic Technol
Makkonen L, Stallabrass JR (1984) Ice accretion on cylinders and wires. National Research Council of Canada, NRC Tech. Report, TR-LT-005
Makkonen L, Stallabrass JR (1987) Experiments on the cloud droplet collision efficiency of cylinders. J Climate Appl Meteor 26: 1406–1411
Makkonen L, Oleskiw M (1997) Small-scale experiments on rime icing. Cold Regions Sci Technol 25: 173–182
Makkonen L, Laakso T, Marjaniemi M, Finstad KJ (2001) Modelling and prevention of ice accretion on wind turbines. Wind Engineering 25: 3–21
Marjaniemi M, Peltola E (1996) A blade heating element design method. In: Proc Wind Energy Production in Cold Climates, BOREAS III, Finnish Meteorological Institute: 340–354
Matsuo T, Sasyo Y (1981) Melting of snowflakes below freezing level in the atmosphere. J Meteor Soc Japan 59: 10–26
Mazin IP (1957) Physical principles of aircraft icing. Gidrometeoizdat, Leningrad (in Russian)
McKay GA, Thompson HA (1969) Estimating the hazard of ice accretion in Canada from climatological data. J Appl Meteor 8: 927–935
McComber P (1984) Numerical simulation of cable twisting due to icing. Cold Regions Sci Technol 8: 253–259
McComber P (1990) Effects of cable twisting on atmospheric ice shedding. In: Proc 5th International Workshop on Atmospheric Icing of Structures: A2–8
McComber P, Bouchard G (1986) The numerical calculation of the wind force on two dimensional iced structures. In: Proc 3rd International Workshop on Atmospheric Icing of Structures: 353–361
McComber P, Touzot G (1981) Calculation of the impingement of cloud droplets on a cylinder by the finite-element method. Journal of the Atmospheric Sciences 38: 1027–1036

McComber P, Druez J, Savadjiev K (1995) Cable twisting due to atmospheric icing. In: Proc 5th International Offhore and Polar Engineering Conference: 461–468
Messinger BL (1953) Equilibrium temperature of an unheated icing surface as a function of air speed. J Aeronaut Sci 20: 29–42
Morency F, Tezok F, Paraschivoiu I (1999) Anti-icing systems simulation using CANICE. Journal of Aircraft 36: 999–1006
Mulherin ND (1998) Atmospheric icing and communication tower failure in the United States. Cold Regions Sci Technol 27: 91–104
Myers TG, Charpin JPF, Chapman SJ (2002) The flow and solidification of a thin fluid film on an arbitrary three-dimensional surface. Phys Fluids 14: 2788–2803
Nikiforov EP (1983) Icing related problems, effect of line design and ice load mapping. In: Proc First International Workshop on Atmospheric Icing of Structures: 239–245
Nygaard BEK, Kristjansson JE, Berge E, Makkonen L (2007) Using NWP models to simulate in-cloud atmospheric icing episodes. In: Proc 12th International Workshop on Atmospheric Icing of Structures, Yokohama, CD-ROM
Obreiter E (1987) Physics of the Marine Ice Accretion Process. MSc Thesis, University of Alberta
Oleskiw MM, Lozowski EP (1983) The design and testing of a Lagrangian computer model for simulating time-dependent rime icing on two-dimensional structures. In: Proc First International Workshop on Atmospheric Icing of Structures: 59–66
Olsen W, Walker E (1986) Experimental evidence for modifying the current physical model for ice accretion on aircraft surfaces. In: Proc 3rd International Workshop on Atmospheric Icing of Structures: 193–250
Peltola E, Laakso T, Ronsten G, Tallhaug L, Horbaty R, Baring-Gould I, Lacroix A (2004) Specific recommendations for the development of wind energy projects in cold climate. European Wind Energy Conference, European Wind Energy Association
Personne P, Gayet J-F (1988) Ice accretion on wires and anti-icing induced by Joule effect. J Appl Meteor 27: 101–114
Personne P, Duroure C (1987) Modeling of the structure of soft rime. J de Phys C1: 413–449 (in French)
Personne P, Duroure C, Gayet J-F (1988) Effect of the surface roughness on the ice load characteristics during icing with low airspeeds. In: Proc 4th International Workshop on Atmospheric Icing of Structures: 232–235
Personne P, Duroure C, Isaka H (1990) Theoretical study of air inclusions on rotating cylinders. In: Proc 5th International Workshop on Atmospheric Icing of Structures: A2–6
Phan CL, Laforte J-L (1981) The influence of electro-freezing on ice formation on high- voltage dc transmission lines. Cold Regions Sci Technol 4: 15–25
Poots G (1996) Ice and Snow Accretion on Structures. Research Studies Press
Poots G (1998) Aspects of a model for wet-snow accretion on an overhead line conductor. In: Proc 8th International Workshop on Atmospheric Icing of Structures: 185–189
Poots G, Rodgers GG (1976) The icing of a cable. J Inst Maths Applics 18: 203–217
Poots G, Skelton PLI (1995) Simulation of wet-snow accretion by axial growth on a transmission line. Appl Math Modelling 19: 514–517
Porcu F, Smargiassi E, Prodi F (1995) 2-D and 3-D modelling of low density ice accretion on rotating wires with variable surface irregularities. Atmos Res 36: 233–242
Prusinkiewicz P (2000) Paradigms of pattern formation: towards a computational theory of morphogenesis. In: Pattern Formation in Biology, Vision and Dynamics: 91–95
Reissner J, Rasmussen RM, Bruintjes RT (1998) Explicit forecasting of supercooled liquid water in winter storms using the MM5 mesoscale model. Quart J Roy Meteor Soc 124: 1071–1107
Rink J (1938) The melt water equivalent of rime deposits. Reichsamt fur Wetterdienst, Wissenschaftliche Abhandlungen 5, 26 p (in German)
Rivet J-P, Boon J-P (2001) Lattice Gas Hydrodynamics, Cambridge University Press, Cambridge
Rogers DC, Baumgartner D, Vali G (1983) Determination of supercooled liquid water content by measuring rime rate. J Climate Appl Meteor 22: 153–162

Rudzinski WJ, Farzaneh M, Lozowski EP (2005) Full-scale 3D numerical and laboratory simulations of glaze ice accretion on a non-energized station post insulator. In: Proc 11th International Workshop on Atmospheric Icing of Structures
Sakamoto Y (2000) Snow accretion on overhead wires. Phil Trans R Soc Lond A 358: 2941–2970
Sakamoto Y, Ishihara K (1984) An estimating method of snow load on overhead power lines. In: Proc 2nd International Workshop on Atmospheric Icing of Structures: 137–140
Sakamoto Y, Miura A (1993) Comparative study of wet snow models for estimating snow load on power lines based on general meteorological parameters. In: Proc 6th International Workshop on Atmospheric Icing of Structures: 133–138
Scavuzzo RJ, Chu ML, Ananthaswamy V (1994) Influence of aerodynamic forces in ice shedding. J Aircraft 31: 526–530
Schlichting H (1979) Boundary-layer Theory. McGraw-Hill, New York
Scott JN, Hankey WL, Giessler FJ, Gielda TP (1987) Navier-Stokes solution to the flowfield over ice accretion shapes. J Aircraft 25: 710–716
Shao J (1999) Nowcast of temperature and ice on overhead railway transmission wires. J Appl Meteor
Shin J, Berkowitz B, Ghen HH, Cebeci T (1994) Prediction of ice shapes and their effect on airfoil drag. J Aircraft 31: 263–270
Skelton PLI, Poots G (1991) Snow accretion on overhead line conductors of finite torsional stiffness. Cold Regions Sci Technol 19: 301–316
Smy T, Walkey D, Harris KD, Brett M (2001) Thin film microstructure and thermal transport simulation using 3D-Films. Thin Solid Films 39: 88–100
Snitkovskii AI (1977) Prediction of dangerous weather phenomena and prospects for research in this area. Sov Meteor Hydrol 11: 71–80
Stallabrass JR (1983) Aspects of freezing rain simulation and testing. In: Proc First International Workshop on Atmospheric Icing of Structures: 67–74
Sundin E, Makkonen L (1998) Ice loads on a lattice tower estimated by weather station data. J Appl Meteor 37: 523–529
Szilder K (1992) Simulation of ice accretion on a cylinder due to freezing rain. J Glaciol 40: 180–184
Szilder K (1993) The density and structure of ice accretion predicted by a random walk model. Quart J Roy Meteor Soc 119: 907–924
Szilder K, Lozowski EP (1994) An analytical model of icicle growth. Ann Glaciol 19: 141–145.
Szilder K, Lozowski EP (1995a) A new method of modelling ice accretion on objects of complex geometry. Int J Offshore Polar Engin 5: 37–42
Szilder K, Lozowski EP (1995b) Simulation of icicle growth using a three-dimensional random walk model. Atmos Res 36: 243–249
Szilder K, Lozowski EP (1996) Three-dimensional modelling of ice accretion microstructure. In: Proc 7th International Workshop on Atmospheric Icing of Structures: 60–63
Szilder K, Lozowski EP, Gates EM (1987) Modelling ice accretion on non-rotating cylinders – the incorporation of time dependence and internal heat conduction. Cold Regions Sci Technol 13: 177–191
Szilder K, Lozowski EP (2002) A new discrete approach applied to modelling of in-flight icing. Canadian Aeronautics and Space Journal 48: 181–193
Szilder K, McIlwain S, Lozowski EP (2005) Examining the influence of convective heat transfer on in-flight icing using a morphogenetic model. In: Proc 11th Australian International Aerospace Congress
Tammelin B, Säntti K (1996) Estimation of rime accretion at high altitudes – preliminary results. In: Proc Wind Energy Production in Cold Climates Meeting, BOREAS III: 194–209
Vargas M (2007) Swept wing icing physics studies at NASA Glenn Research Center. SAE 2007 Aircraft & Engine Icing International Conference, Seville, Paper 2007-01-3332
Vassbo T, Kristjansson JE, Fikke SM, Makkonen L (1998) An investigation of the feasibility of predicting icing episodes using numerical weather prediction model output. In: Proc 8th International Workshop on Atmospheric Icing of Structures: 343–347

Waibel K (1956) Meteorological conditions of rime deposition on high voltage lines in the mountains. Arch Met Geoph Biokl Ser B, 7: 74–83 (in German)
Wakahama G, Kuroiwa D, Goto D (1977) Snow accretion on electric wires and its prevention. J Glaciol 19: 479–487
Wang G, Rothmayer AP (2005) Air driven water flow past small scale surface roughness. In: Proc 43rd AIAA Aerospace Sciences Meeting and Exhibit, AIAA-2005-0653
Yano K (1988) Studies of icing and ice-snow accretion in Mt. Zao. In: Proc 4th International Workshop on Atmospheric Icing of Structures: 109–113
Zavarina MV, Glukhov VG, Mytariev MN (1976) A method for the calculation of ice loads on high constructions. Zeitsch Meteor 26: 98–104 (in German)
Zerr RJ (1997) Freezing rain: an observational and theoretical study. J Appl Meteor 36: 1647–1661

Chapter 4
Wet Snow Accretion on Overhead Lines

Pierre Admirat

4.1 Introduction

This chapter firstly describes the rapid metamorphosis of dry snowflakes when they reach a positive air temperature, and analyzes the heat exchanges of the transformation into wet snow, during its accretion on a cylindrical object.

The following part deals with the accretion mechanisms of wet snow observed both on a stranded conductor sample in wind tunnel conditions and on real power lines in natural weather conditions.

The last part puts theory and observations in practice so as to calculate the maximum snow overload from forecasted meteorological data, to demonstrate the efficiency and the limits of a passive prevention method, and to map the snow overload hazard on a given area from historical meteorological databases.

4.2 Microphysics of Wet Snow

> "By crossing air layers in which the air temperature is above 0 °C, a part of the snow particles melt, liquid water adheres to their surface and, during their zigzagging fall, they join together more or less intimately: consequently, at ground level, this kind of snow is formed by large volumes of humid and spongy masses, in which the crystalline shapes are largely disappeared." Professor Chassant (1902), National School of Agriculture, Montpellier, France.

Dry snowflakes and mechanical bonding: Atmospheric ice crystals are formed in the clouds due to nucleation of so-called "freezing nuclei" in the atmosphere below $-12\,^{\circ}$C, or by the freezing of droplets measuring 20–50 μm in diameter at temperatures between -3 and $-35\,^{\circ}$C. The initial crystal shape is hexagonal and it remains in continuous evolution according to the conditions of temperature and humidity existing in the ambient environment. The growth of ice crystals begins firstly by a vapour-deposition process reaching millimetric size, within a few minutes,

P. Admirat
Meteorology Consultant, 96 Chemin des Sept Laux, 38330 Saint Ismier, France
e-mail: tarimda@wanadoo.fr

M. Farzaneh (ed.), *Atmospheric Icing of Power Networks*,

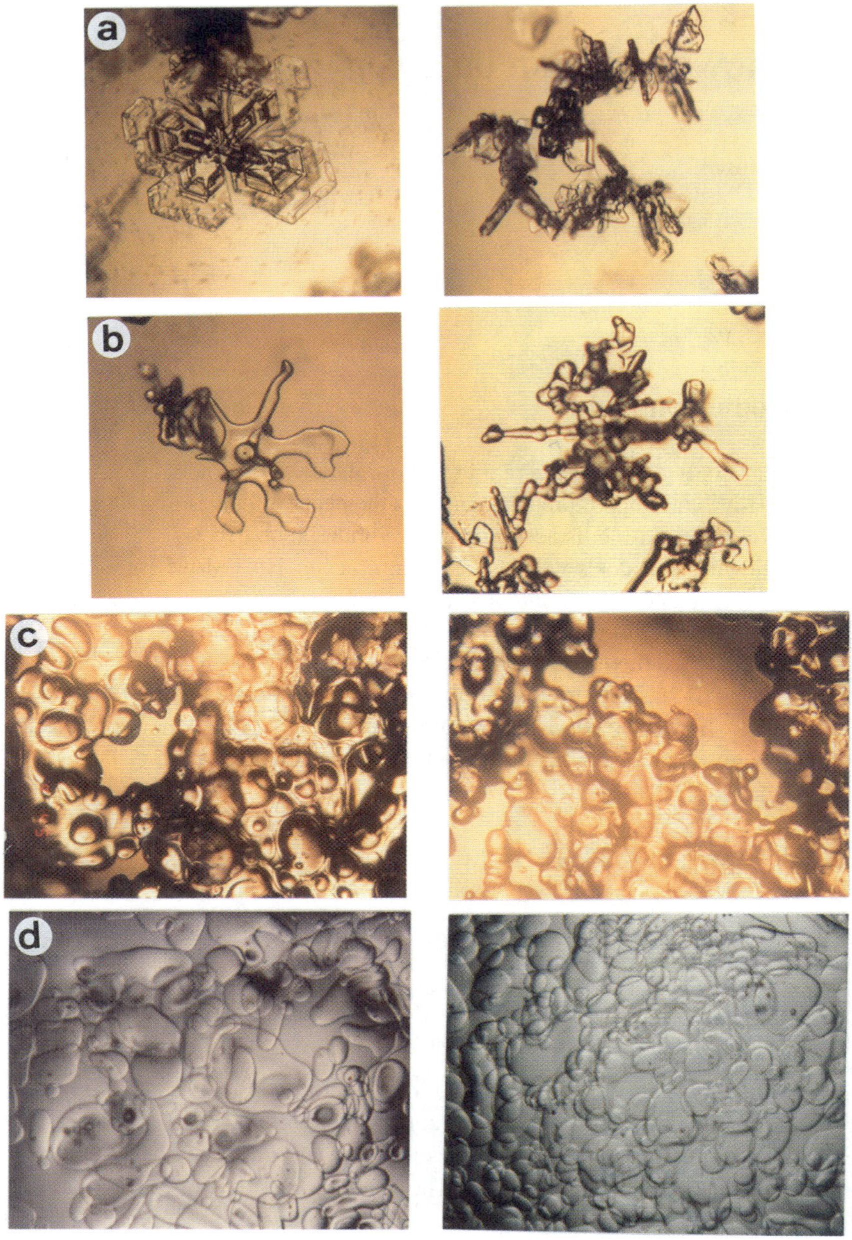

Fig. 4.1 Snow samples at different stages of their metamorphosis at positive air temperature. (**a**) Dry snowflakes (0.5 to 1 cm) falling at negative air temperature: the ice crystals present angular structures; (**b**) First emergence of ice metamorphosis at slightly positive air temperature: crystalline structures are rounding and retracting; a thin water layer covers the ice; (**c**) Typical clusters of wet snow consisting of ice granules (# 100 μm), capillary liquid water (5 to 15%) and air bubbles (often > 50% in volume) strongly joined together by capillary forces; (**d**) Final stage of the ice metamorphosis: the ice platelets are melting and the LWC is over 40%. The adhesive forces are going out and the wet snow material is flowing

followed by a mechanical bonding process with other ice crystals, yielding centimetric snowflakes that fall under gravity (Fig. 4.1a).

Wet snow granules and capillary bonding: In the course of precipitation, if the snowflakes pass through the isotherm 0 °C, they meet air at slightly positive temperature. Then the growth stops and, due to weak heat exchange with the atmosphere, a superficial liquid phase occurs over the ice crystals (Fig. 4.1b). This thin water layer transforms the weak mechanical bonding into strong capillary bonding.

The air-snowflake thermal exchanges accelerate suddenly when, by impacting an obstacle, the snowflakes come to a stop and are exposed to wind at positive temperature. Hence, the metamorphosis of ice the crystals accelerates: the hexagonal multi-branches of snowflakes lose their sharpness, round out, shorten, and turn into agglomerates of sub-millimetric ice granules weltering in their own melting water and surrounded by trapped air bubbles. The first scientific description of this rapid metamorphosis of snow at positive air temperature was published by Wakahama (1965) and theorized by Colbeck (1973).

These agglomerates combine easily with the previous network of agglomerates and adhere readily to all obstacles exposed to the airflow. In this new material, ice granules, liquid water and air micro-bubbles are strongly jointed together by capillary forces. The tensile strengths have already been studied by Wakahama and Mizuno (1979), and Colbeck (1976). The so-called "wet snow", or sometimes "sticky snow" (Fig. 4.1c), was initially called "grain clusters" by Colbeck (1979) as the basic element of wet snow agglomerates. The first photographic documentation of such "grain clusters", presented by Colbeck and Ackley (1982) at the first International Workshop on Atmospheric Icing of Structures, is absolutely identical to that shown in Fig. 4.1c.

The relative ratios between the three components (ice, liquid water, air) are the result of thermodynamic exchanges at the air-snow interface and govern a range of densities of this new material. Ratios and densities are clearly in permanent evolution according to the natural variations of weather conditions. For example, if the heat exchanges are restrained (low wind speed, air temperature close to 0 °C, high snowfall intensity), the wet snow material will contain half-transformed ice, with only a little liquid water and a lot of air bubbles. This sticky-snow material is opaque, very porous, with a light density lower than 100 kg m^{-3}. Conversely, if the heat exchanges become more active (higher wind speed, air temperature close to 2 °C, lower snowfall intensity), the wet snow material will contain fully transformed ice into granules, liquid water content (LWC) can reach 40% or higher with few air-bubbles, due to high wind pressure. This sticky-snow material is less opaque, of hard consistency, with a high density probably over 500 kg m^{-3}. Finally, when the heat exchanges are excessive, the LWC increases over 40% and the ice granules change into ice plates (called frazil) just before they melt completely (Fig. 4.1d).

4.3 Thermodynamic Analysis of Heat Exchanges

Most of the observations of snow accretions on overhead lines during a wet snow event show perfectly cylindrical snow sleeves enclosing the conductors. We shall see later that the accretion starts with a thin deposit of snow on the windward side of the conductors as a transitory state. The full coating of the conductors appears very quickly and the resulting snow sleeves are cylindrical; they continue to grow as long as the weather conditions are favourable. Then, all heat exchanges are strictly localized at the air-snow interface, except in the particular case of Joule effect induced at the conductor-snow interface. These are the thermodynamic criteria that govern the beginning, growth and break-up of the snow cylinders. There are three successive states, clearly defined:

A transitory state begins with the impacting of the first snowflakes on the conductor, itself at positive temperature. The snow then melts and, when in sufficient quantity, rapidly cools the conductor to as low as 0 °C. At that time, the solid state persists, transforms into wet snow and a snow deposit occurs along the conductor.

A permanent growth state of the snow sleeve persists as long as the main thermodynamic conditions of the ambient environment remain favourable (air temperature, wind speed, intensity of snowfall). Except for Joule effect induction, the heat exchanges, by convection, latent heat of condensation or evaporation, latent heat of snow melting, radiance, are permanent and variable at the air-snow interface (Fig. 4.2).

A gradual breaking up state of the snow sleeve for thermodynamic reasons such as interruption of the snowfall, warming, sublimation of ice, excess liquid water, or other, all factors causing a reduction of cohesive forces within the sleeve and partial or total discharges. In addition, in the case of total freezing,

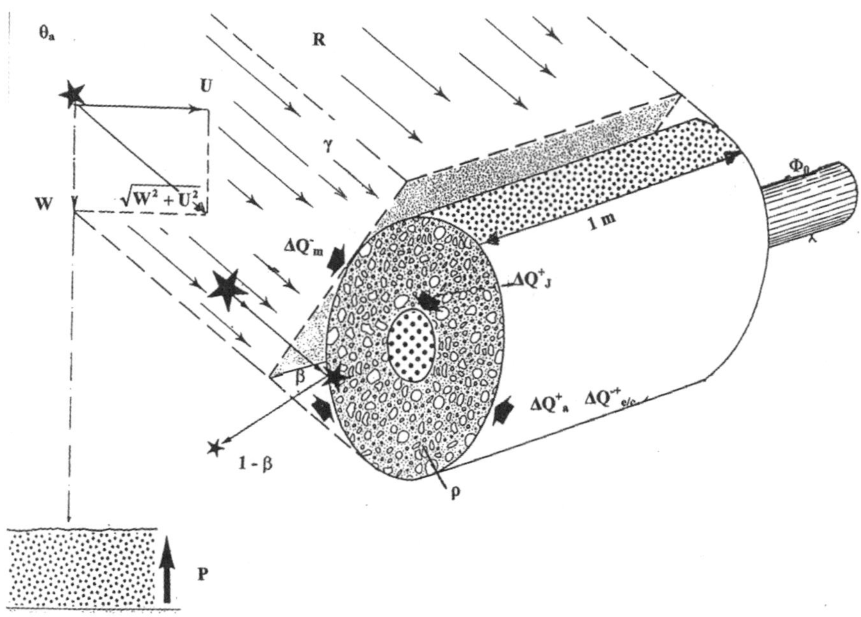

Fig. 4.2 Scheme of main heat exchanges at the air-snow interface and conductor-snow interface

the breaking up is due to mechanical forces from wind and torsional twisting of conductors.

The analysis of heat exchanges has been presented previously (Grenier et al. 1985, 1986; Admirat and Sakamoto 1988).

4.3.1 Heat Exchanges by Convection (ΔQ_a^+)

Convective heat exchanges between air at temperature θ_a and a cylinder of surface S at temperature θ_c can be described as $\Delta Q_a^+ = h\,(\theta_a - \theta_c)\,S$, where h is the exchange coefficient of conduction. Let us suppose a smooth cylinder of diameter Φ perpendicular to an airflow expressed by: $h = \lambda \frac{Nu}{\Phi}$ with λ air thermal conductivity. The Nusselt number Nu itself is a function of the Reynolds number Re of the turbulent airflow around the cylinder: $Nu = 0.64Re^{0.2} + 0.2Re^{0.61}$ (with $Re = \frac{U\Phi}{\nu}$ and ν the air cinematic viscosity). This relationship can be applied to a turbulent airflow because $5,000 < Re < 150,000$ (lowest values of $Re = 4,500$ with $\Phi_0 = 0.03$ m and $U = 2\,\mathrm{m\,s^{-1}}$; extreme values of $Re = 150,000$ with $\Phi_1 = 0.20$ m and $U = 10\,\mathrm{m\,s^{-1}}$).

Now, the first term of the Nu expression soon becomes insignificant when Re exceeds 10,000 and allows the following numerical expression: $Nu \approx 0.2Re^{0.61}$

Moreover, the roughness of a stranded conductor does not correspond to the definition of the smooth cylinder. However, the effects of the roughness disappear as soon as the conductor is fully covered with snow. Under this condition, the exchanges by convection can be calculated between the ambient environment and the surface $S = \pi\,\Phi$ corresponding to a snow cylinder of diameter Φ and of unitary length 1 m. The general equation is as follows

$$\Delta Q_a^+ = 0.2\frac{\pi\lambda}{\nu^{0.61}}\,(U\,\Phi)^{0.61}\,(\theta_a - \theta_c) \tag{4.1}$$

It is quite obvious that the wind speed U is the main parameter governing the convective exchanges (Fig. 4.3a). The numerical expression of this convective exchange is as follows (by taking $\nu = 1.35 \times 10^{-5}\,\mathrm{m^2\,s^{-1}}$, $\lambda = 2.42 \times 10^{-2}\,\mathrm{W\,m^{-1}\,{}^\circ C^{-1}}$, and $\theta_c = 0°C$ temperature of the mixing of ice and water) with Φ m, U m s^{-1}, θ_a°C and $\Delta Q^+{}_a$ W m^{-1}

$$\Delta Q_a^+ = 14.2\,(U\,\Phi)^{0.61}\,\theta_a \tag{4.2}$$

4.3.2 Heat Exchanges by Melting of Part α of Accreted Snow (ΔQ_m^-)

The flow R of the incident snow passing through the surface S (unitary length 1 m; and width Φ m) perpendicular to the air flow just in front of the conductor, is a

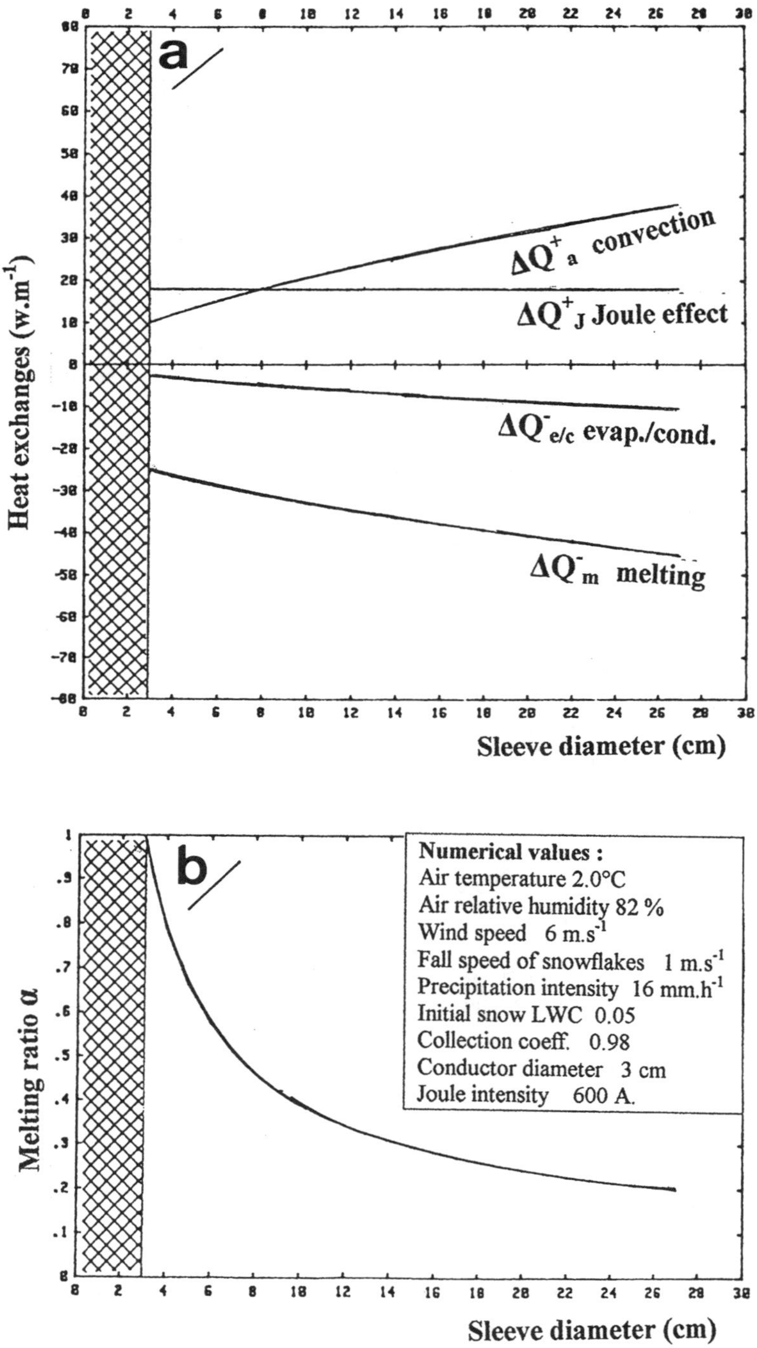

Fig. 4.3 Theoretical variations of the 4 main thermal fluxes and of the melting ratio α as a function of the cylindrical snow sleeve diameter, under constant climatic conditions

function of P (water equivalent of precipitation component at ground level, in mm h^{-1}), of U (horizontal wind speed) and of W (vertical fall velocity of snowflakes, taken to equal 1 m s^{-1})

$$R = \frac{P}{3.6 \times 10^3} \frac{\sqrt{W^2 + U^2}}{W} S$$

Introduction of a collection coefficient β: Among the snowflakes crossing surface S, only a part $\beta(0 < \beta < 1)$ is accreted for aerodynamic reasons (ejection of the smallest snowflakes by the curvature of air streams around the cylinder) and mechanical reasons (breaking up of the largest snowflakes impacting and ejection of their fragments). This accretion coefficient decreases mainly with wind speed and secondarily depends of snowflake dimension, shape, and initial liquid water content. So, the accreted snow is not R but rather βR.

Melting of a part α **of the wet snow**: Just before impact at slightly positive air temperature, the snowflakes have an initial LWC of $\gamma\%$. After impact, the strong heat exchanges with the air flux produce the melting of a part $\boldsymbol{\alpha}$ of the amount $(1 - \gamma)$ of the accreted snow. The heat consumed by the melting is $\Delta Q_m^- = L_m \alpha\ (1 - \gamma)\, \beta\, R\, \Phi$ (Fig. 4.3a) or

$$\Delta Q_m^- = L_m \alpha\ (1 - \gamma)\ \beta \frac{P}{3.6 \times 10^3} \sqrt{1 + \frac{U^2}{W^2}}\ \Phi \tag{4.3}$$

with $L_m = 335$ kJ kg^{-1} of water, P in mm h^{-1}, and $\Delta Q^-{}_m$ in W m^{-1}.

4.3.3 Heat Exchanges by Evaporation or Condensation of Water Vapour ($\Delta Q^{-+}{}_{e/c}$)

Depending on the vapour pressure differences Δp_e between that over-condensed water at 0 °C on the surface of the snow sleeve $P_{es}(0°)$, and that of the atmospheric water vapour pressure at θ_a °C P_e *(air)*, the airflow around the cylinder could produce a mass transfer either by evaporation or condensation of the water vapour molecules at the air-snow interface. The latent heat of the water evaporation/condensation on the surface is as follows

$$\Delta Q^{\pm}{}_{e/c} = \frac{Lv}{C_p} \left(\frac{P_r}{S_c}\right)^{0.63} h\, \Delta r\, S \tag{4.4}$$

L_v is the latent heat of water evaporation
C_p is the heat capacity of the air at constant pressure
Δr is the difference in the mixing ratio $\Delta r = r_{(air)} - r_s(\theta_c)$ or
$\Delta r = 0.622 \frac{P_e^{(air)} - P_{es}^{(\theta_c)}}{P}$, according to the differences in vapour pressures.

This difference Δp_e between the vapour pressure of atmospheric water vapour $P_e(air)$ and the pressure of saturated vapour of the condensed phases on the surface of the snow cylinder $P_{es}(0°)$ governs the type of exchange:

- At 0 °C, $P_e(air) = P_{es}(0°)$ then $\Delta p_e = 0$ there is no thermal exchange.
- If $P_e(air) < P_{es}(0°)$ then Δp_e is < 0, the air flux is under-saturated at θ_a °C. The heat exchange $\Delta Q^{-}{}_{e}$ is < 0; there will be an evaporation process on the surface of the snow cylinder and a cooling effect (Fig. 4.3a).
- If $P_e(air) > P_{es}(0°)$ then Δp_e is > 0, the air flux is saturated at θ_a °C and oversaturated at 0 °C; the heat exchange $\Delta Q^{+}{}_{c}$ is > 0; there will be a condensation process of water vapour on the surface of the snow cylinder and a warming effect.

These three situations could succeed each other according to the natural weather variations. The quantity of heat absorbed by the evaporation/condensation process of the condensed phases of snow sleeve is as follows

$$\Delta Q^{-}{}_{e/c} = 0.1244\pi \frac{L_v}{C_p} \left(\frac{P_r}{S_c}\right)^{0.63} \frac{\lambda}{\nu^{0.61}} (U\,\Phi)^{0.61} \frac{\Delta P_e}{P} \tag{4.5}$$

The numerical expression of these heat exchanges by evaporation/condensation, using $L_v = 2,500$ kJ.kg^{-1}water, $C_p = 1,007$ kJ.kg^{-1}air, $P_r = \dfrac{C_p\mu}{\lambda} = \dfrac{C_p\nu\rho_a}{\lambda} =$ 0.716, $S_c = 0.605$, and $P = 1,000$ hPa, is as follows (Fig. 4.3a)

$$\Delta Q^{\pm}{}_{e/c} = 24.3\ (U\,\Phi)^{0.61}\ \Delta P_e \tag{4.6}$$

$\Delta Q^{-+}{}_{e/c}$ is then given in Wm^{-1} of the conductor, with diameter Φ in m, wind speed U in m s^{-1} and Δp_e in hPa.

4.3.4 Heat Exchanges by Joule Effect at the Conductor-Snow Interface ($\Delta Q^{+}{}_{J}$)

The heat released by unit of length of the conductor where intensity I passes is (Fig. 4.3a)

$$\Delta Q^{+}{}_{J} = R_0 I^2 \quad \text{W.m}^{-1} \tag{4.7}$$

4.3.5 Heat Exchanges by Radiation Effect ($\Delta Q^{-}{}_{rad}$): A Negligible Flow

Between the snow sleeve and the nearby environment there is a small radiation exchange toward the snow sleeve because of the very small difference in temperatures between the air and the snow sleeve: $Q^{-}{}_{ray} = \sigma[(273+\theta_a)^4 - (273+\theta_c)^4]\,S$. Numerically, the surface energy at the ice-ice interface is $\sigma = 5.67\,10^{-8}\,\mathrm{W\,m^{-2}\,K^{-4}}$. If we consider the existing environmental radiant as the black bodies and the maximum temperature differences of about 3 or 4 °C between the bare conductor ($\Phi_0 = 3\,\mathrm{cm}$) and the ambient environment, in this case the radiation exchange will only be $1.7\,\mathrm{W\,m^{-1}}$, i.e. negligible.

An instantaneous period of sunshine at the end of the event can produce a much larger flow of radiant heat of about 20 Watts per linear metre. This thermal flow increases the LWC of the snow sleeve, which accelerates its stretching and breaking off.

4.3.6 Transformation of Ice Crystals into Ice Granules (E_{ss}): A Negligible Flow

The energy released during the process of transformation of crystals into ice grains is equal to the difference between the surface energies of two crystalline states

$$E_{Surface} = \frac{\sigma_{ss}}{I'_g}$$

with $\sigma_{ss} = 6 \times 10^{-2}\,\mathrm{J\,m^{-2}}$ as surface energy at the ice-ice interface. So, the energy to convert 50 μm crystals into grains of 500 μm, is:

$$\Delta E_{trans} = 6 \times 10^{-2}\left(\frac{1}{50} - \frac{1}{500}\right)10^6 \cong 10^3\,\mathrm{J.m^{-3}}$$

During a major wet snow event, the mass and volume of accreted snow on a 10-cm sleeve increase respectively by about $0.3 \times 10^{-3}\,\mathrm{kg\,m^{-1}s^{-1}}$ and $2 \times 10^{-6}\,\mathrm{kg\,m^{-1}s^{-1}}$. The energy produced is therefore negligible. These calculations, used for other weather conditions, do not really change the considerations of the thermal exchanges. Consequently, if the exchanges under Joule effect have been deleted (these exchanges are usually negligible under the working conditions of a high voltage wire), the simplified thermal balance amounts mainly to the exchanges between the air and the snow $\Delta Q^{+}{}_{a}$ on a snow sleeve surface, which leads to the melting of a more or less significant part α of the accreted snow. The very fast decrease of the α factor goes from the value 1 (applied to $\Phi = \Phi_0$) to the value 0.5 (applied to $\Phi = 7\,\mathrm{cm}$), corresponding to a 2-cm-thick snow sleeve.

4.3.7 Thermal Balance

The main heat sources and sinks are not emitted evenly over the surface of the snow sleeve and are found in different locations. In fact:

- $\Delta Q^{+}{}_{a}$ and $\Delta Q^{+-}{}_{e/c}$ are distributed over the entire surface of the snow sleeve, but the exchanges are more significant on the windward side or on the back side of the cylinder rather than on the sides situated perpendicular to the air flow.
- $\Delta Q^{-}{}_{m}$ is limited on the windward face of the cylinder exposed to snow precipitation.
- $\Delta Q^{+}{}_{J}$ is produced at the conductor-snow interface. This source of heat, which contributes to increase the average LWC of the snow sleeve, concerns mainly the internal layers of the snow sleeve. However, the liquid phase rapidly migrates to the external layers.

By supposing that the four main thermal flows diffused in the snow sleeves, the instantaneous energy balance of each new snow layer is valid for the entire snow sleeve is

$$\Delta Q^{+}{}_{a} + \Delta Q^{\pm}{}_{e/c} + \Delta Q^{-}{}_{m} + \Delta Q^{+}{}_{J} = 0 \tag{4.8}$$

Usually, during the growth of a snow sleeve, this equation is continuously balanced by the melting rate, α, which allows for determining the exact proportion of melted snow as a function of diameter Φ. The very first snowflakes to land on the conductor at positive temperature melt completely ($\alpha = 1$). When the absorption of the latent heat decreases conductor temperature to 0 °C, melting becomes partial and snow accretion starts, meaning that α decreases (Fig. 4.3b), giving $\Delta Q^{-}{}_{m}$ the appropriate thermal equilibrium value.

In fact, if α remains close to one due to excessive melting at high air temperature, the LWC of the wet snow agglomerates reaches extreme limits and the capillary adhesive strength decreases to a point where accretion becomes impossible.

4.3.8 Calculating the Liquid Water Content (LWC) of Accreted Snow

The total LWC of wet snow layers is made up of the initial LWC γ of the snowflakes before impact, the LWC α of the melting ice mass and (1-γ) of the agglomerates. The LWC of a snow layer can be written as

$$\mathrm{LWC}_{\{\text{snow layer}\}} = \alpha\,(1-\gamma) + \gamma \tag{4.9}$$

The initial water content γ is a simple function of air temperature and atmospheric thermal stratification given by: $\gamma = \zeta\,\theta^2$ (ζ = factor of thermal stratification of the air). Low numerical values of $\gamma < 5\%$ are usually obtained when the air temperature is between 0 and 1 °C. Each elementary snow layer is similar to a small concentric sleeve of diameter Φ and thickness $d\Phi$, having a specific melting ratio $\alpha_{(\text{snow layer})}$, according to heat exchanges, meaning functions of Φ and U. The LWC of the sleeve changes from one snow layer to the next.

The thermal balance is as follows

$$\alpha = \frac{\Delta Q^{+}{}_{a} + \Delta Q^{\pm}{}_{e/c} + \Delta Q^{+}{}_{J}}{0.93\,\beta\,(1-\gamma)\;P\sqrt{1+\frac{U^2}{W^2}}\,\Phi} \tag{4.10}$$

By integrating through a snow sleeve having as total thickness depth $\Phi_1 - \Phi_0$, the averaged value of the ratio α_{av} is: $\alpha_{av} = \frac{\int_{\Phi_0}^{\Phi_1} \alpha.dM}{M}$. The total averaged $\mathrm{LWC_{av}}$ of a snow sleeve is rather complex, depending on 13 parameters

$$\mathrm{L.W.C._{av}} = \frac{2}{\Phi_1^2 - \Phi_0^2}\,\frac{0.62\,a\left(\Phi_1^{1.61} - \Phi_0^{1.61}\right) + b\,(\Phi_1 - \Phi_0)}{0.93\,\beta\,P\sqrt{1+\frac{U^2}{W^2}}} + \gamma \tag{4.11}$$

with $a = (14.2\,\theta_{\mathrm{a}} + 24.3\,\Delta p_e)\,U^{0.61}$ and $b = R_0\,I^2$.

Increases in $\mathrm{LWC_{av}}$ are mainly due to:

- The convective exchanges which control melting, correlated to wind speed and air temperature.
- The electrical current (Joule effect behaves like a liquid water source).

Conversely, decreases in LWC are mainly due to:

- An increase in snowfall intensity.
- An increase of collection coefficient β when wind speed decreases.

In conclusion, putting in equation the thermal exchanges during the snow accretion allows for understanding the influences and interdependences of each parameter.

4.4 Modelling the Cylindrical Growth of Wet Snow Sleeves

The instantaneous variation of the mass dM of the snow cylinder of density ρ_s expressed from the instantaneous variation of its diameter Φ is $dM = \rho_s \frac{\pi\Phi}{2}\,d\Phi$; then

$$\Phi\,d\Phi = \frac{2}{\pi}\frac{1}{\rho_s}\,\frac{dM}{dt}dt$$

and

$$\Phi\frac{d\Phi}{dt} = \frac{2}{\pi\rho_s}\frac{dM}{dt} \tag{4.12}$$

Now, the instantaneous mass *dM* of the snow sleeve is the sum or difference of the snow mass actually accreted and of the mass of liquid water condensed or evaporated during the time interval *dt*

$$\frac{dM}{dt\text{ accreted}} = \beta\, R\, S = \beta\, \frac{P}{3.6 \times 10^3}\sqrt{1 + \frac{U^2}{W^2}}\, \Phi \qquad (4.13)$$

$$\frac{dM}{dt\text{ evap/cond}} = \frac{\Delta Q^{\pm}{}_{e/c}}{L_v} = 24.3\frac{(U\,\Phi)^{0.61}}{L_v}\, |\Delta P_e| \qquad (4.14)$$

It can be shown that this last amount of condensed or evaporated water (Eq. 4.14) will never be greater than 1% of the accreted snow; so it can be neglected from a practical point of view. So Eq. (4.12) becomes

$$\frac{d\Phi}{dt} = \frac{2\beta}{\pi \rho_s}\,\frac{P}{3.6 \times 10^3}\sqrt{1 + \frac{U^2}{W^2}} \quad \text{or} \quad \frac{d\Phi}{dt} = \frac{2}{\pi \rho_s}\text{K}$$

$$\text{with K} = \frac{\beta P}{3.6 \times 10^3}\sqrt{1 + \frac{U^2}{W^2}}$$

If the meteorological conditions stabilize, that means if P, U, W are constant over the time interval, and if β has an averaged and constant value, by integrating it can be written as follows

$$\Phi = \frac{2}{\pi \rho_s}\text{K}t + \Phi_0 \qquad (4.15)$$

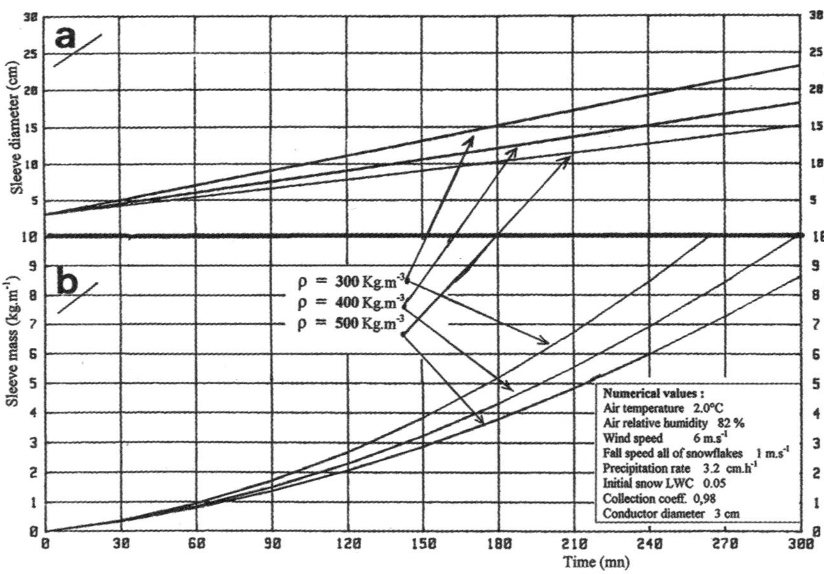

Fig. 4.4 Computed values of the masses and diameters of wet snow sleeves as function of time, according to 3 values of wet snow densities, under constant climatic conditions

In this case, the growth of the snow sleeve diameter is a linear function of time, only depending on averaged ρ_{av} values (Fig. 4.4). Then $\frac{dM}{dt} = \mathrm{K}\frac{2}{\pi\rho_s}\,(\mathrm{K}t + \Phi_0)$ and

$$M = \frac{\mathrm{K}^2}{\pi\rho_s}t^2 + \mathrm{K}\Phi_0\,t + \mathrm{M}_0 \tag{4.16}$$

With the same hypothesis of constant meteorological conditions, the growth of the snow sleeve mass is in proportion to time squared and also depends on averaged ρ_{av} values (Fig. 4.4).

This type of physical model seems well adapted for wind tunnel simulations, where the meteorological conditions are held constant over several hours. By extension, the model must also be representative of snow-sleeve growth in natural and therefore varying conditions, by proceeding with successive short time-intervals (for example 1 hour) during which averaged values of the meteorological parameters can be considered as being constant.

4.5 Simulation of Accretion Mechanisms in Wind Tunnel Conditions

The purpose of wind-tunnel simulations is to reproduce the mechanisms of snow accretion on conductor samples, under controlled aerodynamic conditions, to verify the influence of each variable, according to the previous theory. All equipment has to, in the first place, reproduce the five main meteorological conditions – wind speed, air temperature, relative air humidity, snowfall intensity, microphysical quality of snowflakes – and, secondly, expose a conductor sample to snowfall under controlled torsional stiffness conditions. The aforementioned experimentation was carried out in the only wind tunnel available at the CRIEPI Ishiuchi Laboratory, over three successive winters (Admirat and Dalle 1984, 1985, Admirat et al. 1985).

4.5.1 Technology Used in the Ishiuchi Wind Tunnel

The wind tunnel laboratory is located in the Ishiuchi village where, during winter, heavy snowfalls frequently occur at temperatures below zero.

Air temperature: The wind tunnel operates in an open circuit inside a large shed, where the air temperature is continuously maintained and controlled to remain slightly positive, adjustable between 0 and 2 °C by means of an automatic heating system.

Wind speed: A powerful ventilator blows a somewhat turbulent air flux into a horizontal tunnel (Fig. 4.5a). The average wind speed can be adjusted within the range of 5 ms^{-1} to 15 ms^{-1}, but short wind speed variations of 50% have been measured along the target conductor, inducing some heterogeneousness in the characteristics of the wet snow generated.

Relative air humidity: Due to the permanent presence of a large quantity of snow inside and outside the laboratory, the relative air humidity is controlled between 85 and 94% at air temperature θ_a, between 2.0 and 0.5 °C, which means very close to the 0 °C ice saturation point.

Snowfall is simulated by a continuous injection of an adjustable amount of centimetric snow particles. These particles are produced from small blocks of natural fresh snow brought inside by a conveyor belt and passed through a cylindrical multi-comb rotating at high speed (Fig. 4.5a). This system shreds the snow blocks and injects into the air stream a relatively broad spectrum of snow particles ranging from 3 mm to 1.5 cm in diameter. Because of the sedimentation speed of the different snow particles and of their respective trajectories, the intensity of the snowfall through the test section is, of course, approximately constant but not identical throughout the section (Fig. 4.5b$_3$, 4.5b$_4$). So, the snow flux R throughout the test section must be calibrated according to the adjustable shredded snow mass, to the selected wind speed, and to the selected position of the conductor sample in the test section (Admirat and Dalle 1985).The corresponding precipitation rates were usually between 8 and 30 mm h^{-1}.

Thermal exchanges between the air and the snow particles first begin during the short period of time (less than 1 second) between the injection of snow particles into the airflow and the impact on the target: an examination under microscope shows the beginning of the ice metamorphosis just before impact (Fig. 4.1b). The main and very rapid exchanges occur when, immobilized by impacting the bare conductor or the former snow layer, the snowflakes are suddenly open to the airflow at positive temperature. The rounding of ice crystals and their partial melting immediately occur (Fig. 4.1c); then, the new ice grains and the liquid water become encrusted into the former wet snow layer.

Conductor samples 1.5 m in length are placed horizontally in the test section (Fig. 4.5b), perpendicular to the flow of snow particles. A torsional stiffness system applies variable stiffness to simulate from free-rotating to non-rotating spans.

A computer monitors and records all experimental conditions for each experiment. Qualitative and quantitative results were widely published between 1985 and 2000 (Admirat et al. 1986; Sakamoto et al. 1986, 1988; Lapeyre 1986).

4.5.2 Qualitative Wind-Tunnel Tests: Characteristics of "Wet Snow"

There is no snow accretion on the conductor samples when the test is carried out at negative air temperatures and wind speeds above 2 ms^{-1}. These "dry snow" particles are not subjected to either thermal exchange or metamorphosis. These particles are characterized by low mechanical cohesion; moreover, many snowflakes are broken at the impact point and swept away by the air flux.

Snow accretion on the conductor samples only occurs when the test is carried out at slightly positive air temperatures. The thermal exchanges rapidly induce the metamorphosis of ice crystals, transforming “dry snow” into “wet snow”: in a few seconds, the crystalline ice branches round out, retract, and create ice granules; a thin water layer occurs and the “ice-water granules” adhere to the conductor sample, or the existing wet snow layer, or whatever the obstacle. In this way, these qualitative tests confirm that the cohesiveness of “wet snow” sleeves around conductors is generated by the capillary forces.

Finally, when the ice granules are submitted to excessive thermal exchanges, caused by both positive air temperatures above 4 °C and relatively high wind speeds (# 15 ms^{-1}), they produce excess liquid water and a reduction of the adhesive forces between the granules, leading to complete melting (Fig. 4.1d). No accretion process can develop.

4.5.3 Accretion on Stranded Conductors: Two Possible Mechanisms

When the adequate aerodynamic conditions are fulfilled, the accretion always begins in the same manner: a thin snow layer starts to accrete on the windward face of the conductor sample; this snow deposit grows, gets wider by taking an angular profile according to wind speed: the greater the wind speed, the more angular the profile. The centre of gravity of the conductor sample-snow deposit system becomes more and more eccentric and the torque created induces conductor-sample rotation. Therefore, either an axial or a cylindrical accretion mechanism follows, according to the torsional stiffness of the conductor sample.

Self-limiting axial accretion mechanism on a fixed conductor (Figs. $4.5b_1$ and 4.6): The fixed conductor resists the torque, meaning that the same side of the conductor is always facing the snow precipitation. The snow deposit can only grow in the axial wind direction. However, the snow accretion rate decreases as a function of time, initially by a decrease in accretion coefficient (deviation of both airstream and snowflakes by the more and more angular profile), and also, beyond a certain dimension, the snow deposit itself breaks up under wind pressure, then builds up again, and breaks up, etc. This axial accretion mechanism is therefore clearly self-limiting through a loading-unloading process.

Unlimited cylindrical accretion mechanism on a free-rotation conductor (Figs. $4.5b_2$ and 4.7): In this case, the conductor sample does not resist the torque and rotational motion is generated. A new area of the conductor now faces windward, on which a complementary snow layer begins to form. After a one-half-turn rotation, the successive snow layers entirely cover the conductor, thus forming a perfectly cylindrical snow sleeve. A special test was conducted until a snow overload up to 20 kg m^{-1} (Fig. 4.5c) and more than 10 complete rotations were achieved (Fig. 4.8a). This cylindrical accretion mechanism is clearly unlimited and continues for as long as the simulated meteorological conditions are present (at least within the limits of wind-tunnel capacities).

Unlimited cylindrical accretion on a smooth conductor: Wet snow accretion on a smooth cylindrical conductor (metallic or sheathed cable) will always be

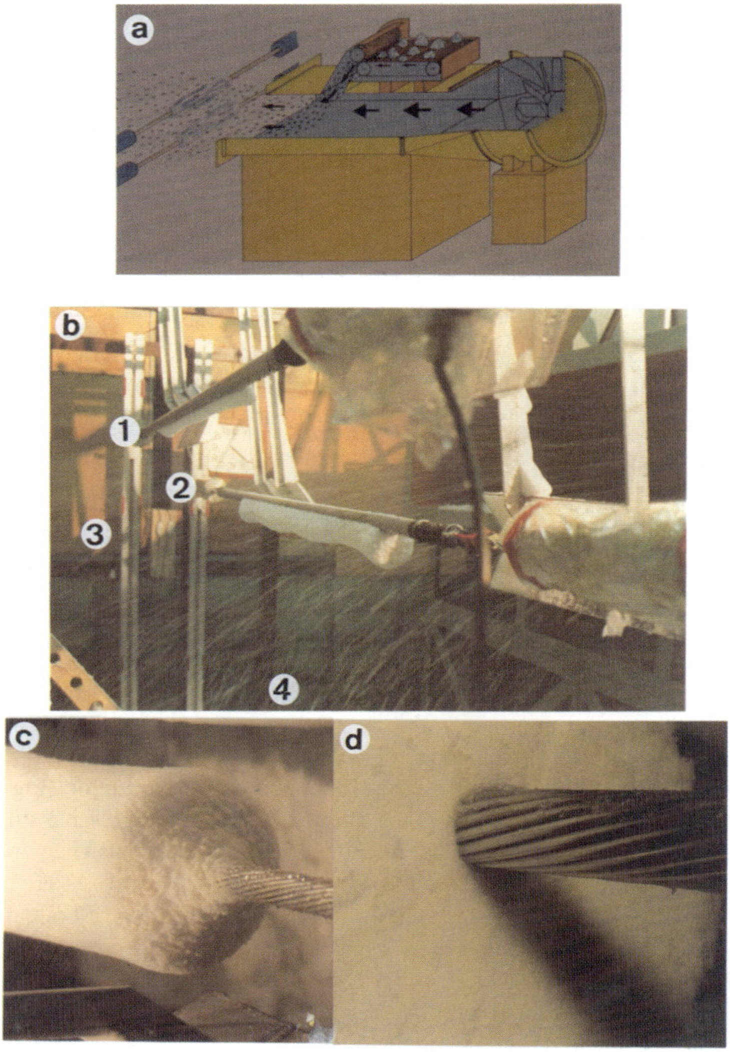

Fig. 4.5 Ishiuchi wind tunnel testing the two snow accretion mechanisms on a fixed and on a free rotating conductor samples. (**a**) Schematic view of the simulating system; (**b**) Appearance of the two accretion mechanisms 20 minutes after starting time: (b_1) axial accretion on the fixed conductor sample, (b_2) cylindrical accretion on the free-rotating conductor sample, (b_3) trajectories of the smallest snowflakes, (b_4) trajectories of the largest snowflakes; (c) Snow sleeve 2 hours after starting time and nine complete rotations: Wind speed $5\,\mathrm{m\,s^{-1}}$, Sleeve diameter 45 cm, Air temperature $+1.5\,°\mathrm{C}$ Snow overload $20\,\mathrm{kg\,m^{-1}}$, Precipitation $19\,\mathrm{mm\,h^{-1}}$, Averaged LWC 12%, Averaged density $150\,\mathrm{kg\,m^3}$; (d) Section of the sleeve prepared for wet snow sampling

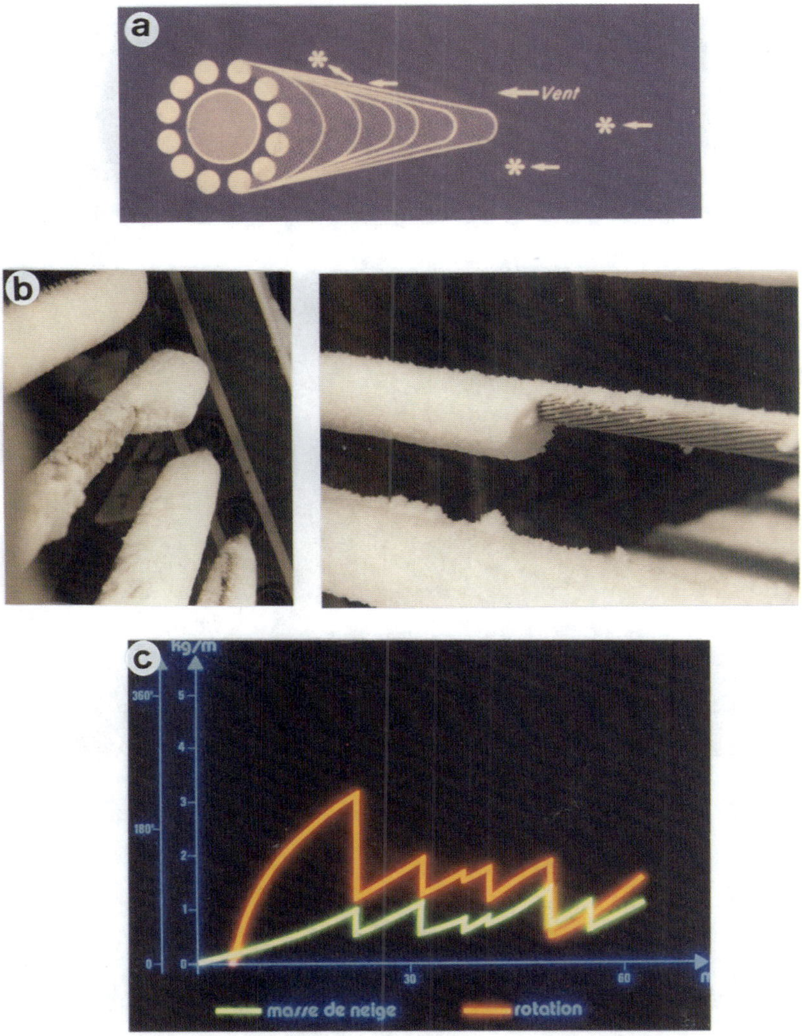

Fig. 4.6 Schematic concept and experimental tests of the axial snow accretion on a fixed conductor and evidence of its self-limited growth. **(a)** Fixed conductor: snow accretion can only occur on the axial windward face. The accretion rate decreases due to a lower accretion coefficient and the deviation of snowflakes on the more and more angular profile of the snow deposit. Beyond a certain dimension, the snow deposit breaks up under wind pressure or excessive LWC; **(b)** Axial accretion test in wind tunnel conditions showing the breaking up of the snow deposits and their angular profiles; **(c)** Schematic concept of both the self-limiting snow accretion and conductor rotation under a successive loading- unloading process

cylindrical. The sleeve formation is the result of permanent and rapid sliding of the snow on the smooth conductor-snow interface (a thin water layer enhances the sliding). Conductor stiffness has no effect in this case, and this type of cylindrical accretion is also theoretically unlimited.

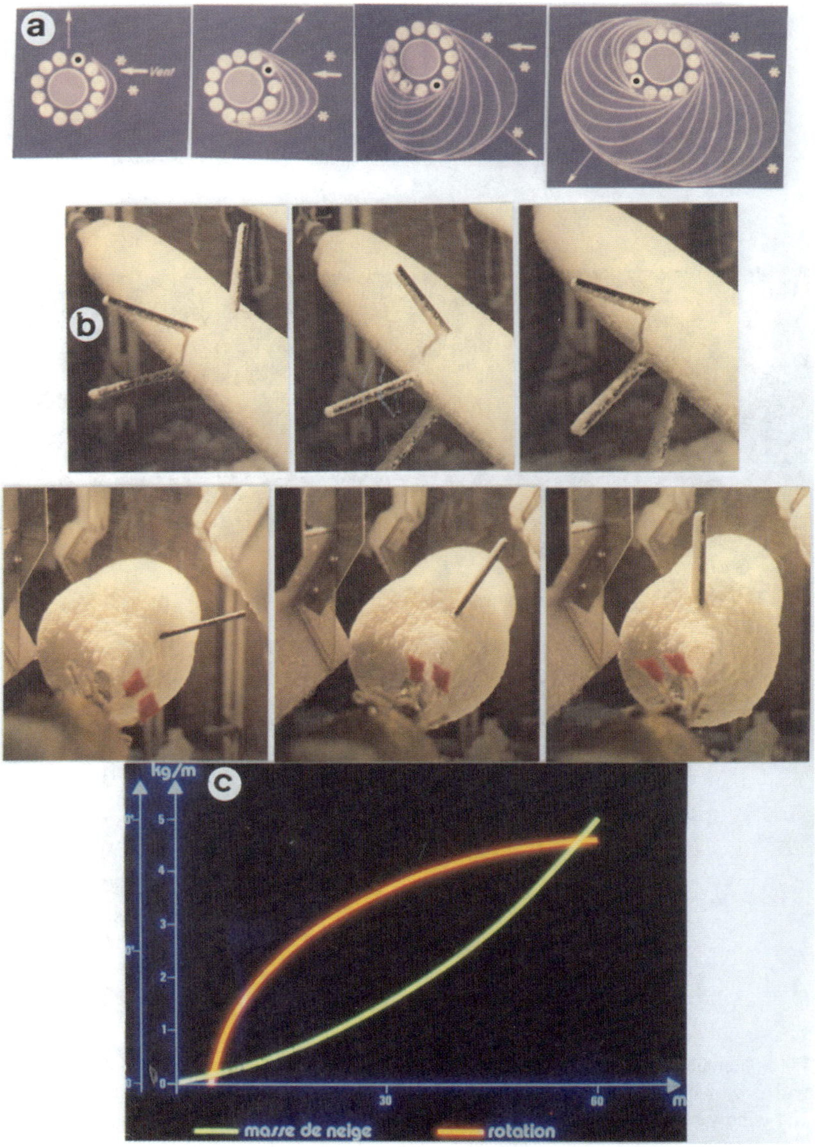

Fig. 4.7 Schematic concept and experimental test of the cylindrical snow accretion on a free-rotating conductor and evidence of its unlimited growth. (**a**) Free-rotating conductor: the snow accretion allows a slow and permanent rotation and a permanent accretion. After a one-half-turn rotation, the successive snow layers entirely cover the conductor, thus forming a perfectly cylindrical snow sleeve; (**b**) Cylindrical accretion test in wind tunnel conditions showing both the unlimited growth of the same snow sleeve at two different stages and the permanent conductor rotation; (**c**) Schematic concept of both the unlimited snow accretion and the conductor rotation under a permanent loading process

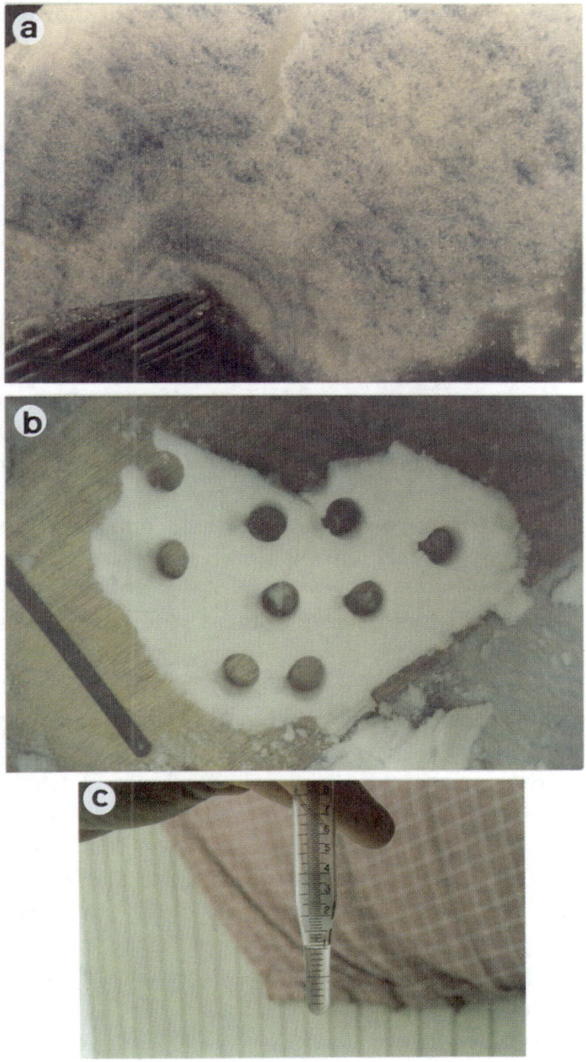

Fig. 4.8 Wind tunnel facilities to study: (**a**) Conductor rotations and snow layers (more than 10); (**b**) Multi-sampling for snow density variations; (**c**) LWC measurements

4.5.4 Quantitative Tests: Wet Snow Characteristics

Different tests have made it possible to examine a wide range of snowflakes just before and just after their impact on the snow sleeve. A range of numerical values have thus been obtained to describe the microphysical features of wet snow sleeves under the influence of the main parameters governing the heat exchanges: wind speed, air temperature and snowfall intensity. An example of experimental conditions has been published and the results can be summarized as follows:

Observation of Liquid Water Content of snowflakes just before impact: The metamorphosis of snowflakes is clearly observed just before impacting, mainly due to the rounded shape of the ice crystals (Fig. 4.1b), but the existence of a thin layer of capillary water is more supposed than actually observed in most of the tests, except for some specific tests at air temperatures above 2.5 °C.

Averaged densities ρ of wet snow sleeves: The measurements of ρ inside a sleeve section are relatively easy to make by weighting a well-known volume of wet snow sample (Fig. 4.8 and methodology Fig. 4.18). Because the wet snow material is made of ice granules, liquid water and air bubbles, its density fundamentally depends of the air content, itself depending mainly of wind pressure. Effectively, in constant meteorological conditions, the measurements made at different points of a same sleeve section (Fig. 4.8b) did not clearly detect small variations of ρ from one snow layer to another. But by comparing two sleeve sections grown in different meteorological conditions, the measurements of ρ confirm a close relationship to applied wind speed during sleeve formation: a strong wind pressure makes the sleeves more compact and the measured ρ values increase from 100 to 400 kg m^{-3} when wind speed increases from 3 m s^{-1} to 10 m s^{-1}. An exceptional ρ value close to 500 kg m^{-3} has been measured in a specific test at a wind speed of 15 m s^{-1}.

Average Liquid Water Content of the wet snow sleeves: The LWC measurements can be made by centrifuging a calibrated wet snow sample, taken at different points of a sleeve section (Fig. 4.8c and methodology Fig. 4.19). Inside each sleeve section, the LWC values are theoretically different from the central part until the periphery, according to the different accreted snow layers (Fig. 4.8a), but experimental measurements cannot clearly detect such small variations. Conversely, the experimental LWC values show a close relationship to heat exchange intensity: by comparing two sleeves having the same diameter, formed at the same air temperature and precipitation rate, but the first at a wind speed of 3 m s^{-1} and the second at 10 m s^{-1}, the former clearly produced a lower LWC. Most of the averaged LWC values obtained from many tests vary from 7 to 15% inside all sleeves having a hard consistency (perforating the sleeves with a pencil is rather difficult).

By increasing the initial LWC of snowflakes by a wide margin just before impact, by means of IR radiation (Fig. 4.9) or by additional water spray at the air-sleeve interface, or by Joule effect at the conductor-snow interface, some LWC values greater than 35% have been obtained in sleeves having a spongy consistency (perforating the sleeves with a pencil is very easy). Then, with LWC contents above 40% (Fig. 4.8c), the capillary forces become null and induce flowing and stretching of the snow, causing snow chunks to fall away (Fig. 4.9b,c).

Estimating the accretion coefficient β: Basically, β cannot remain constant during an accretion test due to the increases in sleeve diameter and Reynolds number. Besides, experimental measurements of β during a simulation tests are practically impossible. However, by comparing both the measured and calculated values of the sleeve diameter and mass, using the same constant meteorological values and the

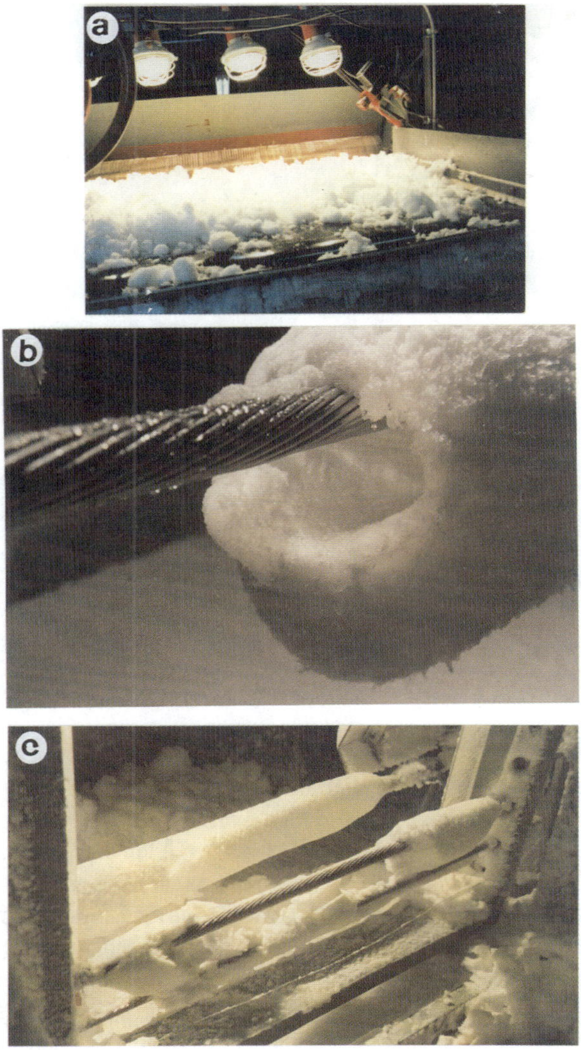

Fig. 4.9 By strongly increasing the initial LWC (IR radiation) of the snow before impact, the capillary forces decrease and induce flowing, stretching and falling down of the snow sleeve (LWC # 43%, in the sleeve reported in Fig. 4.8)

same ρ value, agreement between the two methods depends only of the numerical value given to β. In the case of an isolated test, such an adjustable numerical β value is not so difficult to find (Fig. 4.10). Applied to a series of tests, this method provided a succession of numerical values for β, growing from 0.1 to 0.4 for wind speeds decreasing from 10 to $3\,\mathrm{m\,s^{-1}}$. Finally, this method shows that β decreases when the wind speed increases and provides the first empirical relationship: $\beta = 1/U$ (with $U > 1\,\mathrm{m\,s^{-1}}$).

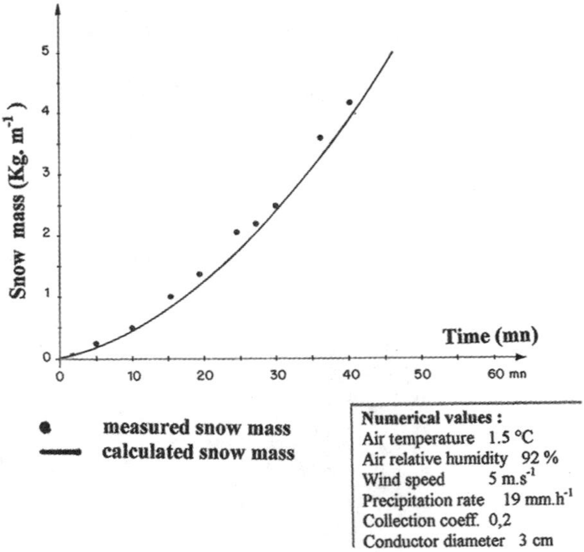

Fig. 4.10 Calibration of the β collection coefficient in wind tunnel: the best agreement between calculated and measured values is obtained with $\beta = 0.2$ (wind speed $= 5\,\mathrm{m\,s^{-1}}$)

4.6 Observation of Accretion Mechanisms in Natural Climatic Conditions

4.6.1 Occurrence of Wet Snow Events

Frequency and intensity of the events: Low-intensity wet snow events are the most frequent result of usual winter atmospheric disturbances. These events last 3 to 5 hours with precipitation intensities ranging from 2 to $5\,\mathrm{mm\,h^{-1}}$ (less than 15 cm of snow accumulation on the ground). They produce snow sleeves of less than 10 cm in diameter and less than $1\,\mathrm{kg\,m^{-1}}$ in overload, causing no or little damage to overhead lines.

The more serious events occur less frequently. Stormy weather conditions during winter cause the heaviest damage to overhead lines. These stormy conditions last 15 to 24 hours with precipitation intensities ranging from 5 to $10\,\mathrm{mm\,h^{-1}}$ (up to 50 cm of snow accumulation on the ground). They produce snow sleeves greater than 20 cm in diameter and up to $10\,\mathrm{kg\,m^{-1}}$ in overload, which often exceeds the mechanical resistance of electrical equipment, conductors and supports.

Atmospheric localisation of wet snow air layer: From a meteorological point of view, we know that all dry snowflakes falling at negative air temperatures from high altitudes, and crossing the 0 °C isotherm, turn into adhesive clusters of wet snow at temperatures between 0 and 2 °C; that is to say in an atmospheric air layer

Table 4.1 Heavy wet snow damage on power lines on Nov. 26, 1982 (EDF data, Central France)

		Length of damaged lines (km)	Number of damaged supports
Distribution Lines	Low voltage	2,650	6,500
	Medium voltage	3,050	2,700
Transmission Lines	High voltage	50	164
	Very high voltage	10	15

only 300-m deep at most. With warmer air temperatures, the LWC of each ice cluster increases up to 100% and rain drops are formed.

Let us suppose an observer submitted to rainfall at relatively high positive temperature (for example: altitude 100 m and air temperature 5 °C). If he climbs up a mountain road, he will inevitably meet the successive kinds of precipitations, according to the usual air temperature gradient of 0.6 °C/100 m:

- From 5.0 (alt. 100 m) to 3.5 °C (alt. 350 m), he will be submitted to rainfall.
- From 3.5 up to 2.0 °C (alt. 600 m)) he will observe a gradual change from rainfall to snowfall.
- Between 2.0 and 0 °C (from alt. 600 m up to alt. 950 m), he will be in the typical wet snow atmospheric layer. Here, he can see the wet snow adhere to all aerial obstacles with strong capillary bonding.
- Above 950 m in altitude, the observer enters a "dry snow" air layer at negative temperature. He sees dry snow deposits adhering to aerial obstacles with weak mechanical bonding.

The most serious damage to overhead lines results from the coincidence of topographical, aerological and infrastructure factors: such a severe event has a high potential risk when the atmospheric wet snow layer is at the same altitude as a large horizontal area, where a large number of activities are affected. This same severe event, when it happens in a mountainous area, presents a low potential risk, because it concerns a restricted area and affects a moderate number of activities.

Many northern countries, including France, have reported severe damage to power lines at successive IWAIS meetings (Table 4.1).

4.6.2 *Observation Systems Adapted to Topography*

Depending on the topography, there exist different observation systems:

Observation zones (0–1,500 m of altitude): Observations can be made by climbing a mountain road in an equipped car until the altitude of wet snow layer is found. This kind of observation system is not very expensive, and is relatively easy to operate. Identified overhead lines are necessarily required near the road for observations and measurements to be made.

Observation sites: Flat landscapes are ideal for numerous in situ observations, with observers living nearby to report on the occurrence, intensity, air temperature, wind speed, snow accumulation; to take reference photographs of snow sleeves along, with estimations of maximum sleeve diameter, and to make measurements of accreted snow density, using a small metallic rack for in situ sampling (Fig. 4.11). This observation system is not very expensive, but it needs special atmospheric coincidences between the right air temperature at the right altitude of the observation zone, and of course, trained observers.

Fig. 4.11 Example of structures supporting a smooth metallic cylinder 1.5 m in length, available for in situ wet snow sampling. About 30 of themselves have been installed in France over different areas

Testing station: In an area where wet snow events occur frequently, an experimental observation station can be set up with all the necessary equipment to record the meteorological and mechanical variables and carry out video surveys. In such a station, there should be several spans, provided with various mechanisms as a referring span, spans with different torsional stiffness, with mechanical prevention systems, spans with Joule effect, etc. (Lapeyre et al. 1986; Tournadre 1986, Eliasson and Thorsteins 2000). When a wet snow event is expected, a team can be sent on location to make specific observations: measurement and documentation of accretion parameters, conductor rotation, snow samples, density and LWC measurements, sleeve weight and diameter measurements, quality control of prevention systems, etc. Of course, this kind of equipment is expensive (construction and maintenance).

4.6.3 Results of Observations

At the beginning of the event, an observer notices a snow deposit on the windward side along the span. Some hours later, he will see a very homogenous cylindrical snow sleeve along the span, except at its two extremities, near the anchoring system, where a simple, more or less fragmented deposit is observed (Admirat et al. 1984, 1985, 1986, 1987, 1988).

The axial accretion mechanism: is observed at the anchoring points where conductor stiffness is infinite and over a short distance of high torsional stiffness, making conductor rotation impossible. The axial accretion only covers a distance of 15–20 m on each side of the anchoring points of the span (Fig. 4.12 and 4.13). These observations in natural meteorological conditions fully confirm and elucidate the equivalent tests in wind tunnel conditions.

The cylindrical accretion mechanism: takes place on the rest of the span length having low torsional stiffness, where the conductors can rotate and generate a continuous, cylindrical and homogenous snow sleeve (Fig. 4.14). This is also true for wind-tunnel conditions.

A cylindrical snow accretion mechanism is also observed on all smooth metallic wires, due to the permanent sliding of the snow on the metallic surface. This mechanism can be applied to the 1.5-m long smooth metallic rack used for in situ sampling (Fig. 4.11).

Changing the accretion mechanisms by increasing the torsional stiffness of a span: A conclusive test was carried out at the Luchon testing station, during the March 5, 1988 event, where the accretion mechanisms on four conductors with different torsional stiffnesses were compared. The main result consists of an axial accretion on a 166-m length of a span, equipped with four anti-gyration counterweights (Fig. 4.15).

This test represents the first demonstration in natural conditions of the limited effect of a single counterweight on span stiffness and the need for several counterweights along a span to produce enough torsional stiffness to yield the self-limiting axial accretion mechanism only. As previously described for service spans, the

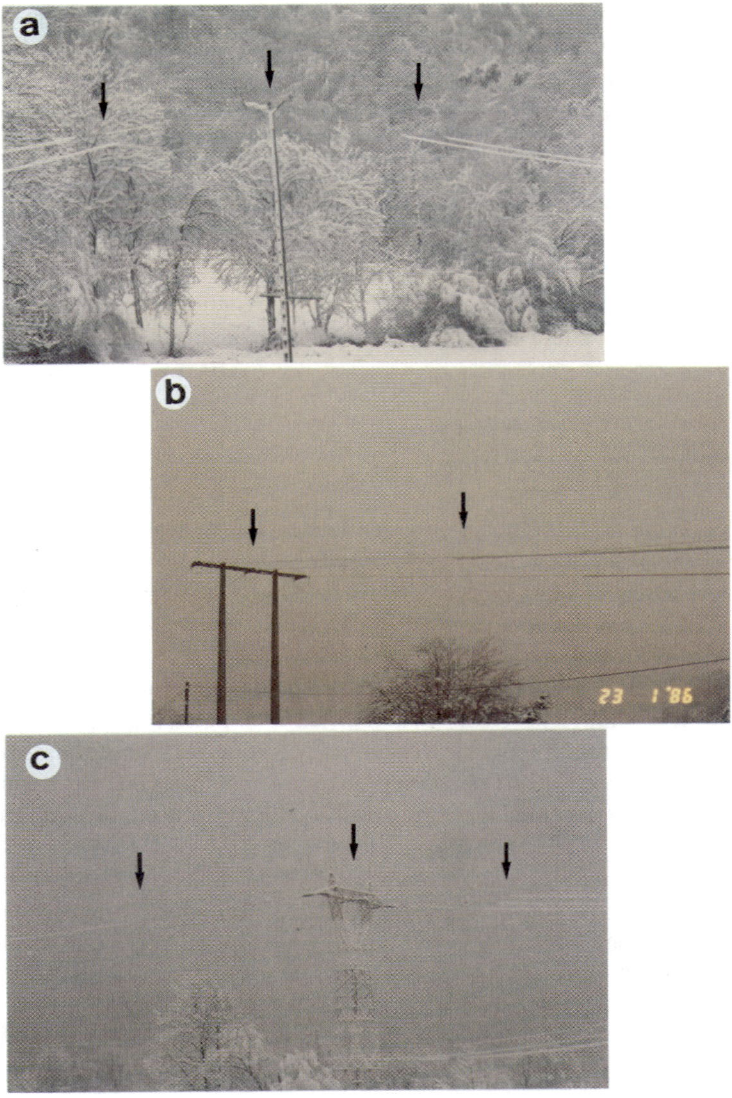

Fig. 4.12 Usual observation of the two accretion mechanisms on overhead lines: axial accretion (i.e. snow deposits) on each part of the anchoring points and cylindrical accretion (i.e. snow sleeve) beyond this distance. (**a**) and (**b**) On these 20 kV lines, there is no accretion on either side of the anchoring points, except for some fragmented snow deposits (axial snow accretion), while snow sleeves (cylindrical accretion) are clearly identified 10–15 m beyond each anchoring point; (**c**) Same observations on this 150 kV line: there is no visible snow accretion along 20–25 m on either side of the anchoring point (axial accretion) in the zone of high torsional stiffness, while snow sleeves (cylindrical accretion) are visible beyond this distance, as soon as the torsional stiffness adequately decreases to allow for conductor rotation

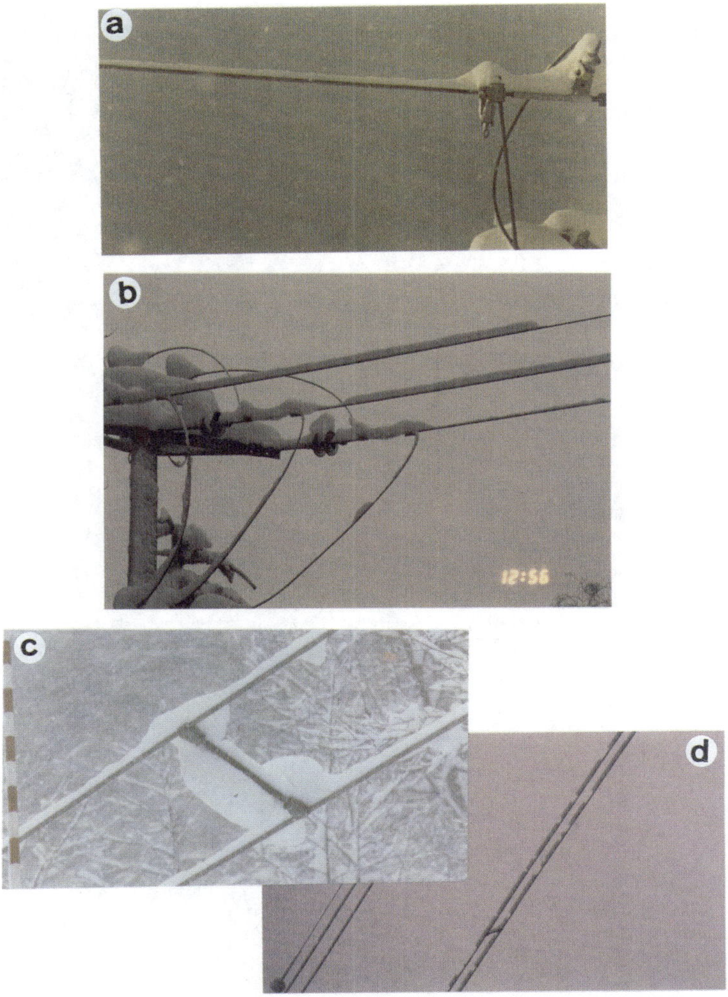

Fig. 4.13 Detailed examples of axial accretions (i.e. more or less fragmented snow deposits) along the high stiffness sections of the spans. (**a**) Zoom showing a typical axial accretion (snow deposit) near the anchoring point; (**b**) Zoom of axial accretion (fragmented deposits) on the three conductors; (**c**) Zoom of axial accretion on each side of the spacer of this HV line; (**d**) Fragmented snow deposits along 10–15 m because of forced rigidity due to the spacer

strong torsional stiffness of each side of an anchoring point decreases rapidly and limits the axial accretion mechanism to a short length of the span. The theoretical torsional stiffness (Gosselin and Lapeyre 1985; Admirat and Lapeyre 1988) of stranded conductors is clearly distinct from the low stiffness observed on spans subjected to snow accretion (Fig. 4.16). Therefore, to increase the torsional stiffness of a span, it is necessary to artificially increase the number of rigidity points by means of a successive anti-gyration counterweights.

Fig. 4.14 Examples of cylindrical accretions (cylindrical snow sleeves) along the low stiffness sections of the spans. (**a**) Precision work by a Japanese technician removing a snow sample from a sleeve 25 cm in diameter; (**b**) Usual view of the 3 similar cylindrical wet snow sleeves on a 20 kV line; (**c**) The snow sleeves begin to break into small fragments about 1 to 2 m, shortly after the event, probably under the influence of solar radiation; (**d**) Usual view showing the two types of mechanisms on the same span (135 m, 20 kV line): axial accretion from anchoring points up to a short distance (10–15 m), followed by a continuous cylindrical accretion along the spans, about 100 m, itself followed by the second axial accretion by approaching the second anchoring point

Observations of sleeve discharge: Usually, partial discharges occur a few hours following the event, if the air temperature remains positive. In this case, the excess heat increase the LWC of the sleeve, which causes stretching until some fragments of 1 to 2 m in length fall off (as in Fig. 4.14b). Sometimes the air temperature falls

Fig. 4.15 Observation of the close relationship between the accretion mechanisms and the torsional stiffness of the conductors (EDF testing station of Luchon, March 5, 1988). **Cable 1** (reference cable): Cylindrical accretion along 145 m caused by a 270-degree cable rotation (as indicated by the three red marks). Axial accretion on the last 10 m approaching the anchoring points at the two extremities of the span. The averaged overload measured by tensiometric instrumentation is 3.2 kg m^{-1}; **Cables 2 and 4** (1 counterweight at 83 m at mid-span): Cylindrical accretions along 2 × 65 m caused by a 100-degree rotation of the two mid-spans (as indicated by the three red marks on cable 2). Axial accretions along 2 × 8 m on each part of the counterweight (rotation of only 35 degrees) and on the last 2 × 10 m approaching the anchoring points. The average overload is 2.5 kg m^{-1}, a reduction of 20%; **Cable 3** (4 counterweights and 5 intervals of 33 m): Permanent axial accretion on the 166 m of the span length. The span rigidity is equivalent to the one of a fixed cable. The average overload is 1.2 kg m^{-1}, a reduction of 60%

below zero when cold air follows the atmospheric disturbance and the cooling effect causes the liquid water to freeze quickly, transforming the wet snow sleeve into a frozen snow sleeve, which can remain on the span for a more or less extended period (Fig. 4.17).

Measurements of snow sleeve densities: The measurement of ρ is more difficult in natural conditions than in wind tunnel conditions. However, more than 200 measurements were made during the four 1985–1988 winter periods at the Villefort and Luchon EDF testing stations, and sometimes at different sites, thanks to the availability of metallic mini-racks (Fig. 4.11), and using the same technical method (Fig. 4.18). The extreme values of ρ were found to be between 90 and 300 kg m^{-3}, with 50% of the values being between 150 and 250 kg m^{-3}. These values are lower than those obtained in wind tunnel experiments. In addition, the dependence between densities and wind speeds is not as clear as in wind tunnel conditions, mainly due to the lack in wind-speed measurements at the exact position of the density measurements.

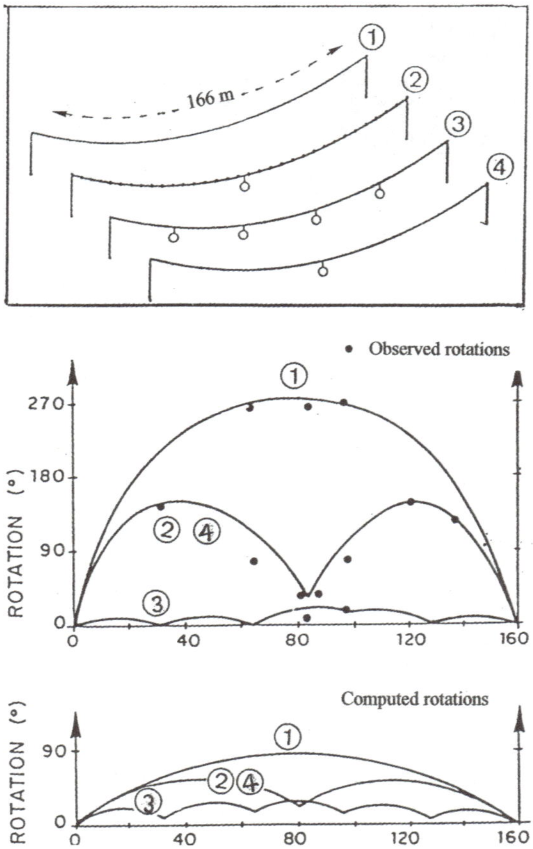

Fig. 4.16 Calculated and observed rotation angles (or torsional stiffness) of an ASTER 570 conductor according to the experimental device of the EDF testing station of Luchon (applied to the wet snow event March 5, 1988, Fig. 4.15). *High:* Schematic view of the EDF testing station showing the 4 spans, (AAAC 228 conductor), each 166-m long; *middle:* observed conductor rotation (see Fig. 4.15); *bottom:* computed conductor rotation using a mechanical model

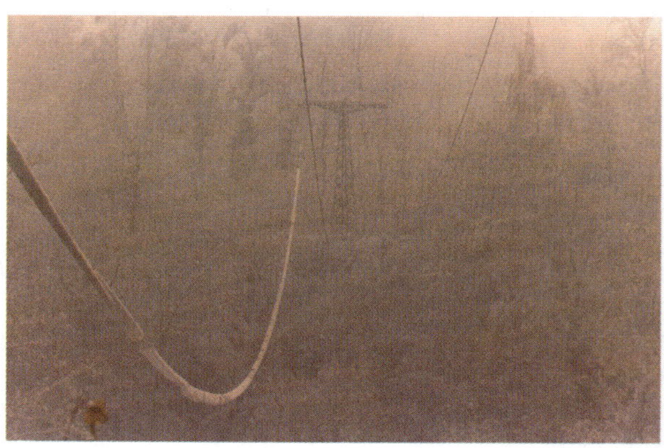

Fig. 4.17 Single frozen snow sleeve remaining and twisting on a 150 kV span, followed few times later by an abrupt unloading

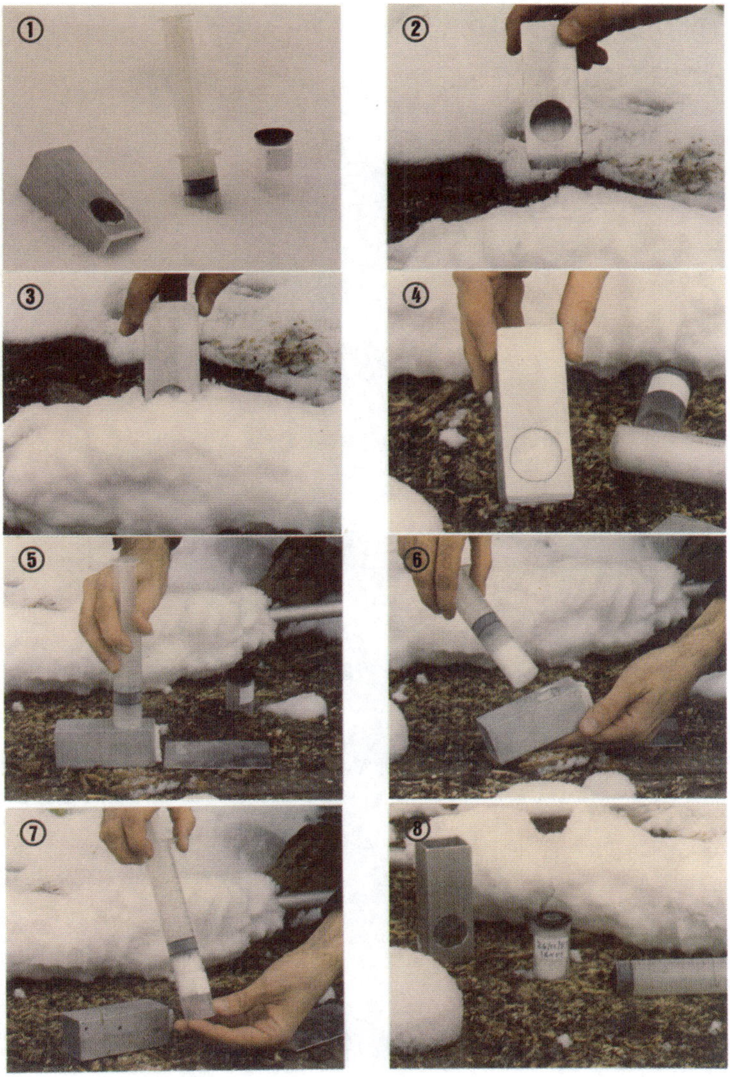

Fig. 4.18 In situ measurements of a wet snow sleeve density. (**1**) Cooling the equipment: mini-core, syringe, storage box; (**2**) The mini-core; (**3, 4**) Taking and extracting the wet snow sample; (**5, 6**) Taking and extracting the calibrated snow volume into the syringe; (**7, 8**) Stocking the calibrated sample into the storage box, waiting for total melting to measure the equivalent water volume (snow mass). This practical work takes one minute to perform

Measurements of snow-sleeve Liquid Water Content: Only 25 LWC measurements of wet snow samples were successful, due to several technical difficulties in natural conditions (Fig. 4.19), the values of which were found to be between 2 and 12%. These results are lower than those obtained under the stable and optimal conditions of the wind tunnel.

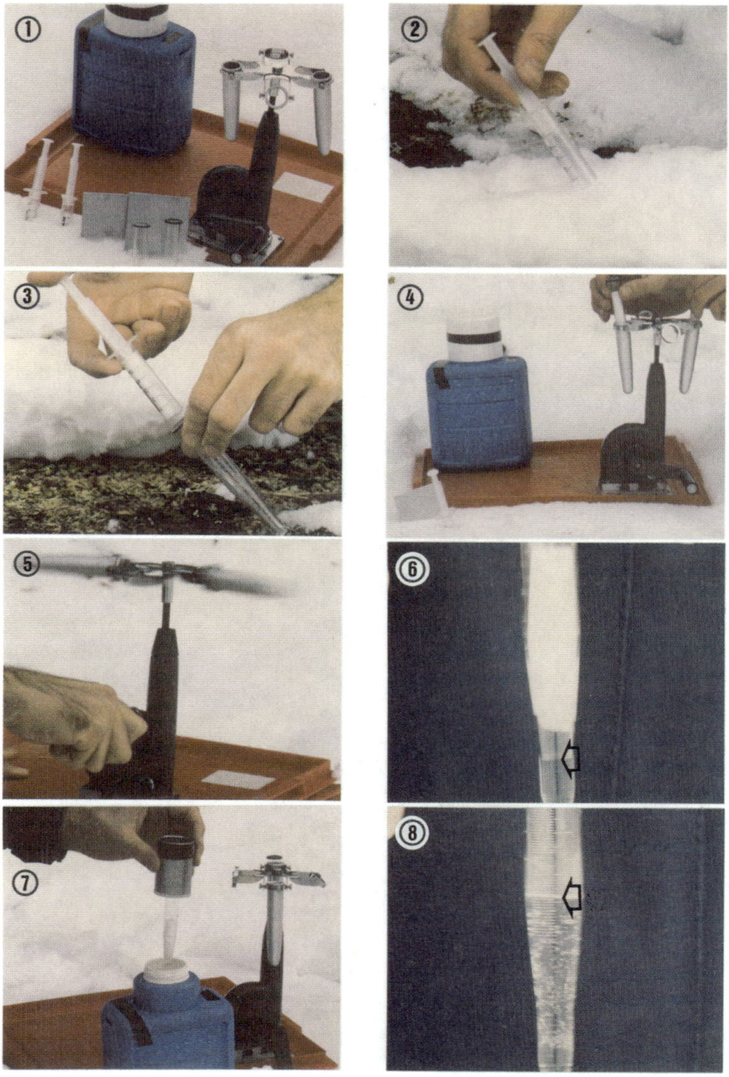

Fig. 4.19 In situ measurements of a wet snow sleeve LWC. (**1**) Cooling the equipment: double syringes, double test tubes, centrifugal machine, warm water container; (**2**) Taking two wet snow samples of any volume; (**3**) Introducing the samples into the two test tubes at 0 °C; (**4, 5**) Centrifuging at 0 °C; (**6**) Measuring the volume (the mass) of liquid water extracted from the sample at 0 °C; (**7**) Melting the sample by introducing into the warm bath; (**8**) Measuring the total volume (the total mass) of liquid water and obtaining the ratio of the two volumes. This practical work takes five minutes to perform

4.6.4 Final Calibration of the Cylindrical Accretion Model

An empirical calibration method consists in finding the appropriate values of β and ρ, allowing for the best agreement between observed and calculated data at the observation time of each event. For example, applied to the Japanese Irihirose wet snow event, the best agreement (however still with a high correlation coefficient) was found with $\beta = 0.7$ and $\rho = 400\,\text{kg m}^{-3}$. In this case, the model overestimates the observations (Table 4.2).

Applied to other 13 documented events in Japan, the results are as follows (Table 4.3):

In spite of this empirical method, the β and ρ values however show a remarkable correlation with the average daily wind speed U, as shown in Fig. 4.20 with the following expressions of $\beta = 0.88\ U^{-0.88}$ (if $U > 1\,\text{m s}^{-1}$; $r = 0.95$) and of $\rho = 333.0 + 16.5\ U$ ($r = 0.80$)

In the end, the resulting practical calibrated values proposed for the Japanese events are as follows

Table 4.2 Comparisons between meteorological data, in situ observations and calculated data for the Irihirose wet snow event on April 9, 1982 (Admirat et al. 1986)

Hours	U m s^{-1}	θ air	P mm h^{-1}	LWC%	Φ cm	M kg m^{-1}
20	1.0	5.0	1.0			
21	2.0	5.0	4.0			
22	1.0	5.0	3.0			
23	1.0	4.0	3.0			
24	1.0	3.5	2.0			
01	1.0	3.3	4.0			
02	1.0	2.0	5.0	60	4	0.3
03	1.0	0.5	7.0	28	6	0.8
04	1.0	0.4	7.0	19	8	1.5
05	1.0	0.5	6.0	16	10	2.3
06	1.0	0.4	6.0	13	12	3.2
07	1.0	0.3	7.0	11	14	4.5
08	1.0	0.2	7.0	9	16	5.9
09	3.0	0.2	5.0	8	17	6.8
10	3.0	0.2	4.0	8	17	7.5
11	3.0	0.4	4.0	8	18	8.3
12	2.0	0.4	3.0	9	19	8.9
13	3.0	0.6	2.0	9	19	9.3
14	2.0	0.7	1.0	10	19	9.5

In situ observation data	Calculated data ($\beta = 0.7$; $\rho = 400\,\text{kg m}^{-3}$)
Duration of the snow event 12 h	
Sleeve diameter at 11 h: 12–13 cm	Sleeve diameter at 11 h = 18 cm
Incidents on MV lines at 13 h	Incidents on MV line at 7 h ($M > 4\,\text{kg m}^{-1}$)
No incident on HV lines	Incidents on HV line at 9 h ($M > 6\,\text{kg m}^{-1}$)

Table 4.3 Climatic data and optimal β and ρ values giving the best agreements between both the observed and calculated snow sleeve diameters or masses (Admirat et al. 1986, 1988)

Date	Observ. Site	U m s^{-1}	P mm h^{-1}	β	ρ
Feb. 12, 1972	Sapporo	5.0	5.2	0.25	350
Feb. 26, 1972	Kucchan	10.8	4.1	0.15	500
	Muroran	9.9	2.7	0.15	500
Nov. 30, 1972	Wakanai	14.2	7.8	0.05	600
	Obihiro	1.2	4.8	0.85	350
Dec. 23, 1980	Fukushima	1.8	3.9	0.4	350
	Koriyama	2.0	3.6	0.4	400
	Sendai	5.7	4.3	0.15	500
Jan. 3, 1981	Toyama	4.6	1.7	0.3	350
Apr. 9, 1982	Irihirose	1.4	5.2	0.7	400
	Koide	2.2	3.2	0.4	350
	Nagaoka	11.5	6.7	0.1	500
	Tokamachi	3.5	4.8	0.35	400

$$\beta = 1/U \tag{4.17}$$

$$\rho = 300 + 20\,U \tag{4.18}$$

with $1 < U < 10\,\mathrm{m\,s^{-1}}$.

Applying this empirical calibration to 22 documented wet sow events in France, such as the Alsace event on January 1, 1962 (Admirat et al. 1987) or the Roussillon event on January 31, 1986 (Maccagnan et al. 1988), and several events at the Luchon station, the main result has been to induce an appreciable reduction in the former ρ value to yield these two final equations

$$\beta = 1/U \tag{4.19}$$

$$\rho = 200 + 20\,U \tag{4.20}$$

with $1 < U < 10\,\mathrm{m\,s^{-1}}$.

Quality-control of the calibrated model was carried out during a snow event at the Luchon station, where the meteorological and mechanical data were recorded as a function of time. By including the successive hourly data and these new equations for β and ρ into the model, the comparison between measured and calculated hourly snow masses is in general very acceptable (Fig. 4.21).

Naturally, there exist some empirical or statistical models for calculating wet snow loads (Makkonen 1989; Shackleton et al. 1993; Sakamoto and Miura 1993a,b). There are also numerical expressions of wet snow densities (Eliasson et al. 2000; Sakamoto et al. 2005), which require large meteorological data bases. The main advantage of the physical model presented in Section 4.4 in association with the equations for β and ρ, is that it can be applied to any wet snow event in real time, past time and future time, if only three meteorological data are known or forecasted:

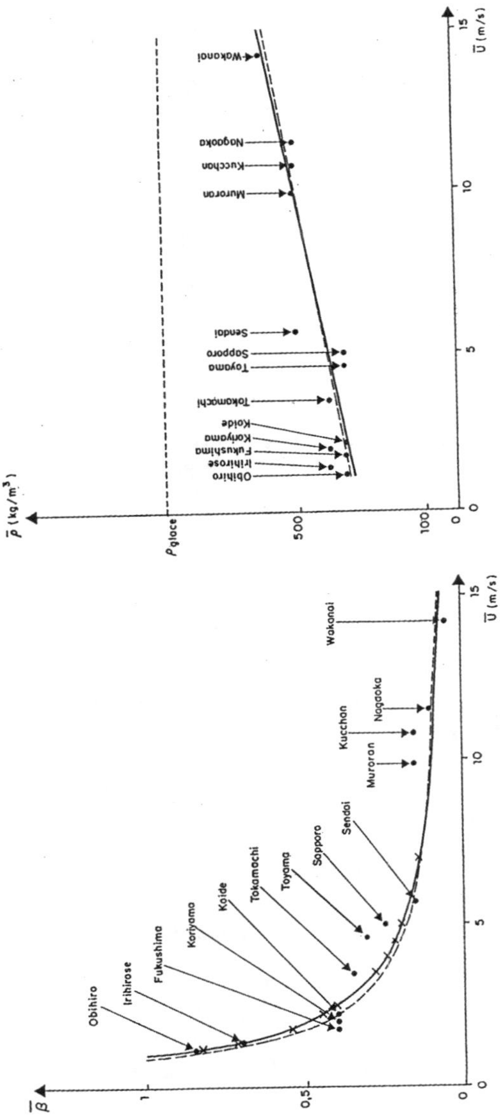

$$\beta = 0{,}88 \ U^{-0.88}$$

$$\rho = 16.5 \ U + 333$$

Fig. 4.20 Classified values of β and ρ versus corresponding average wind speed (Table 1)

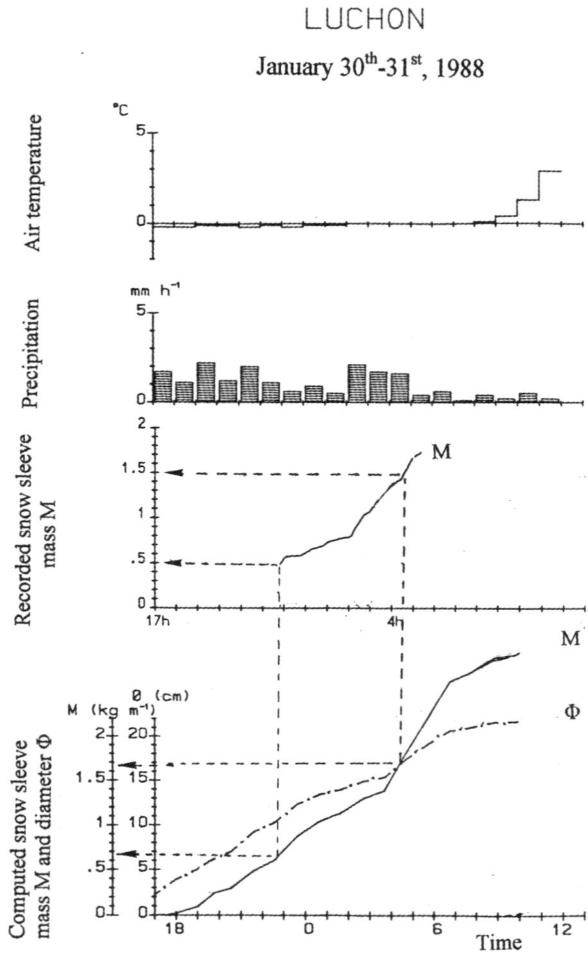

Fig. 4.21 Exceptional quality-control of the calibrated model: computed values 1.5 and 0.5 kg m^{-1} and corresponding measured values 1.7 and 0.7 kg m^{-1} (Luchon testing station, January 31, 1988)

wind speed, air temperature, amount of precipitation (Admirat et al. 1990). Three applications are presented in the next section.

4.7 Applications to Forecasting, Preventing, and Mapping the Wet Snow Overload Hazard

4.7.1 Forecasting the Wet Snow Overload Hazard

The first application of the previous study is the forecasting of a wet snow event occurrence and of the maximum overloads expected on a target area. Such information concerns the electrical power management staff by helping, if necessary, to anticipate the best prevention countermeasures.

This application needs short-range forecasted meteorological data on the atmospheric disturbances to localize the maximum hazard in space and time, and to compute the maximum snow overloads on the target area (Gland and Admirat 1986; Lapeyre and Gland 1987). The essential data are the altitudes of the 0 °C isotherm before and after the atmospheric disturbance, cumulative height of the expected precipitation, maximum duration of the precipitation, and wind speed at ground level.

A continuous weather survey: is necessary to monitor the on-going situation (satellite, radar, weather observations) and the short-range forecasted meteorological situation at different scales in time and space. When a forecasted atmospheric disturbance is approaching the target area at medium-range in time (72 or 48 hours), the survey process is activated.

Assistance of two selection criteria: First is the forecasted altitude of the 0 °C isotherm, which localizes the two extreme altitudes of the 300-m-deep atmospheric wet snow layer, and immediately informs if the target area will be included or not in this layer. The second criterion is the forecasted water equivalent height of the precipitation, which immediately informs about the hazard level: a cumulative height less than 10 mm cannot produce a snow overload greater than 1 kg m^{-1}, and the expected wet snow event cannot cause damage to the power network. These two selection criteria are repeated with the short-range forecasted data at 36, 24, 12 and 6 hours. When these two criteria are positive (hazard level $>$ 1 kg m^{-1} at any point of the target area), an alert-alarm process is engaged to meet the expected hazard.

Forecasted Hazard Warning (FHW): An alert report is established and sent to the management staff of the target area. It contains a large amount of information on the expected event in terms of maximum hazard level, potential damage, surface area affected, time and duration of the event.

The data entered into the FHW system are the following:

- Geographical identification of the target area and corresponding altimetric database.
- Forecasted altitudes of the 0 °C isotherm, before and after the atmospheric disturbance.
- Forecasted cumulative heights (water equivalent) at certain location in the area.
- Forecasted transit time of the wet snow event over the target area.
- Forecasted wind speed before and after the event.

The FHW system firstly organizes the meteorological variations for each 100 m in altitude into the atmospheric wet snow layer and computes the expected characteristics of the corresponding snow sleeve as a function of time (Table 4.4). Then, the FHW translates the geographical-meteorological database into appropriate terms for the users, according to both an altitude-duration time diagram and a geographical-overloads diagram (another example set in Fig. 4.22) which indicates:

- The maximum snow overloads each 100 m in altitude.
- Altitude asl. concerned by the wet snow atmospheric layer.

Table 4.4 Meteorological data processing and calculation of the expected snow sleeve characteristics as a function of time, at altitude 500 m

Forecasted meteorological data (example data)	
Maximum precipitation height	68 mm
Duration time:	24 hours
Average wind speed:	2,6 m s^{-1}
Altitudes of the 0 °C isotherm:	600 m coming down to 500 m

Hours	U m s^{-1}	θ air	P mm h^{-1}	LWC%	Φ cm	M kg m^{-1}
19	2.6	0.6	0.3			
20	2.6	0.5	0.5			
21	2.6	0.5	0.8			
22	2.6	0.5	1.0			
23	2.6	0.5	1.2			
24	2.6	0.4	1.7			
1	2.6	0.4	2.3			
2	2.6	0.4	3.0			
3	2.6	0.4	3.7	42	5	0.2
4	2.6	0.3	4.4	32	7	0.5
5	2.6	0.3	5.1	25	10	1.0
6	2.6	0.3	5.5	21	12	1.7
7	2.6	0.3	5.8	18	15	2.6
8	2.6	0.3	5.8	16	17	3.5
9	2.6	0.2	5.5	14	20	4.5
10	2.6	0.2	5.1	13	22	5.5
11	2.6	0.2	4.4	13	23	6.4
12	2.6	0.2	3.7	12	25	7.2
13	2.6	0.1	3.0	12	26	7.8
14	2.6	0.1	2.3	13	26	7.8
15	2.6	0.1	1.7	15	26	7.8
16	2.6	0.1	1.2	16	26	7.8
17	2.6	0.0	0.8	17	26	7.8
18	2.6	0.0	0.3	18	26	7.8

- Duration time before the overloads of 2, 3 ... 6 kg m^{-1}.
- Location of the hazard area $>$ 3 kg m^{-1} (or $>$ 4 kg m^{-1}) where damage is expected.
- The geographical area where the maximum wet snow overload is expected.

Waiting for the imminent real event, the FHW system updates the information twice a day.

Really Hazard Warning (RHW): An alarm report is established when the forecasted event becomes a real event. It is sent to the management staff just after verifying by local observers. Then, the RHW system re-updates the same kind of information, by taking into account the most elaborated meteorological data.

This double presentation makes it possible for the users to immediately identify the space-time importance of the real event and to save the residual time (about 10–15 hours) for preparing the appropriated counter-measures and facing the risk of

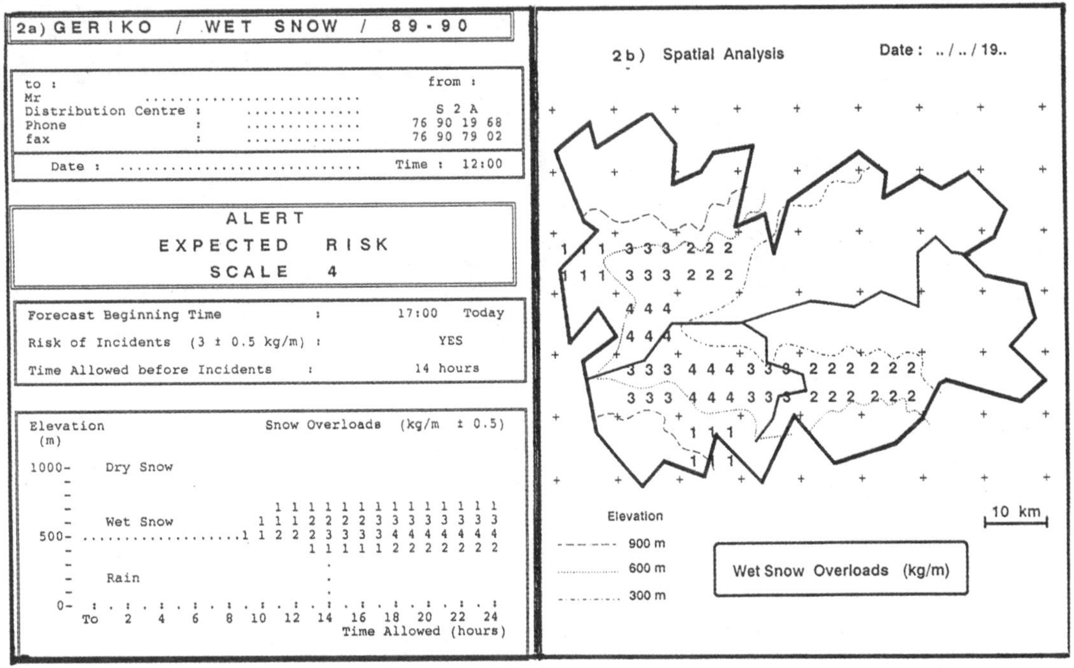

Fig. 4.22 Warning messages before a forecasted wet snow event: Maximum snow overloads expected on an altitude-duration diagram, Maximum snow overloads expected on a spatial-altitude diagram. (**a**) The atmospheric wet snow layer will be about 700 and 400 m in altitude: The maximum snow overloads will be of 4 kg m^{-1} at an altitude of 500 m, Dry snowfall at altitudes above 700 m; rainfall at altitudes below 300 m, The maximum duration of the event is 24 hours, Time 0 = is the starting time of the event, Overloads > 2, 3, 4 kg m^{-1} are expected at 11, 15, and 18 hours after starting time, The duration time between starting time and first damages to power lines is 14 hours, Damage is expected between 500 and 600 m asl (snow overload criterion > 3 kg m^{-1}); (**b**) Corresponding surface area, with the maximum overloads at the end of the event

damage to the power network. This method was tested in France from 1986–1988 by some EDF Centres (Admirat et al. 1984, 1985, 1986, 1987, 1988; Admirat and Maccagnan 1986) and has been widely used since1989 (GERIKO licence, EDF patent).

4.7.2 Preventing the Wet Snow Overload Hazard

The second application of the aforementioned study is to propose a controlled method for preventing excessive overloads on spans. As seen in wind tunnel experiments (Section 4.5.3) and in natural conditions (Section 4.6.3), increasing the torsional stiffness of the conductors changes the accretion mechanism from unlimited cylindrical to self-limiting axial. The problem lies in determining how many counterweights are needed on a given span to obtain an efficient prevention method.

Limited practical efficiency of a single anti-gyration counterweight: As already demonstrated at the Luchon testing station, a single counterweight placed at mid-span (i.e. 83 m from the anchoring points) induces an axial accretion only over a very short-distance (8–10 m) on each side of the counterweight, itself rotating 100 degrees (cables 2 and 4, Fig. 4.15). By comparing with the average snow overload on the reference cable, the reducing effect is approximately 20%. This reduction factor cannot be greater due to the preponderant length of the cylindrical accretion mechanism on these low-stiffness cables. So, the 83-m interval must be reduced.

Conversely, the four counterweights placed every 33 m render the 166-m span quasi-rigid and induce axial snow accretion along the entire span (cable 3, Fig. 4.12). The reducing effect is about 60% in this experiment and could be greater if the event had continued because the growth on the cable would have remained self-limiting, compared to the potentially unlimited growth of the sleeve on the reference cable. The best demonstration of the reducing effect of a quasi-rigid span is found in the records of the April 24, 1991 event, at the Luchon testing station, with a reduction rate of 78% (Fig. 4.23). In addition, a reduction rate of the same order of magnitude was registered on the 166-m, two-conductor bundle with high torsional stiffness induced by four spacers (and two anchoring points), having the equivalent action as four counterweights. The resulting conclusion is that a set of counterweights, placed at about 40-m intervals, is absolutely required to maintain strong and continuous torsional stiffness along a span and to obtain a weighable reduction of snow overloads.

Number of counterweights to be installed on a span to obtain total prevention: The records made at the testing stations of Luchon and Villefort have been further documented thanks to many other observations on standard power lines (Figs. 4.24 and 4.25).

Therefore, based on these observations as a whole, a highly efficient prevention method by increasing the torsional stiffness of the conductors requires installing counterweights, at about 40-m intervals, along the span. This 40-m length l (i.e. 2×20 m, on each side of each counterweight) is the maximum protected span length against the cylindrical accretion mechanism.

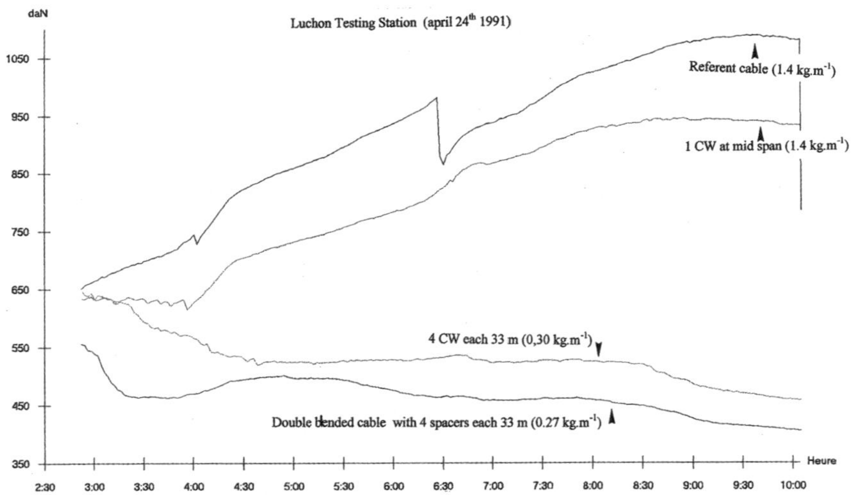

Fig. 4.23 Evaluation of snow overload reductions on 2 spans of complete high torsional stiffness during the nocturnal event on April 24, 1991 (registered measurements of Luchon testing station). **Cable 1** Usual growth of the sleeve on this reference ASTER 570 cable (with a small unloading at 06h30). The final tensiometric measurement reached 1,100 decaNewton, equivalent to 1.4 kg m^{-1} of snow overload, under the probable continuous cylindrical accretion along the cable, excepted in the usual anchoring zone, as in Fig. 4.15; **Cable 2** (1 single counterweight at mid-span): The final tensiometric measurement yields 950 daN or a snow overload 1.1 kg m^{-1} (a reduction of 20%); same kind of accretion than on cables 2 and 4 in Fig. 4.15; **Cable 3** (4 counterweights each 33 m): The final tensiometric measurement yields 450 daN or a snow overload of 0.3 kg m^{-1} (a reduction of 78%) due to the continuous axial accretion mechanism along the span, as cable 3 on Fig. 4.15; **Cable 4** (double bundle cable with 4 spacers): The final tensiometric measurement yields 410 daN or a snow overload of 0.27 kg m^{-1} also induced by a double continuous axial accretion

If

n is the number of counterweights to be installed along the span length L
s_x is the maximum overload taken into account by span design
S_x is the maximum snow overload expected on the span for preventive conditions
l is the maximum protected span length ($l = 40$ m for ASTER 570 cable)

then

$L\, s_x$ is the maximum overload allowable before breaking
$L\, S_x$ is the maximum overload induces by snow on an unprotected span
$2\,l\,(n+1)$ is the total protected span length induces by n counterweights (placed at 40-m intervals) and the two anchoring points
$2\,l\,(n+1)\,S_x$ is the residual snow overload on the span length where the cylindrical accretion mechanism is still possible, because of residual length of low torsional stiffness (lack of counterweights, excessive intervals, etc.)

Fig. 4.24 Cylindrical accretion along the conductor 2 of standard torsional stiffness and axial accretion along conductors 1 and 3 of high torsional stiffness (20 kV line). This 20 kV line is equipped on phases 1 and 3 with counterweights every 40 m, inducing axial accretions visible up to the anchoring point (two black lines of the downward face of the conductors), while on phase 2, without counterweights, a continuous cylindrical accretion of 12 cm in diameter is visible along the span, except of course very close to the anchoring point; The snow overload reduction is 1.7 kg m^{-1} to 0.15 kg m^{-1} (90%)

The "total prevention equation" can be obtained by writing that the residual snow overload must be lower than the yield point of the span before breaking: $[L - 2\,l\,(n + 1\,)]\,S_x < L s_x$, or $2\,l\,(n + 1\,)\,S_x > L\,(S_x - s_x)$ and:

$$n > [(S_x - s_x) - 1)]\,L/40 \qquad (4.21)$$

According to the selected S_x, a double choice is given to the designer for both technical and economical considerations:

- Either to reconstruct the overhead line by taking S_x as the new mechanical constraint.

Fig. 4.25 Limited preventive effect of a single counterweight on a 225 kV line. The 3 conductors are equipped with only a single counterweight placed 80 m from the tower; an axial accretion is visible (*three black lines*) 15–20 m on each side of the 3 counterweights, followed by a 50-m-long cylindrical accretion, 15 cm in diameter, on the 3 conductors, itself followed by an axial accretion of 10–12-m long (no visible on the figure) up to the 3 anchoring points

- Either to install a number of the counterweights corresponding to a partial or total prevention.

Such a prevention method has been studied from experimental and theoretical points of view (Admirat and Lapeyre 1988; Saotome et al. 1988; Yamaoka et al. 1988; Poots and Skelton 1988, 1993; Skelton and Poots 1991; Hardy et al. 2005). It has been introduced into EDF Centres since 1987, and several hundreds of kilometres of medium and high voltage lines have been equipped with anti-gyration counterweights.

Using Joule effect as an active prevention method: The heat released by Joule effect at the conductor-snow interface turns the snow into liquid water, which spreads inside the snow sleeve. The increase in LWC, up to 35–40%, causes the flowing of snow and the stretching of the sleeves, up to tearing out some fragments of 1 to 2 m in length. The heat intensity and the corresponding increase in LWC have been studied from theoretical and experimental points of view in different meteorological conditions. In severe conditions, a Joule effect of about 30–40 $W\,m^{-1}$ (I # 700–800 amperes) is adequate for a total melting of the snow impacting the conductor, while only 15–25 $W\,m^{-1}$ (I # 500–600 amperes) is enough to increase the LWC by up to 40–50%. However, this active prevention method, very easy to operate in wind tunnel experiments (Colbeck 1976; Lapeyre and Hiriart 1984; Admirat et al. 1988; Zsolt et al. 2005), is really difficult to bring into actual operation. It seems more interesting for rime ice prevention (Prud'Homme et al. 2005) than for wet snow prevention.

4.7.3 Mapping the Maximum Snow Overload Hazard

The third application of the calibrated model refers to both the mapping of a single wet snow event and the mapping of the historical hazard calculated from adequate meteorological data over a historical period, with events having really existed (Meteo-France or EDF archives or newspaper clippings) or supposed to have existed.

Mapping the spatial-time-development of a single wet snow episode: With hourly weather data from meteorological stations, the development of a snow event can be recreated by modelling the snow accretion at different space-time developments (Admirat et al. 1987; Maccagnan et al. 1988). At each information point, the mass and diameter of the snow sleeves are calculated hour by hour by taking the average hourly values of wind speed, air temperature and snowfall intensity (Fig. 4.26).

Calculating the maximum "Snow overload hazard" from local meteorological data base: This former method could be applied to a series of events having really or potentially existed over a target area by selecting available days, at the nearest meteorological centres and at many local stations of the French climatic

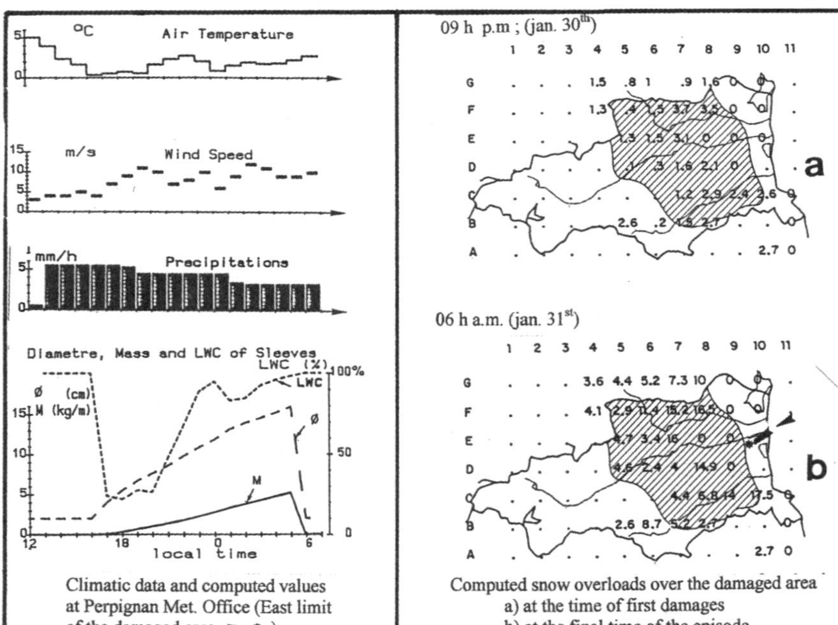

Climatic data and computed values at Perpignan Met. Office (East limit of the damaged area ➤*)

Computed snow overloads over the damaged area
a) at the time of first damages
b) at the final time of the episode

Fig. 4.26 Rebuilding the spatial-time development of a major wet snow event (January 30–31, 1986, Perpignan, France). The average climatic data, the corresponding computed values of the diameter, the mass and the LWC of the snow sleeves as a function of time at Perpignan city on January 30–31, 1986 (Fig. 19). Two spatial recreations of the snow overloads at 09h00 p.m. (time of the first damage) and at 06h00 a.m. at the final stage of the event network, where the air temperature and amount of precipitation are measured daily over a 20- or 30-year historical period. The operating procedure is as follows:

- At each station, a "wet snow day" is defined as any day having a daily precipitation greater than 15 mm and extreme air temperatures between −2 and 4 °C. Then, a database of the 10–15 largest "wet snow days" is established, documented with 3 parameters:

Snow Overload (Sm)

Nice-Alpes d'Azur — Period: 1971-2000 — 30 years

Clans (331m)

Date	Tn	Tx	Pmm	Vm	Sx	Sm	AF Annual Frequency	RP Return Period
Feb. 11, 1975	-0.7	5.9	64.5	4.0	6.1	4.9	0.03	30.0
Dec. 8, 1977	-3.5	8.2	68.0	5.0	5.8	4.6	0.07	15.0
Dec. 24, 2000	0.0	5.0	53.2	3.7	4.2	3.4	0.10	10.0
Jan. 28, 1978	-0.6	2.2	59.6	7.1	3.8	3.0	0.13	7.5
Dec. 21, 1983	-0.7	6.8	45.7	4.5	3.0	2.4	0.17	6.0
Jan. 10, 1999	-3.3	2.0	42.7	4.3	2.6	2.1	0.20	5.0
Jan. 5, 1994	-2.0	7.5	37.2	3.1	2.5	2.0	0.23	4.3
Jan. 18, 1995	-0.5	3.0	39.0	3.2	2.5	2.0	0.27	3.8
Dec. 28, 1981	-2.1	3.2	39.5	4.3	2.3	1.8	0.30	3.3
Dec. 8, 1990	-6.0	3.0	36.2	5.1	1.8	1.4	0.33	3.0

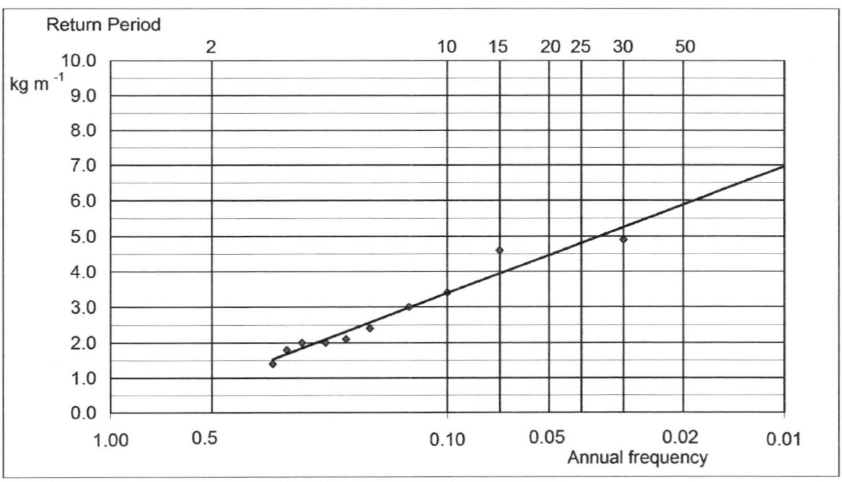

Average: 2.8 — Standard deviation: 1.19

Return period (year)	2	10	15	20	25	30
Hazard references (kg m^{-1})		3.4	4	4.5		

Fig. 4.27 Selected "wet snow days" and classified "snow overload hazards" from meteorological database of the Clans station (alt. 331 m) over a 30-year period. Tn and Tx = minimum and maximum air temperatures of the day; Pmm = amount of the daily precipitation in mm water-equivalent; Vm = average wind speed of the day; Sx = maximum calculated value of the "snow overload hazard" Sm = hazard taken into account (= Sx reduction by a factor of 0.8); AF = annual frequency; RP = return period

Table 4.5 Hazard references calculated from a "wet snow days" database selected at the station in Clans, over a 30-year historical period (Admirat et al. 1988)

Hazard references / Station in Clans	
Two-year hazard	$1.0\,\text{kg m}^{-1}$
Five-year hazard	$2.5\,\text{kg m}^{-1}$
Ten-year hazard	$3.4\,\text{kg m}^{-1}$
Twenty-year hazard	$4.5\,\text{kg m}^{-1}$

   - Constant positive air temperature of 1 °C,
   - The amount of precipitations P (mm water equivalent),
   - The daily wind speed.

- The model put in order 3 standard duration-precipitation levels (8 hours if $15 < P < 25$ mm; 12 hours if $25 < P < 40$ mm; 18 hours if $P > 40$ mm) and distributes the precipitation as a "bell-curve" with a maximum in the middle of the event. Then, the maximum snow overload is calculated for each of the 10–15 "wet snow day" selected. The results are classified in decreasing order in a semi-logarithmic diagram according to their frequencies over the historical period.
- A linear fitting determines the references of the "snow overload hazard" at any period of time. For example, the database of the station in Clans (Fig. 4.27)

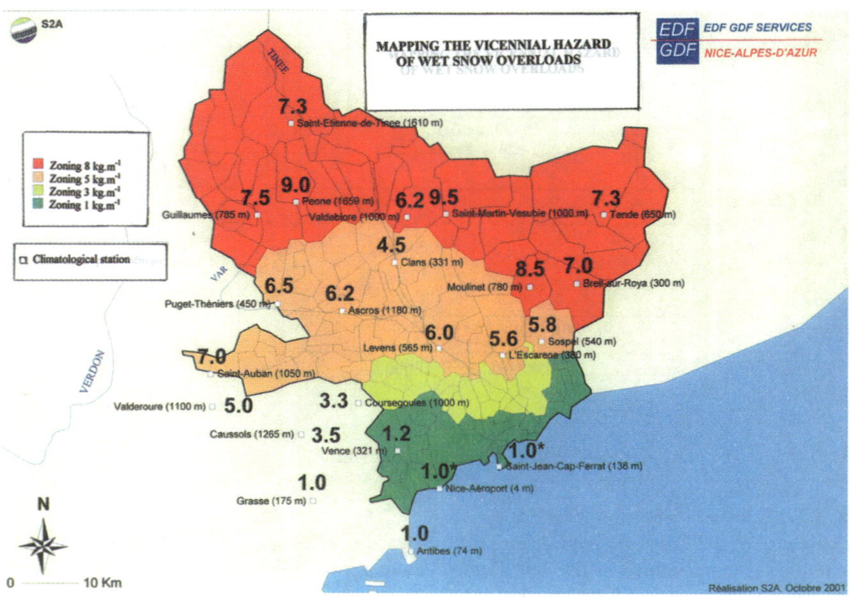

Fig. 4.28 Small-scale mapping of the 20-year hazard in the "Nice-Alpes-d'Azur" EDF area, underlined with four sub-areas of standardized hazard levels

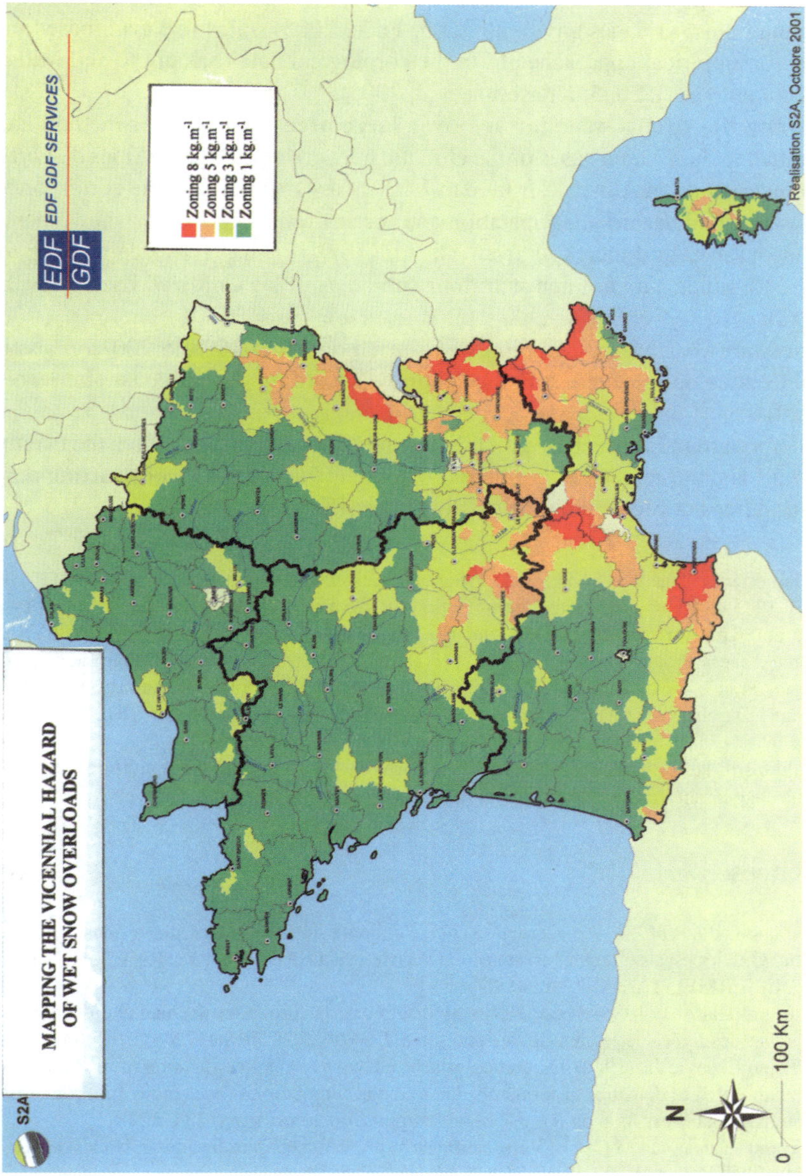

Fig. 4.29 Large-scale mapping of the four 20-year standardized hazard levels in France, based on more than 2,000 meteorological databases

includes 10 selected "wet snow days" over a 30-year historical period. The classification and the linear adjustment yield the hazard references (Table 4.5).

Other procedures to determine extreme values of wet snow loads and corresponding return periods have been used in Norway, based on highest 24-hour winter precipitations (Krishnasamy et al. 2000; Fikke 2005) or, as in Japan, based on statistical or empirical approaches from meteorological data (Sakamoto and Miura 1993; Sakamoto et al. 2005; Kitashima et al. 2005).

Mapping the twenty-year hazard over large areas: The former method has been applied to the 23 data bases of the climatic network of the "Nice Alpes d'Azur" area (French Southern Alps; alt. from 0 to 1,600 m asl.) to calculate the corresponding snow overload hazard at each station and for any return period. As shown here, the twenty-year hazard increases from $1\,\mathrm{kg\,m^{-1}}$ to $9.5\,\mathrm{kg\,m^{-1}}$ from sea level up to the highest altitudes, underlined with four sub-areas of standardized hazard levels (Fig. 4.28).

This twenty-year hazard mapping was carried out from 1995 to 2001 on about 85 EDF Centres up to covering France entirely, according to four constraint levels (Fig. 4.29).

Henceforward in France, based on this new physico-statistical method, the twenty-year hazard is the reference return period to take into account for constructing new distribution lines or reinforcing the existing ones (EDF 1995).

Acknowledgments The Wet Snow R&D Program received the support of ELECTRICITE DE FRANCE, 1983–2001 with the active participation of: Bernard DALLE, Head of the Infrastructure Program, Transmission Lines Department EDF, Clamart, France; Jean Louis LAPEYRE, Head of the Access Program to Distribution EDF Network, Distribution Lines Department, EDF, Courbevoie, France; Jean Claude GRENIER, Professor, Joseph Fourier University, Grenoble, France; and Yukichi SAKAMOTO, Head of the Research Department of the CRIEPI, Tokyo, Japan.

I would also like to thank Prof. Masoud FARZANEH for giving me the opportunity to contribute this chapter and for his unconditional help in finalizing the text.

References

Admirat P and Dalle B (1984) Accrétion de la neige collante sur les conducteurs aériens: synthèse des études et des essais effectués en soufflerie avec le C.R.I.E.P.I. en janvier 1984. Tech Report, EDF/DER/ERMEL/TA/HM/72-5200, 27 p

Admirat P and Dalle B (1985) Théorie et modélisation de la formation des manchons de neige collante sur les lignes aériennes. Tech Report n°8, EDF/INSU/CNRS M72/1B5912, 68 p

Admirat P and Lapeyre JL (1988) Theoretical study and experimental verification of the torsion of cables submitted to densities of moments due to the accumulation of wet snow. In: Proc of the 4th International Workshop on Atmospheric Icing of Structures, Paris: 324–329

Admirat P and Maccagnan M (1986) Vérification de la prévision quantitative des épisodes de neige collante. Tech Report, EDF/DER/ERMEL TA/HM/72-5550

Admirat P and Sakamoto Y (1988) Wet snow on overhead lines: a state of the art. In: Proc of the 4th International Workshop on Atmospheric Icing of Structures, Paris: 7–13

Admirat P et al. (1985) Simulation en soufflerie des mécanismes d'accrétion cylindrique de neige collante. Tech Report EDF/DER/ERMEL/TA/HM/72-5291

Admirat P et al. (1986) Calibration of a wet snow accumulation model with 13 documented episodes in Japan. Tech Report EDF/DER/ERMEL M72/1B 5600

Admirat P et al. (1986) Quantitative results and proposed mechanisms on wet snow accretion in the Ishiuchi wind-tunnel facilities. In: Proc of the 3rd International Workshop on Atmospheric Icing of Structures, Vancouver

Admirat P et al. (1987) Première reconstitution quantitative d'un épisode de neige collante: Mulhouse/Belfort/Colmar 1–2 janvier 1962. Tech Report, EDF/DER/ERMEL/TA/HM 72 – 5612, 30 p

Admirat P et al. (1988) Calibration of a wet snow accumulation model on real cases in Japan and France. In: Proc of the 4th International Workshop on Atmospheric Icing of Structures, Paris, pp 129–133

Admirat P et al. (1988) Influence of Joule effect and of climatic conditions on Liquid Water Content of wet snow accreted on conductors. In: Proc of the 4th International Workshop on Atmospheric Icing of Structures, Paris: 367–371

Admirat P et al. (1990) Synthesis of observations and practical results of the EDF wet snow program 1983–1990. In: Proc of the 5th International Workshop on Atmospheric Icing of Structures, Tokyo

Admirat P et al. (1984) Observations of wet snow episodes on electric lines. Tech Report no 5912, EDF/DER/ERMEL/TA/M72

Admirat P et al. (1985) Observations of wet snow episodes on electric lines. Tech Report no 5287, EDF/DER/ERMEL/TA/M72

Admirat P et al. (1986) Observations of wet snow episodes on electric lines. Tech Report no 5553, EDF/DER/ERMEL/TA/M72

Admirat P et al. (1987) Observations of wet snow episodes on electric lines. Tech Report no 5631, EDF/DER/ERMEL/TA/M72

Admirat P et al. (1988) Observations of wet snow episodes on electric lines. Tech Report no 5699, EDF/DER/ERMEL/TA/M72

Chassant M (1902) Les chutes de neige sous le climat méditerranéen. Ecole Nationale d'Agriculture, Montpellier

Colbeck SC (1973) Theory of wet snow metamorphosis. Research Report 313, CRREL, Hanover

Colbeck SC (1976) Thermodynalical deformation of wet snow. CREEL Report, 74–44

Colbeck SC (1979) Grain clusters in wet snow. Journal of Colloid and Interface Sciences, 72–3: 371–384

Colbeck SC and Ackley SF (1982) Mechanisms for ice bonding in wet snow accretions on power lines. In: Proc of the 1st International Workshop on Atmospheric Icing of Structures, Hanover

EDF (1995) Technical Convention of Electricity – Guideline 175, April, Reims

Eliasson AJ and Thorsteins E (2000) Field measurements of wet snow icing accumulation. In: Proc of the 9th International Workshop on Atmospheric Icing of Structures, Chester

Eliasson AJ et al. (2000) Study of wet snow events on the south coast of Iceland. In: Proc of the 9th International Workshop on Atmospheric Icing of Structures, Chester

Fikke S (2005) Modern Meteorology and atmospheric Icing. In: Proc of the 11th International Workshop on Atmospheric Icing of Structures, Montreal

Gland H and Admirat P (1986) Meteorological conditions for wet snow occurence in France. Calculated and measured results in a recent case study on March 5th, 1985. In: Proc of the 3rd International Workshop on Atmospheric Icing of Structures,Vancouver

Gosselin M and Lapeyre JL (1985) Theoretical study of the twisting of overhead-lines conductors under torque densities caused by the accumulation of wet snow. Tech Report EDF/DER/ERMEL M72/1B 5332

Grenier JC et al. (1985) Théorie et modélisation de la formation des manchons de neige collante. EDF/INSU M72/1B 5912, Report n° 8

Grenier JC et al. (1986) Theoretical study of the heat balance during the growth of wet snow sleeves on electrical conductors. In: Proc of the 3rd International Workshop on Atmospheric Icing of Structures,Vancouver

Hardy C et al. (2005) Theoretical assessment of ice loading of cables as a function of their torsionnal stiffness. In: Proc of the 11th International Workshop on Atmospheric Icing of Structures,Montreal

Kitashima T et al. (2005) A new attempt to estimate wet snow accretion on Overhead wires. In: Proc of the 11th International Workshop on Atmospheric Icing of Structures, Montreal

Krishnasamy S et al. (2000) Estimation of extreme wet snow loads in Southern Norway. In: Proc of the 9th International Workshop on Atmospheric Icing of Structures, Chester

Lapeyre JL (1986) Etude des mécanismes d'accumulation de Neige Collante. CIGRE , Paris

Lapeyre JL and Gland H (1987) Overhead lines faced with climatic overloads. CIRED, Liège

Lapeyre JL and Hiriart A (1984) Protection par effet Joule des lignes aériennes contre les formations de manchons de givre ou de neige collante autour des conducteurs. Tech Report EDF/DER ERMEL, TA/HM/72 – 5133

Lapeyre JL et al. (1986) Fonctionnement de la station de Villefort du 28/01/1986 au 04/04/1987. Etude des surcharges de neige collante sur les portées expérimentales. Tech Report EDF/DER/ERMEL M72/1B 5539

Maccagnan M et al. (1988) Space-time modelisation of a major wet snow episode (Perpignan, January 30–31th, 1986). In: Proc of the 4th International Workshop on Atmospheric Icing of Structures, Paris: 14–18.

Makkonen L (1989) Estimation of wet snow accretion on structures. Cold Regions Sci Technol, vol 17: 83–88

Poots G and Skelton PLI (1988) A theoretical model of ice accretion on an overhead line conductor causing twisting of the conductor. In: Proc of the 4th International Workshop on Atmospheric Icing of Structures, Paris: 219–223

Poots G and Skelton PLI (1993) The effects of includong aerodynamic torque in the model of snow accretion on overhead transmission lines. In: Proc of the 6th International Workshop on Atmospheric Icing of Structures, Budapest

Prud'Homme P et al. (2005) Hydro-Québec TransEnergie line conductor de-icing techniques. In: Proc of the 4th International Workshop on Atmospheric Icing of Structures, Montreal

Sakamoto Y and Miura A (1993a) An estimating method of snow load on overhead power lines. In: Proc of the 6th International Workshop on Atmospheric Icing of Structures, Budapest

Sakamoto Y and Miura A (1993b) Comparative study of wet snow models for estimating snow loads on power lines based on general meteorological parameters. In: Proc of the 6th International Workshop on Atmospheric Icing of Structures, Budapest

Sakamoto Y et al. (1986) Modelling wet snow accretion in a wind-tunnel. In: Proc of the 3rd International Workshop on Atmospheric Icing of Structures, Vancouver

Sakamoto Y et al. (1988) Thermodynamic simulation of wet snow accretion under wind-tunnel conditions. In: Proc of the 4th International Workshop on Atmospheric Icing of Structures, Paris: 180–185

Sakamoto Y et al. (2005) Snow accretion on overhead wires. In: Proc of the 11th International Workshop on Atmospheric Icing of Structures, Montreal

Saotome H et al. (1988) Countermeasures for snow accretion on conductors. In: Proc of the 4th International Workshop on Atmospheric Icing of Structures, Paris: 363–366

Shackleton L et al. (1993) Comparison of growth rates on groups of stranded conductors. EALT contract

Skelton PLI and Poots G (1991) Snow accretion on overhead line conductors of finite torsionnal stiffness. Cold Region Science and Technology, vol 19: 301–316

Tournadre R (1986) Station d'études de la neige collante de Luchon-Vallée du Lys. Tech Report EDF/DER /ERMEL M72/1B 5534

Wakahama G (1965) Metamorphosis of wet snow. Institute of Low Temperature Sciences, series A, 23, Hokkaïdo: 51–66

Wakahama G and Mizuno Y (1979) Studies on tensile strenghtstrength of wet snow. CREEL, Special Report no 185, Hanover

Yamaoka M et al. (1988) Countermeasures to both of the wet snow accretion and the galloping damages on the transmission lines. In: Proc of the 4th International Workshop on Atmospheric Icing of Structures, Paris: 376–380

Zsolt P et al. (2005) Power line conductor icing prevention by the Joule effect: parametric analysis and energy requirements. In: Proc of the 11th International Workshop on Atmospheric Icing of Structures, Montreal

Chapter 5
Effect of Ice and Snow on the Dynamics of Transmission Line Conductors

Pierre Van Dyke, Dave Havard and André Laneville

5.1 Introduction

Transmission line conductors exposed to natural winds are subjected to wind-induced vibrations, including aeolian vibrations and wake-induced oscillations (in the latter case, only on bundles). These vibrations may impair the reliability and lifespan of conductors and their accessories, but can be attenuated to non-damaging levels using damping devices and/or spacers or spacer-dampers when required. However, when ice precipitations accrete on the conductors, the severity of the two phenomena can increase dramatically. Moreover, aeolian vibration of conductors coated with ice may then occur in frequency ranges outside damper capabilities.

Galloping, another wind-induced instability, also occurs on ice-accreted conductors and may result in spectacular conductor displacements. Galloping is a movement-induced excitation and its mechanism will be described on both single and bundle conductors.

This review of the combined effects of ice or wet snow and wind on overhead power lines includes serious loadings and instabilities such as the rebound of conductors following ice drop and the rolling of bundles due to accumulated glaze or rime ice and wind action in exposed spans or in elevated mountain routes.

Direct observation of the effects of icing on overhead power lines is an absolutely necessary preliminary to any understanding of the many phenomena that can occur. Without such observations, unrealistic assumptions about the mechanisms in presence could be made and lead to incorrect or incomplete solutions based on analysis. Since the number of variables such as the overhead line designs, the thickness and density of the ice or snow deposits, and the wind loadings to name a few, is large, an overview of the state of the art of these different phenomena will be presented here. Measurements conducted on transmission lines or test lines with natural or artificial ice are included to better understand and quantify those phenomena.

P. Van Dyke
Hydro-Québec Research Institute – IREQ, 1800, boul. Lionel-Boulet, Varennes (Québec), Canada J3X 1S1
e-mail: van_dyke.pierre@ireq.ca

M. Farzaneh (ed.), *Atmospheric Icing of Power Networks*,

Many of these effects have partial or approximate models supported by field data, to assist in reducing the damage. However, there is a need for more work for complete understanding, including cooperative ventures employing the different resources available to universities and utilities.

5.2 Aeolian Vibrations

5.2.1 Aeolian Vibration Mechanism

Aeolian vibrations are associated with the pressure fluctuations applied by the wind on the surface of conductors as the vortices are shed in the wake. These pressure fluctuations are present whether the conductors are in motion or not. The aerodynamics of these types of motion is usually referenced through the Reynolds number. Reynolds number (Re) is given by $\left(\mathrm{Re} = \frac{Vd}{\nu}\right)$ where V is the wind speed, d is the conductor diameter and ν is the kinematic viscosity of air. Then for example, with a wind velocity of about 20 km/h acting on a 25 mm diameter conductor gives a Reynolds number of the order of 10^4. In the case of a bare conductor with this Reynolds number, the boundary layer remains laminar from the point of stagnation, where the fluid is at rest on the conductor, to the point of separation located at an average angle of 91°. The shear layer proceeding from the point of separation undergoes transition and rolls on itself to form a vortex that will be shed downstream in the wake. Vortex shedding is then a fluctuating flow instability the frequency (f_v) of which is linked, via the Strouhal number (S_n), to the oncoming flow velocity (V) and conductor diameter only (d), as shown in Eq. (5.1) (Strouhal 1878).

$$f_v = \frac{S_n V}{d} \tag{5.1}$$

Aeolian vibrations of the conductor are initiated as the vortex shedding frequency approaches one of the conductor's resonance frequencies. In practice, aeolian vibrations cover a range of wind velocities extending from 80 to 120% the value of the matching Strouhal wind speed. For a circular cylinder, the Strouhal number remains nearly constant with a value of 0.2 within the range of the Reynolds number of interest. The uniform stranding around the conductor causes only minor departures from the smooth cylinder behaviour (Pon et al. 1990). However, as the conductor shape is modified by the ice accretion, the Strouhal number may change.

On transmission line conductors, aeolian vibration takes the form of multiple sinusoidal waves across the span. The frequency range may vary from 3 to 150 Hz and the amplitude may reach the magnitude of the conductor diameter at the antinode of these waves, or 'loops'. In this range of wind speeds, the conductor will vibrate and its steady amplitude of motion will depend upon an energy balance between the wind power input from the flow vortices and the damping ability of the line, mainly through conductor self-damping and by any dampers. The nature of vortex shedding is altered as the amplitude of conductor motion increases. Figure 5.1

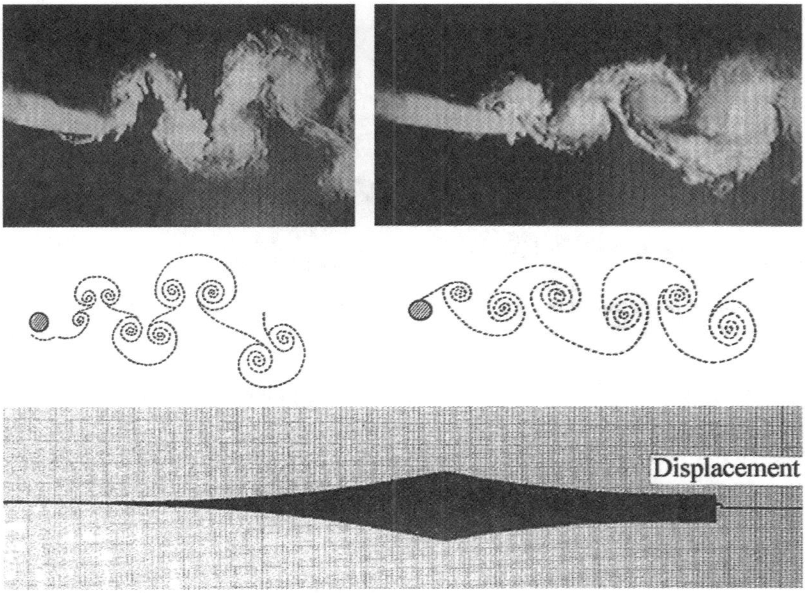

Fig. 5.1 Vortex shedding and transient response at the antinodes of an aeroelastic model of a conductor (Re = 7114) (Brika and Laneville 1993) (Reproduced by permission of Cambridge University Press)

shows two examples of such vortices in the wake of a cylinder in a wind tunnel and the resulting variation of antinode amplitude with time. The amplitude (peak-to-peak /2) will seldom exceed the diameter of the conductor at the antinode. Since conductor internal damping increases with frequency, aeolian vibration amplitudes tend to decrease with the frequency. The wind power (P) transferred from the wind to a vibrating conductor may be expressed in the following general form (Rawlins et al. 1979):

$$P = d^4 f^3 fnc\left(\frac{Y}{d}\right) \tag{5.2}$$

where Y is the vibration amplitude and the function $fnc\left(\frac{Y}{d}\right)$ is determined from wind tunnel tests.

5.2.2 Aeolian Vibrations with Ice Accretion

Ice and/or snow precipitation will affect aeolian vibrations through different mechanisms. A snow cover may smooth terrain obstacles that would normally contribute to wind velocity fluctuations, that is, to reduce the turbulence of the wind. A more

Fig. 5.2 Example of ice accretion on a conductor (Reproduced by permission of Hydro-Québec)

constant wind velocity and azimuth will produce winds that are more propitious to severe aeolian vibrations.

When ice is present, other factors will also contribute to increasing the severity of aeolian vibrations. For example, iced conductors may lock the conductor strands together so that conductor internal damping through strand slippage decreases. Moreover, internal conductor damping depends heavily on the mechanical tension of the conductor. The weight of ice will increase conductor tension, which will also reduce conductor self-damping. An example of ice accretion on a transmission line conductor is shown in Fig. 5.2.

Rawlins has observed on overhead ground wires that the vibration amplitude in the presence of ice was much more severe than what he had measured on a bare wire in the same conditions (Rawlins 1988). Based on Eq. (5.2), he explained that the ice accretion increases the conductor diameter, and at a given frequency, aeolian power increases to about the fourth power of diameter. This is illustrated in Figs. 5.3–5.5, which have been reproduced from his paper. These figures show records of numbers

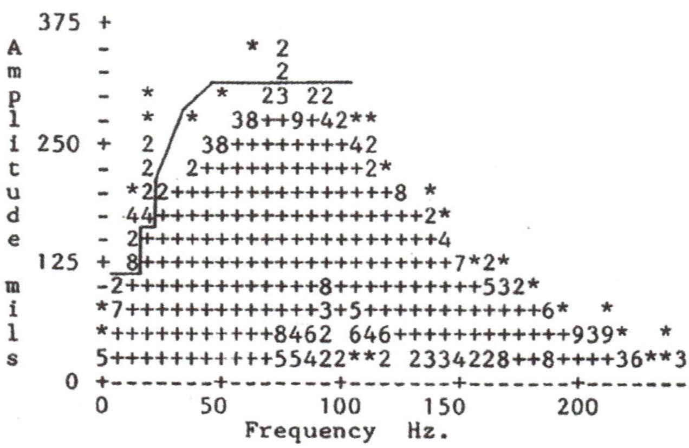

Fig. 5.3 Vibration summary 128 days in winter. 9.1 mm (3/8″) EHS Steel OHGW in 244 m span. Amplitude measured 406 mm from support. (Rawlins 1988) (Reproduced by permission of IEEE) * Only one sample occurred at the frequency-amplitude combination; 2–9 No of occurrences; + More than nine samples occurred; Points located above the *solid line* were recorded during icing periods

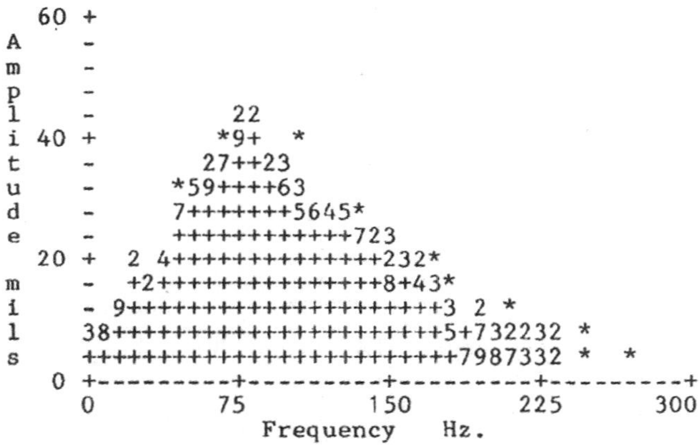

Fig. 5.4 Vibration summary 51-day test. Data from periods with icing deleted. 3 No. 6 AW OHGW in 366-m span. Amplitude measured 89 mm from support (Rawlins, 1988) (Reproduced by permission of IEEE)

of vibration occurrences at each combination of amplitude and frequency for one winter. In Fig. 5.3, the days with icing are distinguished from those with the wire bare, from measurements far from the end of the span. The figure clearly shows that icing periods have produced the most severe vibrations. Comparison of Figs. 5.4 and 5.5, which show comparable measurements without and with ice respectively, on an alumoweld overhead ground wire, leads to the same conclusion.

Rawlins concluded that the increase in aeolian vibration power seems the most likely cause of the episodes of damper fatigue in apparently well-protected lines.

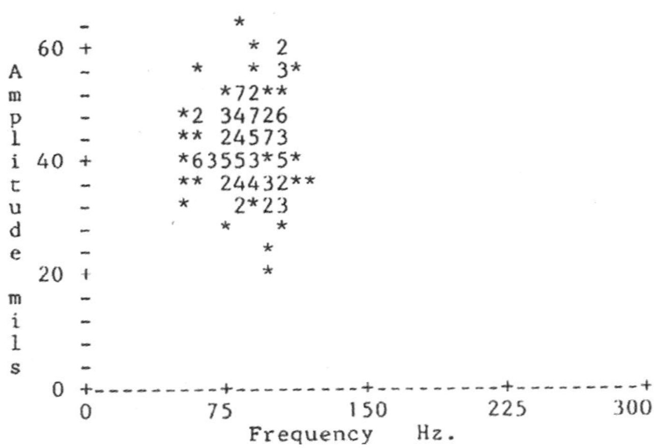

Fig. 5.5 Vibration summary. Same test as Fig. 5.4 iced-conductor data only. 3 No. 6 AW OHGW in 366-m span. Amplitude measured 89 mm from support (Rawlins 1988) (Reproduced by permission of IEEE)

When overpowered, the dampers would permit larger amplitudes, capable of inducing fatigue in the dampers themselves.

He also observed that as the ice builds up and the conductor diameter increases, the conductor frequency decreases, as was expected according to Eq. (5.1), and sometimes remains the same. Also, if the conductor section deviates too much from a circular section, the Strouhal number may also vary as mentioned earlier. The decrease of vibration frequency, compared to that determined from the Strouhal relation based on bare conductor diameter, is illustrated in Fig. 5.6.

Figure 5.7 illustrates a case where conductors carry an ice or wet snow coating and the effective diameter is increased accordingly. The frequencies of aeolian vibration are thus much lower. Consequently, there can be aeolian vibration modes, which occur with frequencies below the lowest frequency at which the damper can absorb the wind energy, and the damper can then be damaged by these modes of vibration. These lower vibration modes are different from galloping, which occurs more commonly when a hard coating of ice is accreted on the conductor and strong winds drive the large amplitude motion. The aeolian vibration occurs under lighter steady winds, and can cause the damage, as illustrated by Fig. 5.8.

Over the years, Hydro-Québec, among other utilities (Loudon 1999), also experienced numerous fatigue failures of various Stockbridge-type dampers on its lines (Fig. 5.9). Studies aimed at understanding the phenomenon have led to the conclusion that aeolian vibrations under icy conditions could cause this fatigue problem. Consequently, it was decided that significant improvement of the dampers' fatigue endurance was needed.

A new type of vibration damper has been developed (Van Dyke et al. 2001) and, since the durability of the aeolian vibration damper was the main concern, its artic-

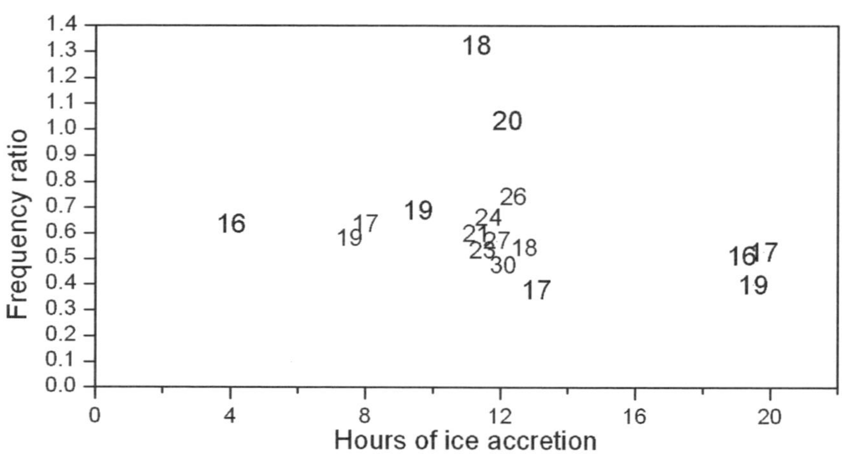

Fig. 5.6 Ratio of recorded frequency to calculated Strouhal frequency, as function of duration of icing, for records in the low-frequency peak of Fig. 5.3 Entries in the graph are equal to one-tenth the amplitudes indicated in Fig. 5.3. Note: several records have been omitted due to doubtful wind data. (Rawlins 1988) (Reproduced by permission of IEEE)

Fig. 5.7 Example of build–up of low density ice accretion on conductors (Reproduced by permission of Tap Seppa)

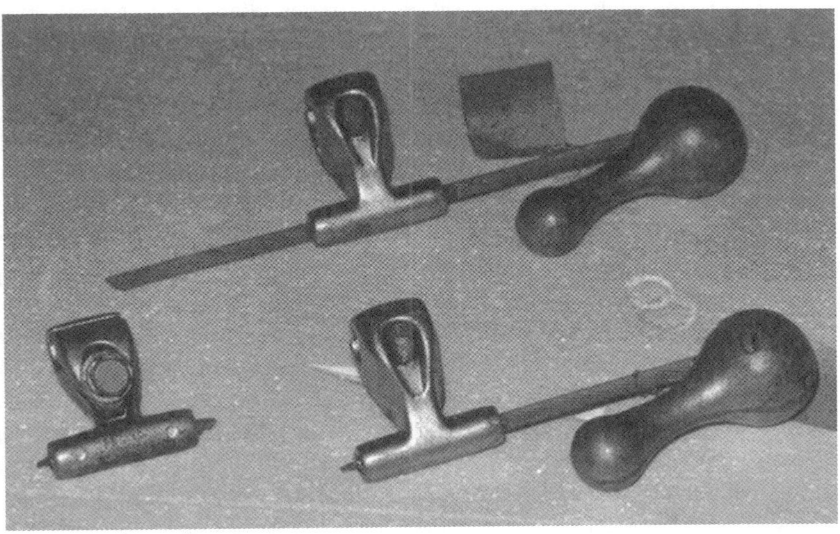

Fig. 5.8 Failed Stockbridge type dampers damaged by low frequency aeolian vibration of conductors covered with hoar frost (Photo: D. Havard)

Fig. 5.9 Stockbridge dampers with missing masses on a transmission line (Reproduced by permission of Hydro-Québec)

ulation is based on a proven technology used for the Hydro-Québec spacer damper. This technology allows mechanical stops to be incorporated in the articulations to avoid damaging dampers under severe ice storm conditions. This articulation is based on elastomeric cylinders located in cavities in such a way that the arm rotation not only produces shear in the elastomer but also compression to minimize any risk of cracking of the elastomer. Performance tests conducted on a laboratory span and full-scale test line, and measurements on a transmission line have shown that the Hydro-Québec damper is as efficient as a Stockbridge damper for controlling aeolian vibrations on a conductor.

Regarding damper endurance, laboratory fatigue tests were conducted and the endurance of the damper was confirmed on an experimental line. During galloping tests conducted at this experimental test line, single Condor conductors were covered with 63-mm D-shape artificial ice to induce conductor galloping. During preliminary tests, the shapes were found to induce severe aeolian vibrations of the order of 50 mm peak-to-peak around 10 Hz. During the first tests, the commercial Stockbridge dampers lasted from a few hours to two weeks. During subsequent tests, the Hydro-Québec dampers were used and they all lasted through eight weeks of testing without any damage (Van Dyke et al. 2001).

These results show that, for areas where icing is severe, it is possible to design and install aeolian vibration dampers that will sustain the increase of power involved.

Fig. 5.10 Stockbridge dampers damaged during the course of galloping tests at IREQ's test line (Reproduced by permission of Hydro-Québec)

5.3 Wake-induced Oscillations

5.3.1 Subspan Oscillation Mechanism

A review of the effects of ice on wake-induced oscillations will be restricted to subspan oscillations since rolling, twisting or vertical galloping (without ice) involves

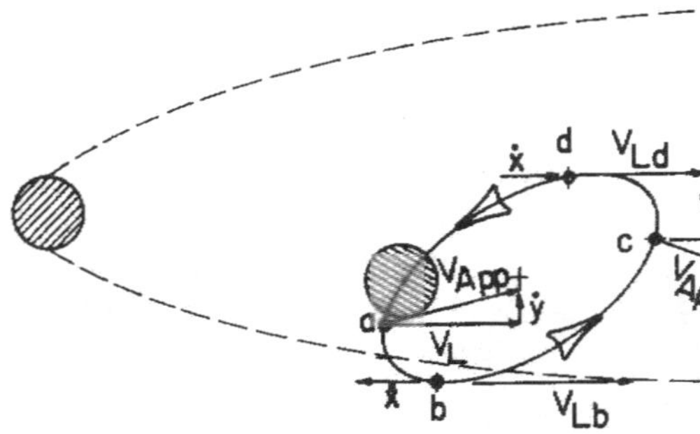

Fig. 5.11 Typical orbit of a circular cylinder in the wake of another cylinder (Rawlins et al. 1979) (Reproduced by permission of EPRI)

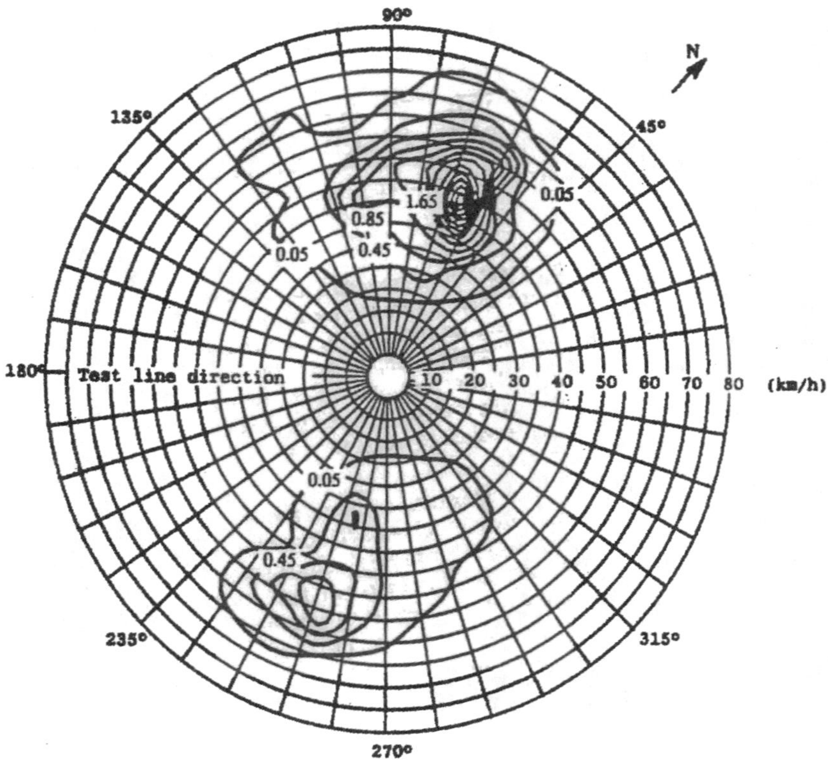

Fig. 5.12 Maximum values of the subspan oscillation instability index (I) of a quad bundle (Hardy and Van Dyke 1995), $I = \frac{l}{60}\sum_i \left(\frac{f_i A_i}{80}\right)^2$, where l is subspan length, A_i is root mean square value of the ith component of displacement (mm) oscillation spectrum at mid-subspan and f_i, the associated frequency (Hz). (Reproduced by permission of Elsevier Limited)

little distortion of the bundle cross-section and is generally less demanding on line hardware. These phenomena occur only at bundle conductors with subconductors arranged one after the other in the direction of the wind. When subspan oscillations appear above a critical wind velocity, the conductor span in the wake of an upstream conductor may then be excited to oscillate, typically in an elliptical orbit (Wardlaw et al. 1973; Price 1975; Simpson 1979; Rawlins et al. 1979; Naudascher and Rockwell 1994; Hardy and Van Dyke 1995). As shown in Fig. 5.11, the leeward conductor acquires its oscillation energy when it moves downstream with the strong flow of the outer wake, and upstream against the weaker flow of the inner wake. The subspan oscillation frequency corresponds to the natural subspan frequencies, which generally range from 1 to 5 Hz. The severity of subspan oscillations is shown as a function of wind velocity and azimuth in Fig. 5.12 for a typical quad bundle. Prolonged subspan oscillation may loosen the spacer clamps, and then produce wear and fatigue of the strands of the conductor in the immediately adjacent sections.

5.3.2 Subspan Oscillations with Ice Accretion

Wake-induced oscillations are mainly influenced by the spacing to diameter ratio (a/d) of the bundle, the subspan length (distance between spacers), the distribution of spacers along the span (mismatched or uniform subspan lengths), and the angle of attack or tilt of the bundle. Two of these parameters may be affected by ice accretion. The first is the ratio a/d, which usually lies between 10 and 20 for transmission lines. For such numbers, there will be no significant aerodynamic coupling leading to the excitation of the windward subconductor (Zdravkovich 1984), but the subconductors are mechanically coupled through the spacers. Experiments on a full-scale test line have shown that the leeward subconductor may experience excitation forces large enough to cause conductor clashing. Moreover, during subspan oscillations, the windward subconductor will be excited through mechanical coupling provided by the spacers staggered along the span. Such occurrences seldom happen anymore on operating lines since subspan oscillations are mitigated by installation of a sufficient number of spacers along the span. However, if a significant amount of ice accretes on the subconductors, the a/d ratio will decrease, which will increase the aerodynamic forces acting on the leeward conductor to a point where significant subspan oscillations may occur, while the phenomenon would have been controlled under normal conditions.

Moreover, it has been shown that for quad bundles, subspan oscillations are more severe for a negative bundle tilt of approximately 10° (Rawlins et al. 1979; Hardy and Van Dyke 1995). A negative bundle tilt means that the leeward conductor is located below the windward one. The conductor in the wake is then located at an optimum position to receive energy through an elliptical oscillation. Unequal ice accretion between the two subconductors may cause a tilt that would be more propitious to subspan oscillations.

Vertical subspan oscillations of the subconductors with ice accretion were observed on 400-kV quadruple bundle lines, lasting for several hours (Kiessling et al. 2003), with the result that the oscillations put extreme stress on the spacers. However, this type of problem happens infrequently compared to those observed during galloping or aeolian vibrations in the presence of ice.

5.4 Galloping Conductors

5.4.1 Introduction

Conductor galloping is a motion-induced excitation occurring when ice accretion on the conductor modifies its circular shape and generates an aerodynamically unstable profile. As the motion of the conductor is initiated by a small perturbation, the aerodynamic forces due to the wind flow at the apparent angle of attack on the unstable ice section feeds energy to the conductor. The first to provide an aerodynamic explanation for the galloping phenomenon was Den Hartog (1932), based on the slope of the curve of the aerodynamic force coefficient of the ice section. There have been a few rare cases of conductor galloping occurring on bare conductors (Davis et al. 1963) and on non-symmetrical conductor profiles, but those cases are not pertinent to the effects of icing, and they will not be covered here.

Conductor galloping is one of the most disturbing phenomena caused by a combination of ice accretion and wind on conductors. This dynamic effect produces very large amplitude vertical motions of conductors when a modest to strong wind blows on conductors covered with ice or wet snow. Following the introduction of vertically oriented double-circuit power lines early in the 1900s, flashovers between adjacent phases during ice storms led to the circuits being forced out of service with resulting power outages. Peak-to-peak galloping amplitudes up to 15 m have been reported. Galloping can result in line outages due to phase to phase contacts and various types of mechanical damage to the conductors, hardware and supporting structures.

5.4.2 Galloping Mechanism

5.4.2.1 Introduction to Some Aerodynamic Concepts

In the following sections, explanation of the galloping mechanism requires the use of aerodynamic terminology such as drag forces, lift forces, moment, angle of attack, azimuth and bluff body.

The drag force acts along the wind direction and is the force resulting from the pressure of the fluid on the body plus the skin friction. The lift force acts in a direction perpendicular to the flow. The lift is induced by the fact that there is a pressure differential between the two sides of the body, because the side with faster flowing fluid has lower pressure than the one with slower moving fluid. These drag and lift forces are generally expressed in terms of coefficients. The relationship between the forces or moment acting on the body and their coefficients are given in the following equations.

$$Drag = \frac{1}{2}\rho V^2 A C_D \tag{5.3}$$

$$Lift = \frac{1}{2}\rho V^2 A C_L \tag{5.4}$$

$$Moment = \frac{1}{2}\rho V^2 A d C_M \tag{5.5}$$

where

A : Projected surface of the body considered in a plane perpendicular to the flow
C_D : Aerodynamic drag coefficient
C_L : Aerodynamic lift coefficient
C_M : Aerodynamic moment coefficient
d : Conductor diameter
V : Wind velocity
ρ : Fluid density

Angle of attack: The angle of attack may be defined as the angle α (Fig. 5.13) between the wind direction and a predetermined reference direction around the conductor which generally lies in the horizontal plane when the conductor is at rest.

Afterbody length: This is the remaining peripheral length of the section located in the wake starting after the separation point.

Azimuth: This is the direction of the wind in a horizontal plane according to a reference which generally corresponds to the conductor direction.

Bluff body: A structure is termed bluff when the flow going across it separates from the cross-section of the structure.

Boundary layer: The boundary layer is the layer of fluid in the immediate vicinity of the bounding surface. It continues to develop downstream as a free shear layer (see Fig. 5.14). Inside the boundary layer, the effect of viscosity is important. Outside it, it is negligible.

Shear layer: The shear layer in a fluid is a region where the velocity of a layer of fluid is different from the velocity of an adjacent layer of fluid (see Fig. 5.14).

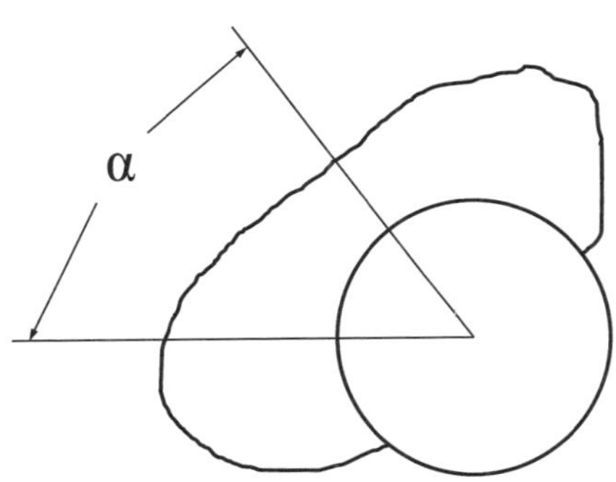

Fig. 5.13 Definition of the angle of attack

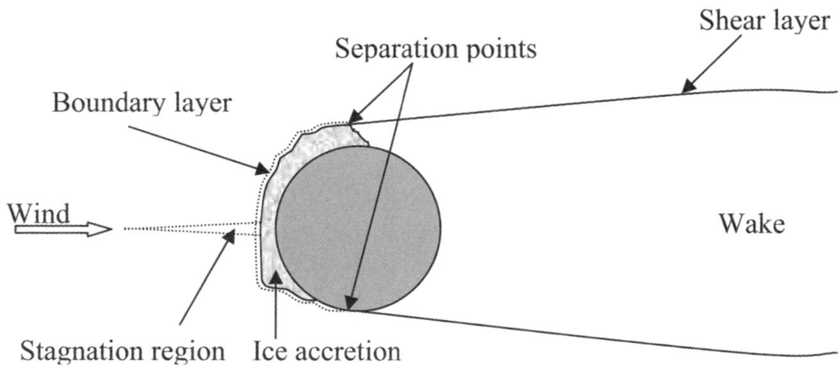

Fig. 5.14 Regions of disturbed flow

Stagnation region: One narrow region of retarded flow on the windward side of the conductor (see Fig. 5.14).

Soft oscillator: A system which oscillates spontaneously from rest with increasing amplitude if the wind conditions are adequate.

Hard oscillator: A system which must have an initial impulse (triggering amplitude) to reach a state where it will oscillate with increasing amplitude if the wind conditions are adequate.

Yaw angle: Direction of the wind in a horizontal plane according to a direction perpendicular to the conductor.

5.4.2.2 Vertical Galloping

Conductor galloping may occur under different possible mechanisms, as will be seen in this section. Den Hartog (1932) was the first to explain conductor galloping based on ice shape drag and lift aerodynamic coefficients.

When ice accretes on the conductor, the resulting shape may have varying aerodynamic coefficients as a function of the angle of attack. Figure 5.15 shows this variation of the lift force acting on the conductor plus the shape of the ice accretion being considered.

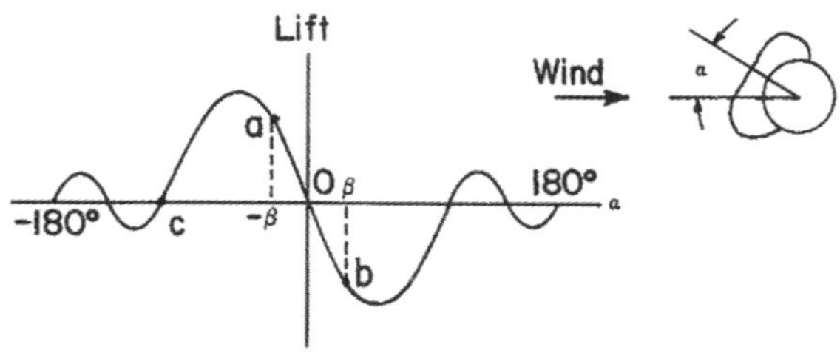

Fig. 5.15 Illustration of variation of lift with angle of attack (Rawlins et al. 1979) (Reproduced by permission of EPRI)

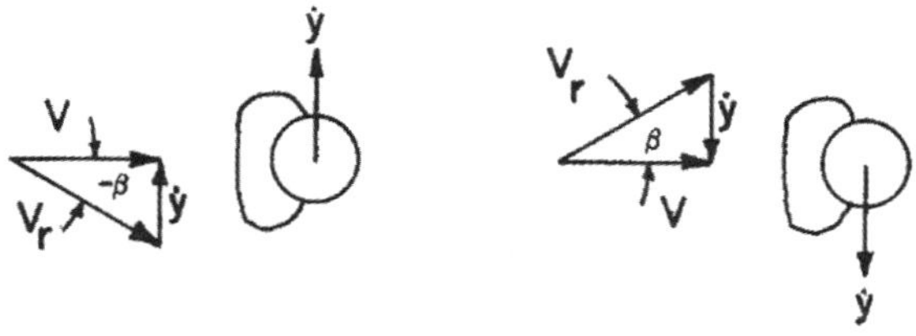

Fig. 5.16 Effect of conductor motion on apparent wind (Rawlins et al. 1979) (Reproduced by permission of EPRI)

An apparent rotation of the conductor relative to the wind direction may result from the conductor velocity itself as shown in Fig. 5.16. When conductor displacement is induced, the apparent variation of angle of attack will bring a variation of the lift force accordingly. If an upward wind velocity coincides with positive lift while a downward wind velocity coincides with negative lift, galloping amplitude will increase. The theory according to Den Hartog was based on this reasoning and lead to the following equations assuming that the fluid force is quasi-steady, which means that the fluid force acting on the conductor is determined solely by the wind velocity and angle of attack. The criterion developed by Den Hartog indicates that the rate of change of the aerodynamic lift coefficient curve, C_L, with respect to angle of attack of the wind, α, must be negative and greater in magnitude than the coefficient of aerodynamic drag, C_D.

$$\text{If } \frac{dC_L}{d\alpha} + C_D < 0 \text{ galloping can build up from small amplitudes} \tag{5.6}$$

$$\text{If } \frac{dC_L}{d\alpha} + C_D > 0 \text{ galloping cannot build up} \tag{5.7}$$

This simplified explanation of galloping requires an ice profile generating a region of negative lift at the normal orientation of the conductor, such as between a and b in Fig. 5.15. However, actual glaze ice shapes observed during galloping of overhead conductors are usually smooth, very thin, crescent profiles without this region of negative lift (Nigol and Clarke 1974). Shapes generated by wet snow accretion are generally thicker and pointed, and thus do exhibit separation points and conform more closely to the Den Hartog model, as exemplified in Fig. 5.17 (Tunstall 1989). The graph of the lift versus angle of attack for this wet snow deposit shows steep negative lift characteristics in the ranges of angle of attack between −100 and −20 and between 20 and 90 degrees.

Further, conductors with both types of ice and wet snow accretion exhibit significant twisting as part of their motion (Edwards and Madeyski 1956). This is most clearly visible during galloping of bundle conductors in the rotation motions of the spacers, but is also present during galloping of single conductors. Thus the angle of attack changes not only due to vertical motion, but also due to the twist of the

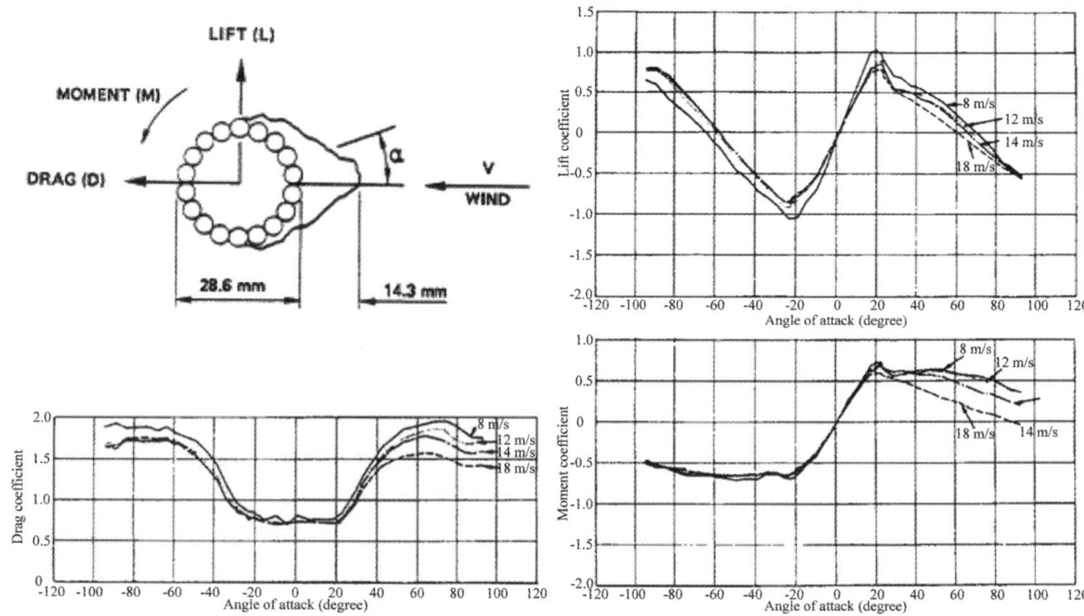

Fig. 5.17 Wet snow shape obtained from a test stand, and aerodynamic lift, drag and moment coefficients versus angle of attack from wind tunnel measurements on replicas of this ice shape (Reproduced by permission of Tunstall 1989)

conductor. The following discussion of bluff body aerodynamic instabilities leads to an understanding of conductor galloping where the accretion creates separation points.

The galloping mechanism may be viewed again with reference to the behaviour of the flow of air around the bluff body. Galloping results generally from an ice accretion on the surface of the conductor that modifies the shape of the conductor's cross-section and causes the flow to separate at some discontinuities of the ice layer. The shear layer proceeding from this imposed point of separation modifies the mean flow and consecutively the mean pressure applied on the surface of the conductor. Since ice accretion remains a time-dependent process, the shape of the conductor "seen by the wind" changes, as well as the location of its center of gravity. The asymmetry of the mean pressure distribution generates a net force on the conductor, in addition to its displacement in the direction of the applied force. If, at the new position, the new mean pressure distribution still generates a net force in the same direction and larger than the combined restoring and damping forces, the conductor motion will persist and might reach very large amplitudes, up to 150d, where d is the conductor diameter. This description fits the case of the galloping of a soft oscillator.

Alternatively, galloping may be initiated by an initial perturbation which, by displacing the shear layer about the conductor, modifies the mean pressure distribution that otherwise would generate a net force in the direction opposite to the motion. This case is defined as a hard oscillator.

In contrast to the aeolian vibrations originating from fluctuating flow instabilities, galloping is then an aeroelastic instability inherently linked to the motion of a bluff body, which in this case is the conductor more or less covered with an irregular layer of ice. The shear layers proceeding from the separation point are then displaced with conductor motion and the entire mean flow field and mean pressure distribution are modified. The details of the interaction between the flow field and a simple moving structure with a separation point follow.

5.4.2.3 From Separated Flow to Galloping: Details

As ice deposits on the surface of the conductor, multiple small peaks may form and become points of separation of the boundary layer from the surface (see Fig. 5.14). Once separated, the boundary layer is transformed in a shear layer that can interact back with the surface from which it originates; this length of interaction is defined as being the afterbody length. In order to examine the details of the flow mechanism of this instability, the simplest shape is the square section with fixed separation points and flat lateral surfaces. The objective is to observe how the motion of the square section (presumed mounted on an elastic system with insufficient damping) and the shear layers interact to generate the instability. The instability is then identified as the underlying cause also of galloping on basically round conductors with an accreted layer of ice or wet snow such as the one shown in Fig. 5.14.

Following Parkinson (1971), Fig. 5.18 shows three instants of a torsionally constrained square prism undergoing galloping: for each instant are indicated the mean

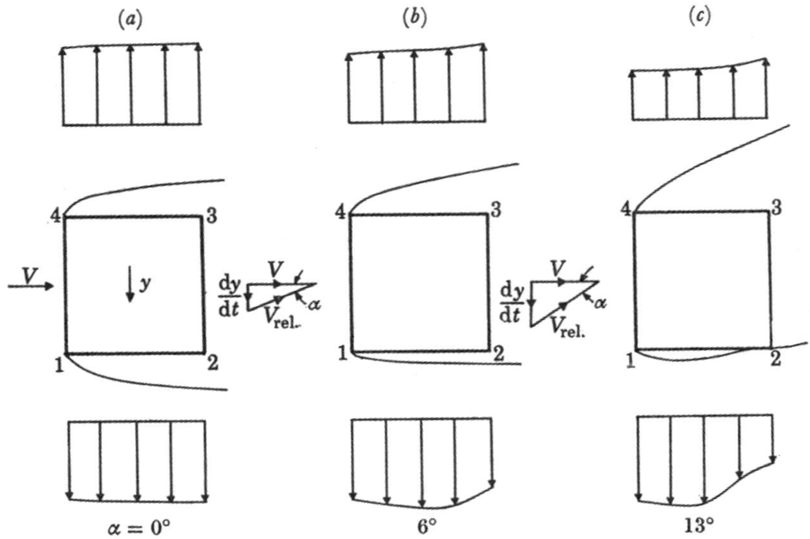

Fig. 5.18 Separated shear layers and pressure distribution on sides during galloping: Case of the square section prism in smooth flow (Parkinson 1971) (Reproduced by permission of the Royal Society of London)

pressure distribution acting on each longitudinal surface as well as the velocity vectors and the equivalent angle of attack.

In Fig. 5.18a, the prism is at zero angle of attack (dy/dt = 0) and the flow separates from corners 1 and 4; the time-averaged positions of the shear layers are symmetrically carried downstream on both sides; the mean aerodynamic force in the transverse direction is then zero.

In order to examine the aeroelastic stability of this shape, a small perturbation in the purely transverse direction will be applied to induce a motion (in practice any small impulse and even the shedding of vortices at any frequency will generate such a perturbation); if the resulting pressure distributions on the longitudinal surface produce a restoring aerodynamic force, the shape of the structure is stable; otherwise the aerodynamic force will be in the direction of the motion, the shape will be unstable and the structure with such a shape will gallop in the transverse direction.

In Fig. 5.18b, the prism is moving down with a velocity of magnitude dy/dt and faces a relative velocity larger than the oncoming wind velocity; in this case the angle of attack is 6°. The flow still separates from corners 1 and 4, but now the shear layers have lost their previous symmetry. The windward shear layer forms closer to side 12 than that the leeward shear layer forms to side 34. Because of the adjacent trapped vorticity, the windward side experiences a pressure distribution leading to a higher suction force than does the leeward one. The net aerodynamic force is then in the same direction as the perturbation (or the motion at dy/dt) and the shape of the structure is aerodynamically unstable. Structural damping will be required to balance this aeroelastic instability. Galloping is sometimes considered a negative aerodynamic damping because of its link to the motion velocity of the structure.

The displacement of transverse galloping will reach its maximum as the net aerodynamic force motion will be the largest. This occurs as the trapped vorticity adjacent to side 12, the windward side, is severely confined while acting over the largest portion of the afterbody length; the windward shear layer approaches the downstream corner and is on the verge to reattach on the side. In Fig. 5.18c, the flow has reattached at corner 2 and the angle of attack is 13°. Higher angles of attack cause the reattachment point to move forward on the windward side and the net suction force on this side to be reduced.

The integration of the pressure distributions, mainly on the lateral surfaces, will yield a force transverse to the flow direction, F_y that can be transformed into C_{Fy}, a dimensionless parameter. This force coefficient can be related to the well known drag and lift coefficients. In the case of the square section, the C_{Fy} data as measured by Smith and Parkinson (Parkinson 1971) with a stationary model in a wind tunnel are shown on the next figure.

Figure 5.19 shows clearly that the transverse force increases as the angle of attack increases in the range from 0 to 13° and then decreases as α continues to be increased. The square section would be unstable and gallop transversely if its motion generates an equivalent angle of attack α in the range $0 < \alpha < 13°$ (positive slope). One condition for the onset of galloping is then:

$$\frac{d\,C_{Fy}}{d\alpha} > 0 \tag{5.8}$$

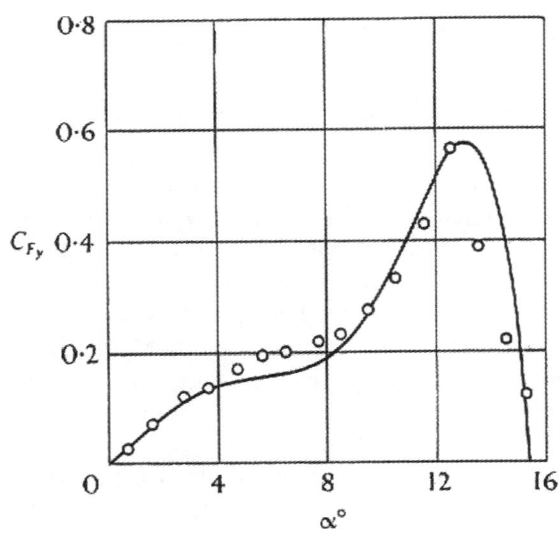

Fig. 5.19 Dimensionless Transverse Force Coefficient as a function of α, the angle of attack (Parkinson and Smith 1964) (Reproduced by permission of Oxford University Press)

And this corresponds to the Den Hartog's criterion already expressed in Eqs. (5.6) and (5.7). In the particular case of the square section, galloping would develop from rest since the criteria is satisfied at $\alpha = 0$. In contrast, such as in the case of the D section, this criterion can be satisfied in a range of angles of attack α, excluding 0, and the onset of instability would therefore need to be induced by a sufficiently large perturbation to reach the triggering amplitude, or angle of attack, required above which the unstable section will oscillate with increasing amplitude.

As an intermediate conclusion, one has observed that galloping of a given shape results from the interaction of the mean flow, more precisely the shear layer emanating from a point of separation, with the afterbody length as this given shape is set in motion. Since galloping is a motion induced instability, the prediction of its onset using the Den Hartog's criterion and of its steady state amplitude will require the knowledge of the aerodynamic transverse force at the different instants of its motion. This is somewhat cumbersome but fortunately, this difficulty can be circumvented using the quasi-steady hypothesis.

5.4.2.4 The Quasi-Steady Hypothesis and a Non-Linear Model

In the case of a single elastic bluff body undergoing galloping, the frequency of oscillation can be sufficiently low relatively to the wind speed: under these conditions, the effect of the body motion will be equivalent to a modification of the relative velocity (vector sum of dy/dt and V) and the angle of attack of this body. The flow "seen" by the body in motion would be equivalent to that "seen" by the same body but stationary, and at the modified angle of attack and relative wind velocity. The lift and drag forces acting on the bluff body in motion are then assumed identical to that of the stationary body under the same flow conditions. Its equivalent angle of attack is simply defined as:

$$\tan(\alpha) = \frac{dy/dt}{V} \tag{5.9}$$

Using a polynomial fit (preferably Tchebichev polynomials for their properties) of the experimental lift and drag data, it is then possible to evaluate the value of the transverse force at any position of the bluff body motion. The equation of motion is then non-linear and the aerodynamic contribution acts as a negative damping. The first non-linear modeling of galloping is the work of G.V. Parkinson and J.D. Smith (1964). This modeling, later extended by M. Novak (1969) was quite successful in predicting double amplitudes and onset wind velocities above which galloping occurs even for conductors covered with ice under smooth and turbulent flows (Novak et al. 1978) as it will be seen.

A. Laneville and G.V. Parkinson (1971) and M. Novak (1971) have shown that the initial hypothesis and resulting non-linear theory can be extended to steady turbulent flows by using the mean oncoming flow velocity in the definition of the angle of attack and by using the stationary force data measured under the identical turbulent flows. These two conditions are prerequisite since the reattachment of the flow is influenced by the turbulence contained in the oncoming flow (Laneville et al. 1975): small-scale turbulence enters the boundary layer via the stagnation point and changes the properties of the shear layer.

The Quasi-Steady hypothesis is not expected to be valid in the case of multiple bluff bodies, such as bundled conductors, because of the additional dynamic effects of the wake that cannot be reproduced under stationary conditions. In the definition of the angle of attack, the oncoming flow velocity is defined as steady; in the case of a conductor vibrating transversely in the wake of an upstream conductor already in motion, the oncoming instantaneous velocity varies significantly as the conductor enters and exits the wake and its position with respect to the wake depends on the motion of the upstream conductor. Y. Nakamura (1980) describes an experimental and analytical study of the galloping of a two-dimensional model of a two-conductor bundle in which ice-accreted conductors were replaced by two identical square prisms: his experimental results show that, as well as galloping type flutter, two other types of instability, namely, torsional and classical type flutter, can occur for bundled conductors. In addition, aerodynamic coupling was observed to cause violent classical flutter to occur when the torsional and the translational resonant frequencies became close to each other.

The use of quasi steady theory to predict the conditions for galloping of iced conductors is supported by theoretical analysis and wind tunnel studies on realistic shapes of iced conductors (Chadha 1974). Those studies included the torsional motions induced through inertial coupling of the eccentricity of the ice mass from the conductor's centroid.

5.4.2.5 Effect of Flow Turbulence on Galloping

Turbulent flows are generally characterized by two criteria, the intensity (symbol i in Fig. 5.20) and the macro-scale of turbulence, the former being defined as the ratio of the r.m.s. velocity fluctuations to the mean oncoming velocity and the latter as the correlation length. The correlation length is a measure of the size of the eddies containing most of the turbulence energy. Figure 5.20, from Parkinson (1989), shows some effects of systematic variation of rectangular afterbody aspect ratio e/d (d/h in the figure).

In Fig. 5.20a, the variation of the base pressure coefficient is shown for the range of afterbody lengths $0 \leq e/d \leq 1$ in a turbulent flow of 12% intensity in addition to its variation and that of the Strouhal number in smooth flow. The intensity of turbulence, as suggested by Gartshore (1973), Laneville, Gartshore and Parkinson (1975) and Hillier and Cherry (1981), creates increased entrainment of fluid by the separated shear layers, thus thickening them and promoting firstly the interference between the shear layer and the trailing edge corner and secondly the reattachment of the shear layer at this trailing edge at lower values of e/d than for smooth flow. From these observations, the galloping behaviour in turbulent flow might be expected to be similar to that in smooth flow, but shifted as this is seen to occur in Fig. 5.20b: as turbulence intensity in the oncoming flow is increased from $i = 0$ to 12%, a soft galloping profile becomes weaker and eventually stable while

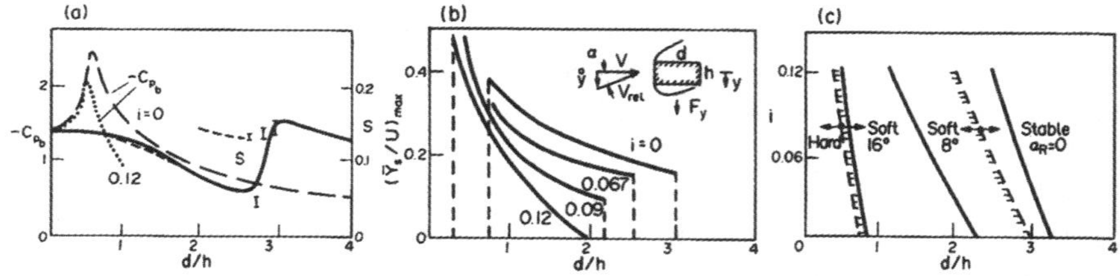

Fig. 5.20 Effects of rectangular afterbody aspect ratio and turbulence intensity; (**a**) base pressure and Strouhal number, (**b**) soft galloping amplitude, (**c**) reattachment and galloping. Experimental data from Brooks (1960), Hoerner (1965), Laneville (1973), Nakamura and Tomonari (1977), Novak and Tanaka (1974), Smith (1962), and Washizu and Ohya (1978)
Source: (Parkinson 1989, page 173) (Reproduced by permission of Elsevier Limited)

a hard galloping profile becomes soft. Other experimental and theoretical studies on models of iced conductors (Chadha and Jaster 1975) suggest that galloping is enhanced by the turbulence intensity and that the range of angle of attack that is unstable is increased by increased turbulence.

Figure 5.20c emphasizes the relationship between galloping, afterbody length, turbulence intensity and reattachment by superimposing curves of constant angles for reattachment (solid lines $\alpha_R = 0°$, $8°$ and $16°$) on an i-e/d map showing galloping zones (bounded by dashed lines). The α_R data are from shadowgraph experiments by Laneville (1973) on rectangular profiles tested in smooth and turbulent flows (two scales and three intensities). The right boundary of soft galloping, at which the profile becomes stable, is close to the curve $\alpha_R = 0°$, but not coincident with it; it would rather coincide with $\alpha_R = 3°$, perhaps indicating that in the case of profiles with $\alpha_R \leq 3°$, the transverse exciting force becomes too weak for galloping to occur, even in a lightly damped structure. The left boundary of soft galloping, at which it becomes hard, so that a sufficient impulse is required to initiate galloping, lies close to the line $\alpha_R = 16°$, the angle of reattachment in the case of a square profile in smooth flow.

The scale, or eddy size, of the incident turbulence does not seem important.

5.4.2.6 Elliptical Trajectories

The theories already discussed in this chapter take into account vertical displacements induced by galloping, while inclined elliptical trajectories are often observed in the field showing the presence of transverse (along-the-wind) motions. The natural frequency of vertical motion is very close to the pendulum or swinging frequency of the span and consequently these motions tend to be sympathetic, and excitation of one mode produces a response in the other. This horizontal displacement superimposed on the vertical one may be explained by the fact that wind velocity generally increases with height. Consequently, the drag force acting on a galloping conductor along the span is lower when the conductor is down then when it is up. Horizontal fluctuating forces are then acting at a frequency corresponding to the vertical displacement and a fluctuating horizontal displacement will then result from this.

A horizontal fluctuating force may also result from the variation of angle of attack of the wind on the ice and conductor profile. Such a variation of angle of attack may bring a variation of the drag coefficient; the drag force acting on the span will then vary accordingly during the cyclic motion and may induce horizontal displacement coupled with the vertical and or torsional displacement already in place.

5.4.3 Galloping on Power Lines

5.4.3.1 Introduction

The need to better understand the galloping phenomenon has led utilities to conduct observation programs on existing and/or test lines. In some cases, artificial ice profiles were used on test lines to obtain galloping under moderate to high winds, but not requiring natural ice accretions. In other cases, existing transmission lines were instrumented and natural galloping events were monitored. Large field observation programs including many utilities were also set up and, through the years, permitted the construction of data banks including many galloping events. These were obtained from Edison Electric Institute, Ontario Hydro, the Electric Power Research Institute (Havard and Pohlman 1984), and the Canadian Electric Association (Pon and Havard 1993; Pon and Havard 1994). This data collection has been pursued by the Cigré Task Force on Galloping (WG11/TFG) through its members who reported galloping case information in their respective country yearly (Lilien et al. 2007). Those data banks were extremely useful to derive semi-empirical curves in order to estimate maximum galloping amplitudes as a function of line parameters as will be seen in the next section. A form was also designed by a CIGRÉ group to make sure that all the pertinent information was collected during those observations (Tunstall et al. 1995). It follows that the following items should be noted:

- Identity of observer.
- Date and time.
- Identity of line, circuit, and phase.
- Voltage.
- Location in the line by tower number.
- Weather conditions, including precipitation, temperature, and wind speed and direction.
- Mode of galloping: standing loops or traveling wave; one mode or several; number of loops; adjacent spans moving synchronously or not; subspan motion in bundles or in spans divided by interphase spacers or other devices.
- Amplitudes of galloping: vertical, horizontal, and torsional, for bundles and for singles fitted with targets; shape of galloping ellipse.
- Support point motions longitudinal and lateral to the line.
- Location in the span where adjacent phases came closest.
- Frequencies of observed modes of motion.
- Shape and thickness of ice coating on conductors.
- Behaviour of nearby circuits.

Nevertheless, the most valuable form of report remains a film of the galloping motion. The observers were encouraged to take all film from a tripod and to expose film that can be scaled to determine amplitudes after the fact. Guidance was provided as to the locations to place the camera for filming to ensure that motions could be scaled relative to known dimensions, such as insulator string length or phase-to-phase separations. When possible, pictures of the ice accretion on the conductor should also be taken at different points along the span.

The most difficult items of requested information pertain to modes, amplitudes, and frequencies. To facilitate identification of galloping modes, a drawing depicting the first three modes, usually corresponding to maximum galloping amplitudes, was made available with the forms (Fig. 5.21). It is also advisable to use films of past galloping episodes in training observers to assess these items. Modes are most easily classified when viewing along the line where the entire span falls within a narrow

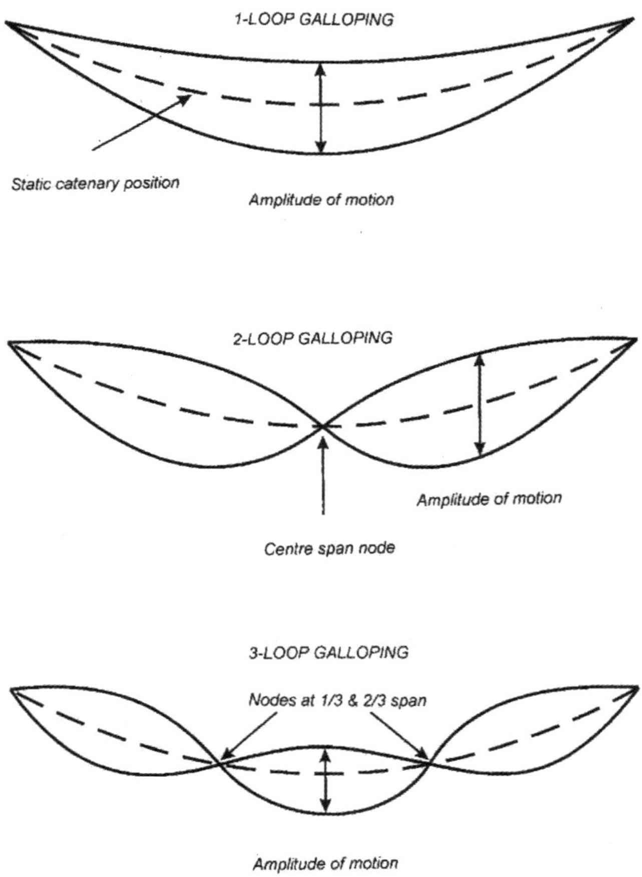

Fig. 5.21 Galloping mode shapes of overhead line galloping (Tunstall et al. 1995) (Reproduced by permission of CIGRE)

field of view, and adjacent phases can be more readily distinguished. Amplitudes are easier to estimate from a broadside position because the middle of the loop can be more accurately located. In addition, known line dimensions, such as insulator string lengths, can be more easily employed because effects of perspective are minimized. One useful technique (Hydro Electric Power Commission of Ontario) is to stand about one span length to the side of the line, hold a pencil vertically at arm's length, and mark off with the thumb a distance on the pencil that corresponds to panel height, insulator string length, or phase spacing. The pencil is then swung, still at arm's length, to line up with the middle of the galloping loop, and amplitude is estimated with reference to the known line dimension.

The mode frequencies of the span considered may be calculated by assuming that the conductor is a flexible string. A wave will travel along the string with a velocity c, regardless of its form.

$$c = \sqrt{\frac{T}{m}} \tag{5.10}$$

where

c: Traveling wave velocity
m: Conductor mass per unit length
T: Conductor tension

It may be shown that the mode frequency is:

$$f = \frac{nc}{2L} \tag{5.11}$$

where

n: Mode number or number of loops on the span
L: Span length

For sagging conductors the longitudinal and vertical in-plane displacements are coupled. It results in a shift to higher values of the symmetric (also called odd) modes. This shift may be substantial for the first mode and decreases for higher odd modes. The mode shape is also affected and the first mode may take the form of a pseudo-fundamental mode with two nodes in the span. Dead-ended spans which are rigidly fixed at each extremity are more affected by that. Their frequency and mode shape may be calculated using formulas derived by Irvine and Caughay (1974).

When suspension spans are involved, mode shape and frequencies may be calculated using a transfer matrix method (Simpson 1966; Rawlins 2001). This method was applied at the Hydro-Québec test line (Van Dyke 2007) which has three suspension spans and two dead-end spans with the following lengths: 150 m; 400 m; 450 m; 425 m; 150 m. Figure 5.22 compares the mode shapes and frequencies of the

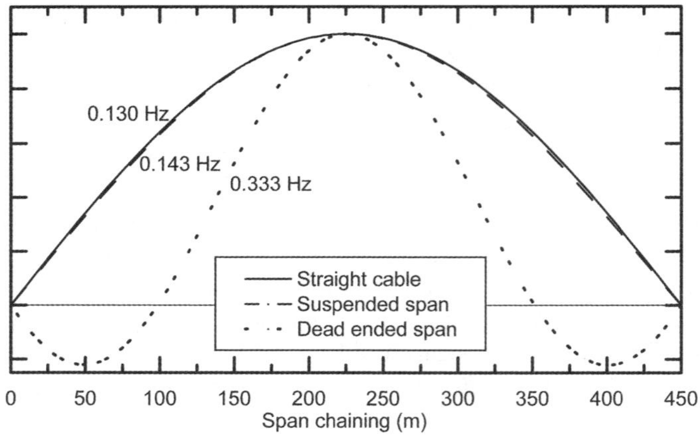

Fig. 5.22 Mode shape and frequency of different 450 m spans (T = 34.5 kN, m = 2.522 kg/m) (Van Dyke 2007) (Reproduced by permission of Hydro-Québec)

middle span of the test line with a 450 m straight cable and a dead-ended span having the same tension and mass per unit length. As expected, the straight cable has the lowest resonant frequency (0.130 Hz). The test line suspension span mode shape is very close to that of the straight cable but its resonant frequency (0.143 Hz) is 10% higher. The first natural frequency of the dead-ended span (0.333 Hz) is almost three times higher than that of the straight cable. Consequently, at 0.333 Hz the natural frequency of its first mode (first symmetric mode) is higher than that of its second mode (first antisymmetric mode), which is 0.260 Hz. In this case, the first symmetric mode is called a pseudo-fundamental mode and its shape is no longer sinusoidal but comprises two nodes. In practice, spans of conductor are neither purely dead-ended or suspension as there is some flexibility in the supporting structure, and along the line swinging of suspension insulators involve adjacent spans with the motion of each span.

5.4.3.2 Galloping on Test Lines Fitted with Artificial Ice Shapes

Generally, research and development tests were conducted with artificial ice profiles since they allow for better control of the test parameters. Moreover, it is no longer required to wait for adequate weather conditions leading to an agglomerate of aerodynamically unstable ice on the conductor; consequently, the data collection is a lot faster. However, questions were raised as to how well those artificial ice profiles represent the real ice accretion. Most anti-galloping systems interact with the galloping mechanism aerodynamically (e.g. airflow spoilers) or mechanically (e.g. detuning pendulums), the only exception being maybe the interphase spacers; consequently, they must be exposed to all kinds of ice types, shapes and quantities combined with wind velocities and directions to assess their efficiency in all conditions. Nevertheless, those experiments were extremely useful to better understand the phenomena, validate mathematical modeling, measure the forces involved in the conductor and at towers, and evaluate the wear and fatigue of components.

Fig. 5.23 D-section (Van Dyke and Laneville 2004) (Reproduced by permission of Hydro-Québec)

D-shapes on single conductors (Fig. 5.23) and modified D-shapes on bundles have been used often to induce galloping since they are known to produce large amplitudes. Figure 5.24 shows artificial ice profiles used to reproduce natural ice accretions during galloping tests (Nigol and Clarke 1974).

Table 5.1 provides an overview of different galloping results obtained on test lines. The galloping amplitudes reached up to 5 m. The first three modes were often excited but other modes were also observed. Most of the time conductor galloping was accompanied with torsion except for four configurations. Those four cases demonstrate that pure vertical galloping or Den Hartog galloping occurs. This latter point is important since some anti-galloping devices are based on the detuning of vertical and torsional modes to reduce galloping amplitudes.

Most of the time, there is only one mode excited, as shown during tests conducted on six single conductors. There was only one mode excited during 40 to 70% of the

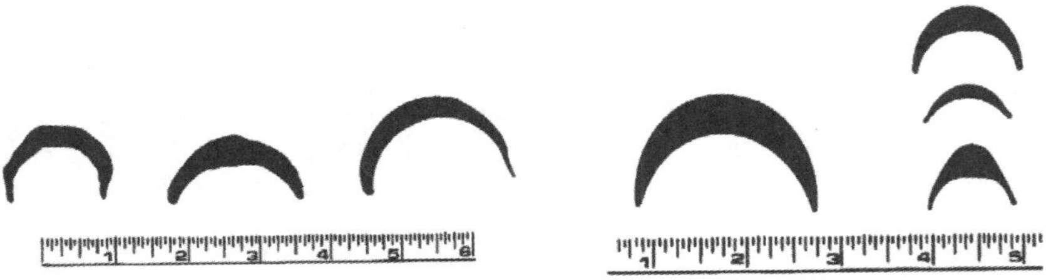

Fig. 5.24 *Left*: Ice sections moulded from conductor ice accretions. *Right*: Artificial ice sections used during test to reproduce natural ice accretions. (Nigol and Clarke 1974) (Reproduced by permission of IEEE)

Table 5.1 Overview of galloping studies on test lines

Ref.	Conductor and configuration	Span length	Ice type and shape	Wind velocity (m/s)	Modes no excited	Vertical displacement (m pp)	Torsion	Dynamic tension (pp)
Stewart, 1937	ACSR 6.4 mm single	32 m	Wax rectangular	i.n.a.	1	1	Over 180°	i.n.a.
Tornquist and Becker, 1947	Copper 3/0 AWG 11.8 mm single	76 m	Wooden 40 mm D section	0.5 to 6 at 40 to 60° from the line direction	2, 4 and 6	0.8	No significant torsion	i.n.a.
	Copper 3/0 AWG 11.8 mm single	4 × 76 m	Wooden 40 mm D section	3 to 5 skew winds	1 and 2	0.3 to 1.3	No significant torsion	i.n.a.
Edwards and Madeyski, 1956	Oriole 18.8 mm single	126 m	Wooden 51 mm D section	6 at 45°	1	1.5	40° pp	0.04 × static tension
Binder, 1962	Copper 300 kcmil 16 mm single	6 × 76 m	Wooden D section	3.7 and more	1	2.4 to 3.6	Coupled with vertical displ.	i.n.a.
	Copper 3/0 AWG 11.8 mm single	6 × 76 m	Wooden D section	5.2 and more	1	2.4 to 3.6	Coupled with vertical displ.	i.n.a.
Ratkowski 1962	Steel ribbon 25.4 mm × 0.25 mm	8.7 m with spring ends	Wooden 54 mm D section	4.5	2, 3 and 4	0.7	No	i.n.a.
Edwards, 1966	Drake 28.1 mm single	9 × 335 m	Polyethylene 63 mm D section	2 and more	2, 3 and 4	3	i.n.a.	i.n.a.

Table 5.1 (continued)

Ref.	Conductor and configuration	Span length	Ice type and shape	Wind velocity (m/s)	Modes no excited	Vertical displacement (m pp)	Torsion	Dynamic tension (pp)
Edwards, 1966	Drake 28.1 mm single	9 × 335 m	63 mm square section	i.n.a.	8 to 16	0.5	i.n.a.	i.n.a.
	Drake 28.1 mm single	9 × 335 m	305 mm square section – lengths of 13.4 m at one and three quarter of the span length	i.n.a.	2	0.6	Torsion with no vertical displacement	i.n.a.
Nigol and Clarke, 1974	Drake 28.1 mm single	3 × 244 m	Crescent shape reproduced from natural ice accretions (Fig. 5.24)	5 to 11	1, 2, 3 and multi-boucle	3.6 (mode 1) 1.8 (mode 2)	20 to 30° pp	i.n.a.
Nigol and Havard, 1978	ACSR 45.7 mm single Drake 28.1 mm single	3 × 244 m	Crescent shape reproduced from natural ice accretions (Fig. 5.24) plus weights to create four ice loadings	i.n.a.	First, stretch and third mode vertical frequencies were measured	i.n.a.	Fundamental torsional frequencies measured	i.n.a.

Table 5.1 (continued)

Ref.	Conductor and configuration	Span length	Ice type and shape	Wind velocity (m/s)	Modes no excited	Vertical displacement (m pp)	Torsion	Dynamic tension (pp)
Tsujimoto et al., 1983	ACSR 28.5 mm quad bundle	459 and 534 m	50 mm D-modified	12	2 and 3	4	Yes	1.2 × static tension
	ACSR 28.5 mm 8-bundle	459 and 534 m	50 mm D-modified	20	1, 2 and 3	3	Lower than for the quad	0.8 × static tension
	ACSR 28.5 mm 8-bundle	353 m	Natural ice accretion	18	1, 2 and 3	3	Lower than for the quad	0.8 × static tension
Morishita et al., 1984	ACSR 28.5 mm Quad bundle	363 + 247 m	Natural hoar frost	12	i.n.a.	4.2	Yes	1.4 × static tension
	ACSR 28.5 mm 6-bundle	363 + 247 m	Natural hoar frost	12	i.n.a.	2.1	Yes	0.9 × static tension
	ACSR 28.5 mm 8-bundle	353 + 230 + 350 m	Natural hoar frost	18	i.n.a.	2.6 4.9 horizontal	i.n.a.	0.6 × static tension
	ACSR 38.4 mm 8-bundle	353 + 230 + 350 m	Natural hoar frost	i.n.a.	i.n.a.	i.n.a.	i.n.a.	1.1 × static tension
	ACSR 28.5 mm 6-bundle	363 + 247 m	Crescent	20	i.n.a.	4.5	i.n.a.	i.n.a.
	ACSR 28.5 mm 6-bundle	363 + 247 m	Crescent	20	i.n.a.	4.5	i.n.a.	i.n.a.

Table 5.1 (continued)

Ref.	Conductor and configuration	Span length	Ice type and shape	Wind velocity (m/s)	Modes no excited	Vertical displacement (m pp)	Torsion	Dynamic tension (pp)
Morishita et al., 1984	ACSR 38.4 mm 10-bundle	230 + 190 m	Crescent	15	i.n.a.	3.5	i.n.a.	i.n.a.
Oura et al., 1995	ACSR 23.1 mm twin bundle	162 m	63 mm D-modified	15s	i.n.a.	i.n.a.	i.n.a.	2.6 × static tension
	ACSR 23.1 mm twin bundle + one interphase spacer	162 m	63 mm D-modified	15	i.n.a.	i.n.a.	i.n.a.	2.0 × static tension
Van Dyke et al. 2001	Bersfort 35.6 mm twin bundle	450 m suspension span	100 mm D-modified	Up to 11	i.n.a.	5	i.n.a.	i.n.a.
Van Dyke 2007	Condor 27.8 mm single	450 m suspension span	75 mm D-shape 1 and 3 kg/m (Fig. 5.23)	Up to 9 and 15	2 to 9	2.2 to 3.5	yes	0.09 to 0.2 × static tension
	Condor 27.8 mm 3 singles with 4 interphase spacers per span	450 m suspension span	75 mm D-shape 1 and 3 kg/m (Fig. 5.23)	Up to 8 and 11	2 and 4	4.5 to 9	Yes with 1 kg/m D and no with 3 kg/m D	0.17 to 0.24 × static tension

n.a. : Not applicable
i.n.a.: Information not available

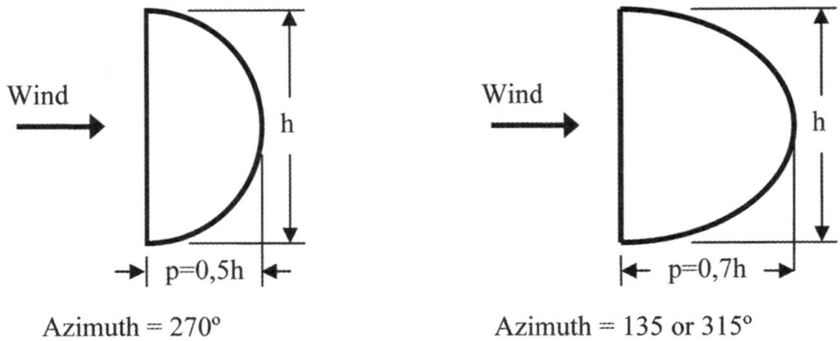

Fig. 5.25 Apparent depth of a D-section as a function of wind azimuth (Van Dyke 2007) (Reproduced by permission of Hydro-Québec)

galloping cases, depending on the configuration and there were two modes excited simultaneously for 10 to 22% of the galloping cases (Van Dyke 2007). During the same tests where D-profiles were used to induce galloping, it was noticed that for conductor galloping under wind directions that are not perpendicular to the conductor, the wind flows around a D-section with effectively a different aspect ratio (apparent section depth over height) (see Fig. 5.25). For example, for a direction of about 50° from the perpendicular to the line, the apparent aspect ratio of the D-section becomes 0.78 instead of 0.5. Nakamura and Tomonari (1981) measured the aerodynamic characteristics of D-sections with different aspect ratios in a turbulent flow. They showed that D-sections with aspect ratios between 0.73 and 1.5 will experience soft galloping (a galloping that starts spontaneously from a resting state). This result emphasizes the fact that a mathematical model based on aerodynamic coefficients corresponding only to a direction perpendicular to the section considered will not provide adequate results for different wind directions since the aerodynamic coefficients vary with azimuth. Moreover, as shown in the case with interphase spacers, non-perpendicular winds may be the most severe (Van Dyke and Laneville 2004).

The other main limitation of galloping modeling is that the applicable ice accretion properties, such as density, shape and aerodynamic coefficients over the conductor are not known and may vary along the span, on each conductor of the same span, and from span to span. However, even with those limitations, it is still useful for understanding the phenomenon and for predicting conductor galloping behaviour under known conditions.

5.4.3.3 Field Observations

A summary of 192 verbal reports on galloping from 28 countries over 20 meetings of the Cigré task force on galloping prepared by Hearnshaw (Lilien et al. 2007) showed that most galloping occurs at wind velocities between 3 and 14 m/s, however galloping has been observed at wind velocities up to 25 m/s and amplitudes up to 15 m have been reported. The minimum wind velocity required to cause galloping depends on system damping and wind turbulence as affected by the smoothness of the upstream terrain, while maximum wind velocity depends on the range of angle of attack where the ice shape is unstable. Novak and Tanaka (1974) showed that galloping ceases above a certain wind velocity because the span tilt angle brings the wind angle of attack outside the range of instability of the ice accretion.

Looking at those verbal reports, it is clear that the first three modes are clearly favoured where high galloping amplitudes have been observed. Travelling waves have also been observed and may induce high stresses on line hardware when the wave is steep.

Galloping is more frequent at temperatures ranging from -5 to $+2^\circ$ C since those temperatures are more propitious to wet snow or freezing rain precipitations. However, it may also occur at lower temperatures when the accretion remains on the conductor while the temperature is dropping or when galloping is associated with hoar frost. As an example, galloping was observed at -50° C in NW Siberia (Lilien et al. 2007).

Regarding ice accretion, any kind of ice accretion that modifies the conductor section may be prone to galloping. However, most reported cases are associated with wet snow or freezing rain. Galloping associated with freezing rain seems to be more common in North America while in Asia and Europe wet snow is generally the predominant cause. However, Japan has reported many galloping cases with hoar frost. A few millimetres of ice that are sometimes difficult to distinguish from the ground are enough to produce galloping. Thicker accretions are also prone to galloping except that, at some point, when the ice accretion increases on one side of the conductor, the conductor will continuously rotate under the gravity effect and may result eventually in a cylindrical shape that will no longer be prone to galloping. Thus the ice shape depends not only on the environmental conditions, but also on the torsional rigidity of the conductor. Bundles, having a higher torsional stiffness, are believed to produce ice shapes more prone to galloping than single conductors.

Galloping has occurred on ground wires, single conductors, bundles and even guy wires.

Morishita et al. (1984) measured tension variations during galloping associated with hoar frost at a test site high up in a mountain range in Japan. They observed that it was less severe for quad bundles than for bundles of six or eight conductors because the torsional natural frequency of large bundles does not coincide with its vertical natural frequency, in contrast to the quad. They also observed that for an ice accretion above 1 kg/m the galloping severity decreased with a further increase of ice accretion. They explained it by the fact that the wind velocity is usually higher at the beginning of the hoar frost formation. Since galloping severity generally increases with wind velocity, severe hoar frost accretion being associated with low wind velocities, would produce less severe galloping. While galloping is generally associated with a predominant vertical displacement, they also observed large horizontal displacements of up to 4.9 m on an 8-bundle, while the vertical displacement was 2.6 m. It may be significant that in mountainous sites such as this the wind can have components of velocity in the vertical direction.

A line was monitored in Norway in a remote area using video cameras (Halsan et al. 1998) and one of the conclusions of the study was that many galloping events take place without any immediate impact on the operation of the lines, especially on a horizontally aligned circuit. The dynamic forces applied during such events, though not sufficient to cause flashovers, may be contributing to progressive deterioration due to wear and fatigue of the conductor hardware.

Similar results were reported from Detroit Edison (Brokenshire 1979), where load cells installed on suspension structures of four transmission lines were monitored over a three-year period. The study illustrated that there were many more unseen and unreported galloping events than expected. These data also showed more events on steel pole structures than on lattice steel structures in the same area, indicating a relationship between galloping propensity and tower flexibility.

During galloping observations, Van Dyke observed galloping on a multispan 120 kV line built on agricultural lands. Some spans of the line passed through a wooded area covered with trees having a maximum height of 10 m. The spans in agricultural areas before and after the wooded area were galloping while the four spans in the wooded area did not. The wind was masked by the trees in that area and its velocity was not sufficient to induce galloping. This observation demonstrates the importance of environment on galloping and the possible effect of cutting trees or any other modification around transmission lines on conductor galloping. The reverse is true when farm lands are turned over to residential housing as part of urban growth.

Different modes may be excited through galloping. Why some modes are preferably excited is not known for certain. Different factors may be involved. The ice accretion shape varies along the span and this variation may play a role in the mode excited since the nodes would coincide with the locations where the ice shape is not unstable. Wind turbulence might also play a role if its main frequency coincides with one of the span mode. The closeness of torsional mode frequencies with the vertical mode frequencies is certainly in many cases. Another factor is the stiffness of the end span supports and the possibility of energy flow between adjacent spans. Frequently different modes are observed on different conductors within the same span of a transmission line.

Higher modes are less prone to high amplitudes since their higher resonant frequencies produce a higher conductor velocity for a given displacement. Thus the angle of attack (see Figs. 5.13 and 5.16) will be higher for those modes and the conductor plus ice section will reach its maximum angle of instability at a lower displacement.

5.4.4 Effects of Galloping

5.4.4.1 Flashovers

Conductor clashing or flashovers are the most common problem caused by galloping. When there are repeated flashovers, the automatic protection systems will open the circuits until the cause of the fault is identified and repaired. If there is no alternate route to bring the energy, this may cause power interruption to the customers which may last hours with the resulting loss of revenue. Repeated power interruption has an impact on the quality of service delivered by the utility, as perceived by the customers. Conductor damage may also result from conductor clashing or flashovers. If a line is left in service while clashing occurs during galloping, the circuit breakers can be seriously damaged by repeated reclosing operations.

5.4.4.2 Dynamic Loads

Besides conductor clashing or flashovers, galloping induces dynamic loads in the conductor itself which are transmitted to the towers through the suspension hardware. These loads are applied repeatedly to the support structures during each galloping event. A recent paper (Havard 2003) included a survey of field measurement work, plus some simplified analysis on this topic. That survey of field measurements revealed that the vertical loads applied to the tower arms and to the support hardware and insulators was up to 2.0 times the static vertical load with ice on the conductor. The survey also showed that the horizontal loads applied to dead end or strain structure could be as high as 2.8 times the static tension in the conductors. These loads are not normally included in the design of the structures, and are generally assumed to be covered by other extreme loadings such as heavy ice loads and maximum wind loads. Furthermore, the loads applied during galloping events are repeated many times, and need to be referred to endurance against fatigue, as opposed to the static strength of the components. The most common effects of these dynamic loads are loosened tower bolts, and occasionally fatigued bracing members. Rare occurrences of tower arm and main member failures have also been reported.

Galloping mode shapes and frequencies may be estimated for line sections using the transfer matrix method (Simpson 1966; Rawlins 2001). With this method, knowing the galloping amplitude and mode excited, it is possible to calculate approximate values for the dynamic forces transmitted to towers, as well as suspension clamp longitudinal displacement for a given mode and antinode amplitude. However, for even modes, dynamic tension of the conductor occurs at twice the galloping frequency and is non-linear. Thus it cannot be predicted using a linear method such as the transfer matrix method (Van Dyke 2007). More complete modeling has been done using the finite-element method where the ice accretion aerodynamic coefficients and wind velocity are used as input and the software is used to estimate the resulting modes excited and amplitudes reached (Chan et al. 1992; Shimizu et al. 1998; Keutgen 1999).

5.4.4.3 Hardware Wear and Fatigue

On distribution lines where pin-type insulators are used, tie wire failures due to galloping happen occasionally. Where clamp-top insulators are used, cement failure of the porcelain insulators due to the bending moment applied on it during galloping has also been observed. Finally, damages to insulators have been observed during galloping events when the insulator string was clashing on the tower arm. In one case, the insulator security clip was broken during galloping and the pin went out with the resulting failure of the insulator string. However, those damages are not related to a fatigue phenomenon. When long twinned or multiple suspension strings are used, most often on lines with large diameter conductors or bundles at high voltages, the insulator strings can undergo violent oscillations within their length. The strings can then clash together and break the porcelain or glass sheds.

In areas subjected to ice loads, suspension insulator everyday vertical loads are normally very low compared to their tensile strength; the ratio being of the order of one to nine. Consequently, the effect of galloping cyclic loads is normally of no consequence. For instance, fatigue tests conducted on porcelain insulators (Matsuura et al. 1990) have shown that they can withstand two million cycles at the following loads: average load of 25% of their M&E rating (mechanical and electrical rating), with a peak-to-peak load of 10% M&E. Their fatigue results showed also that the allowable number of cycles decreases with the dynamic loads. In areas experiencing only sporadic ice loads, the ratio between insulator everyday loads and their tensile strength may not be as high and the utility engineer should refer to the insulator manufacturer for fatigue data and compare with the dynamic loads and number of galloping cycles expected. Dynamic tests have also been conducted on composite line post insulators (de Tourreil and Kuffel 1998) and showed also that the allowable number of cycles decreases with the dynamic loads. In fact, up to now, field observations have shown that tower or tower arms fail before insulators.

Some Stockbridge damper failures have been attributed to the low frequency conductor galloping which produced large displacement of the larger damper mass with resulting messenger wire failures. Steep travelling waves may also damage dampers.

5.4.4.4 Conductor Fatigue

The effect of galloping on conductor fatigue has also been a topic of discussion on many occasions and a study was carried out to better understand its effects. Galloping tests were performed at the Hydro-Québec test line in Varennes on a span of single Condor ACSR conductor (Fig. 5.26) where the performance of different suspension clamps was compared regarding conductor fatigue (Van Dyke and Laneville 2005). A "hinged armour wire support" (HAWS) clamp (made of helical rods, an elastomer insert and a clamp housing) was installed at one end, and standard metal-to-metal clamps were installed on the same conductor at the other end of the middle span where galloping was induced (Fig. 5.27).

The severity of vibrations was assessed using the fy_{max} values where f is the frequency (Hz) and y_{max} (m/s peak) is the maximum amplitude reached along the span. This product is frequently used to show bending stress levels at the end of the span. The first test on the conductor, strung at 41% RTS (rated tensile strength), lasted 38 days, during which time the wind conditions provided about 57 hours of galloping. There was no conductor damage at either end of the span during this test and the fy_{max} values recorded are shown in Fig. 5.28. For the second test, the D-section

Fig. 5.26 Hydro-Québec test line (Reproduced by permission of Hydro-Québec)

Fig. 5.27 HAWS clamp (*left*) and metal-to-metal clamp (*right*) (Reproduced by permission of Hydro-Québec)

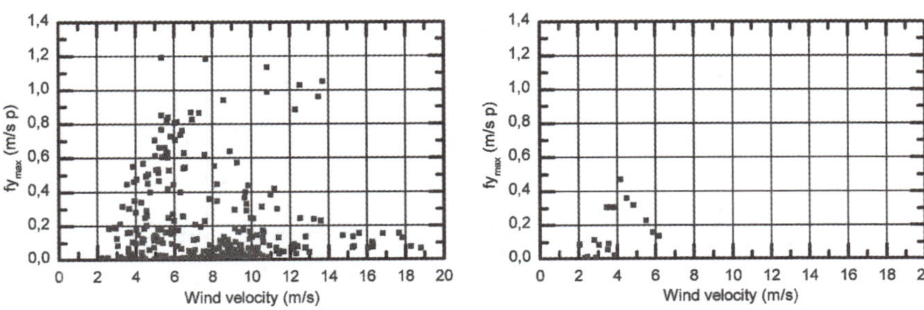

Fig. 5.28 Conductor dynamic stress in terms of fy_{max} at 41% RTS after 38 days of the first galloping test (*left*) and at 55% RTS after six days of the second galloping test (*right*) (Reproduced by permission of Hydro-Québec)

Fig. 5.29 Damaged conductor under the metal-to-metal clamp in the 450-m span after six days of the second galloping test (Reproduced by permission of Hydro-Québec)

mass was increased from 1.0 to 3.0 kg/m and, consequently, the conductor tension increased to 55% RTS. After six days of testing, a visual inspection revealed that the first aluminum layer of the conductor was broken at the outlet of the metal-to-metal clamp in the middle span (Fig. 5.29). There was no apparent damage at the HAWS clamp. The fy_{max} values recorded during those first six days are shown in Fig. 5.28.

The conductor was cut on each side of the metal-to-metal clamp and a 20-m section was replaced with a new conductor using compression joints before resuming the test. The fy_{max} values recorded for the remaining 46 days of the test are shown in Fig. 5.30. At the end of the test, the clamps were removed and six broken wires were found under the metal-to-metal clamp (Fig. 5.31), while there was no damage under the HAWS clamp.

The results obtained during the two tests showed that the clamp/conductor combination was severely stressed and it was made very clear that the HAWS clamp could handle the kind of excitation involved while the metal-to-metal clamp led to conductor fatigue at the same level of excitation. However, at this point, this result does not mean that all suspension clamps must be replaced where galloping has

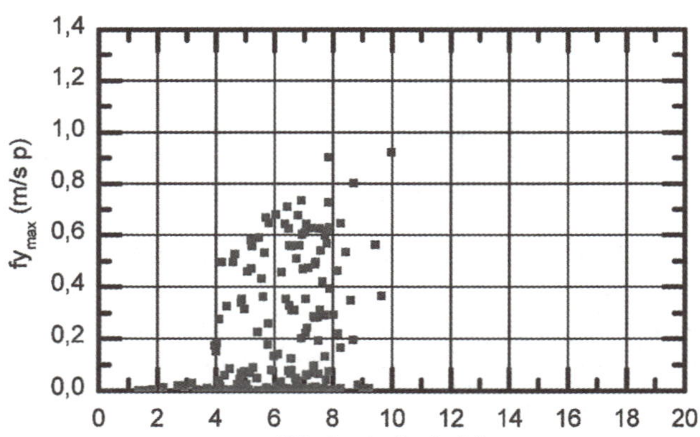

Fig. 5.30 Conductor dynamic stress in terms of fy_{max} at 55% RTS for the 46 remaining days of the galloping test (Reproduced by permission of Hydro-Québec)

Fig. 5.31 Damaged conductor under the metal-to-metal clamp in the 450-m span at the end of the second test (Reproduced by permission of Hydro-Québec)

been observed. Those results must be weighted by the number of galloping events expected on the lines where galloping has been observed to determine whether it is necessary to replace the suspension clamps at these locations.

Fatigue tests were conducted to evaluate conductor lifespan at the amplitude levels observed during galloping. Using Palmgren-Miner's rule (Palmgren 1924; Miner 1945), the cumulative damage was calculated, which consists of adding the ratio of number of cycles at a given stress over the numbers of cycle before fatigue failure at the same stress level. Conductor failures in the metal-to-metal clamp corresponded to cumulative damages of 0.19 to 0.45. Conductor failures were caused by a fretting-fatigue phenomenon.

5.5 Protection Methods for Galloping

5.5.1 Overview

A review of galloping control methods was conducted by a CIGRÉ Task Force in 2000 (Wolfs et al. 2000). The conclusions which were drawn are reproduced here:

- The complexity of galloping is such that control techniques cannot be adequately tested in the laboratory and must be evaluated in the field on trial lines. This testing takes years and may be inconclusive.
- Analytical tools and field test lines with artificial ice are useful in evaluation of galloping risk and appropriate design methods.
- No control method can guarantee it will prevent galloping under all conditions.

- Interphase spacers virtually ensure galloping faults will not occur, but do not necessarily prevent galloping. Their usage is growing and their design is undergoing further development.
- Mechanical dampers to stop vertical motion are still being pursued but to only a very limited extent.
- Torsional devices, which either detune or increase torsional damping or both, are being pursued and actively evaluated.
- Techniques which disrupt either the uniformity of ice accretion by presenting a varying conductor cross-section or the uniformity of the aerodynamics by inducing conductor rotation are being actively pursued.
- Methods of ice removal or prevention are not widely used as specific anti-galloping practices.
- Despacering or using rotating-clamp spacers is still used extensively in a number of parts of Europe subject to wet snow accretions.
- For bundled conductors, the influence of the design of suspension and anchoring dead-end arrangements on the torsional characteristics of the bundle and on the occurrence of vertical/torsional flutter type galloping has been recognized.

Increased clearance between phases and between phases and ground wires including sometimes a horizontal shift of the middle phase of one to two meters on vertical arrangements is also currently used to prevent flashovers during galloping. In the next sections, more details are given about detuning pendulums and interphase spacers, the latter being the most widely used anti-galloping device.

5.5.2 Detuning Pendulums

Analysis of the dynamics of large pure vertical motions, notably by Den Hartog (1932), presented a rationale for the energy flow into the moving conductor and the sustained galloping motions. Later, measurements were made of the motions of a number of single conductor spans during galloping by Edwards and Madeyski (1956). They used targets placed on to the conductors during the galloping events to reveal the torsional component of the motions, and recorded both vertical and torsional modes using movie cameras. Frame by frame analysis of these video films showed that there was dynamic torsional motion simultaneous and synchronized with the very visible vertical motions. This is illustrated by the sample record shown in Fig. 5.32. The torsional motions of single conductors are hard to see from the ground due to the distance and small thickness and transparency of the ice coating.

This torsional motion is also present when bundle conductors gallop and is more visible through the oscillations of the spacers. These torsional motions were the focus of further studies at Ontario Hydro (Nigol and Clarke 1974; Havard and Nigol 1978), and led to development of first torsional dampers, and later torsional detuning pendulums, Fig. 5.33, to modify the motions during galloping.

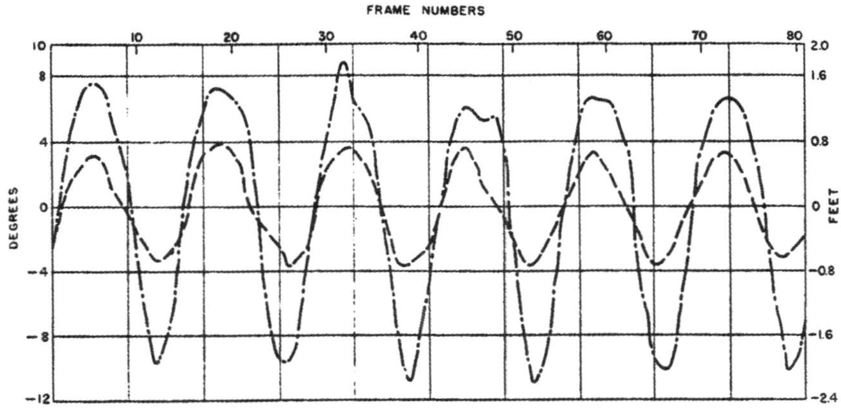

Fig. 5.32 Vertical and torsional galloping motions from frame by frame analysis of movie film of a single conductor span with midspan target. Horizontal scale 0.5 s intervals. – – – – vertical — — — torsional (Edwards and Madeyski, 1956) (Reproduced by permission of IEEE)

Extensive field trials were carried out on operating power lines, mainly in North America, included systematic observation of motions of the overhead conductors during galloping occurrences. The field sites were set up to include identical spans of conductors with and without the galloping controls subject to the same conditions of ice or wet snow and wind. The program generated an extensive database on

Fig. 5.33 Detuning pendulums for galloping control installed on a 345-kV line

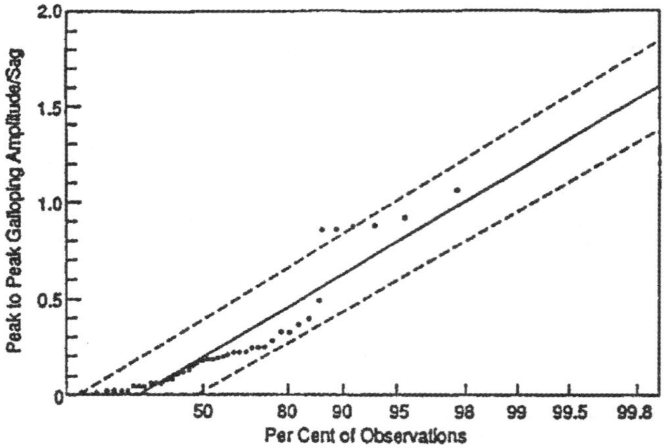

Fig. 5.34 Peak-to-peak galloping amplitudes of large single conductors without galloping controls from field observations during 43 different galloping events. Values normalized by dividing by sag (Havard, 1996)

galloping motions with and without the control devices. These observations led to standards for the amount of detuning pendulum mass and arm lengths needed, and statistically supportable conclusions on the effectiveness of the detuning pendulums (Havard and Pohlman 1984; Havard 1996).

Figures 5.34 and 5.35 show the peak-to-peak galloping amplitudes observed on large single conductor lines with and without detuning pendulums respectively, from 43 different galloping events. The amplitude values are scaled by dividing by the sag, to normalize data from different span lengths. The figures show that the maximum amplitudes observed on the single conductors can exceed the sag of the span,

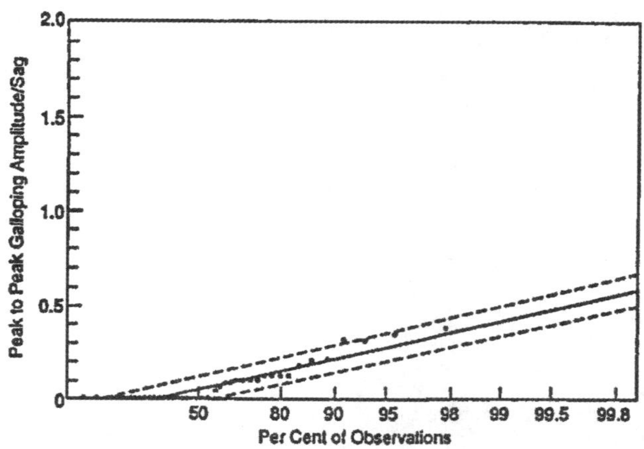

Fig. 5.35 Peak-to-peak galloping amplitudes of large single conductors with detuning pendulums from field observations during 43 different galloping events. Values normalized by dividing by sag (Havard, 1996)

and that the detuning pendulums reduce the motions to about one third. Slightly greater proportional reductions in galloping amplitudes were observed on two-and four-conductor bundle conductor lines based on a statistically significant number of independent field observations.

5.5.3 Interphase Spacers

Interesting information regarding interphase spacers performance was obtained by the CIGRÉ Task Force on Galloping through a questionnaire that was filled by 32 utilities from 13 different countries (Berg et al. 1992). Those are in use on lines ranging from 11 to 425 kV. The main conclusions are that interphase spacers prevent flashovers due to galloping and ice loads and that there are still galloping

Fig. 5.36 Interphase spacer in use at Hydro-Québec (Reproduced by permission of Hydro-Québec)

displacements and the resulting dynamic loads but large amplitude motions appear to have been eliminated. Galloping amplitudes on conductors with interphase spacers were reduced by 27% compared with untreated conductors during field trials conducted at Ontario Hydro (Pon et al. 1982).

Early interphase spacers were made of porcelain insulators, but nowadays they have been replaced by lighter polymer insulators. While such insulators are very strong in tension, care must be taken to make sure that they will not buckle under the compression strain. Figure 5.36 shows the interphase spacer made of a polymer insulator and an aluminum extension tube currently used at Hydro-Québec. Two or four interphase spacers depending on the span length are normally installed at one third and one quarter of the span length to maximize their efficiency over the first modes. Sometimes they are installed at mid-span but while their efficiency is maximized at the first mode, they are located at the second mode node where they have no efficiency to prevent the conductors from coming together at quarter span.

5.6 Galloping Amplitudes

The data obtained in the preceding field studies were also used to develop empirical relationships between maximum galloping amplitude and span properties. Simple relationships describing peak-to-peak galloping amplitude versus span length and peak-to-peak galloping amplitude divided by sag versus span length, for single conductors are shown in Fig. 5.37 (Havard 1998). The equivalent observations for bundle conductors are from a more limited range of span lengths, but the maximum observed values conform to the envelope around the data.

A more sophisticated set of relationships was developed in (Lilien and Havard 2000). Better correspondence with the data was obtained when the peak-to-peak galloping amplitude was divided by conductor diameter, and the span length was represented by a conductor span parameter. This conductor span parameter was given by:

$$100\frac{T.d}{mg.L^2} = \frac{100.d}{8S} \tag{5.12}$$

This parameter has values in the range of 0.015 to 1.1, and generally the value is in reverse order to the span length. Figure 5.38 shows the resulting relationship for single conductor observations and Fig. 5.39 for two-, three-, and four-conductor bundle data. These curves were found to be in agreement with numerical simulations of single and bundle conductor galloping using a finite element based approach (Wang and Lilien 1998). These are full 3 dimensional simulations, including torsional freedom, and they model a full line section. They use a database of aerodynamic lift, drag and moment properties of ice coatings from many different sources. The chosen ice shape was relatively eccentric given high values of lift, drag and aerodynamic moment. This detail about the chosen ice shape serves as a reminder of another aspect of the galloping phenomenon that needs to be studied more care-

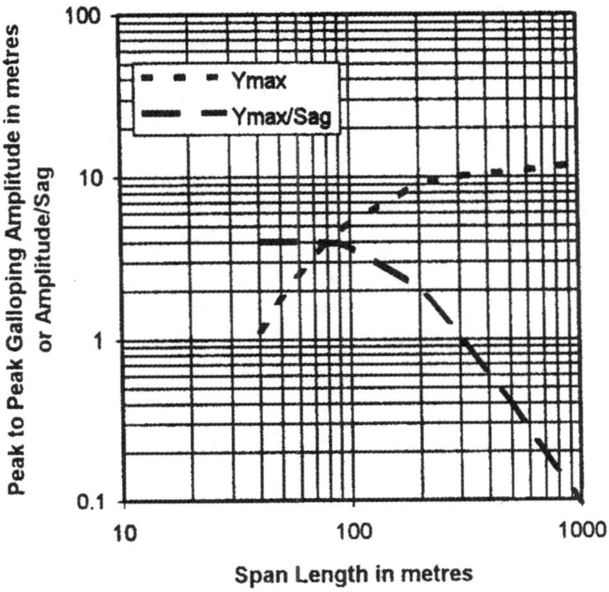

Fig. 5.37 Peak-to-peak galloping amplitude and peak-to-peak galloping amplitude divided by sag as a functions of span length for single conductors (Havard, 1998)

fully. There are a number of theoretical approaches to describe the behaviour of the conductor under ice and wind action, but none of these fully account for the ability of conductors to gallop with virtually no ice in place. There are many records of galloping in which the ice thickness is less than 3 mm, and the changes to the section properties, and to the aerodynamic lift, drag and moment, appear to be insufficient to generate the forces required to drive the motion. It appears that the change in roughness of the conductors from one part of the surface to the other is enough to

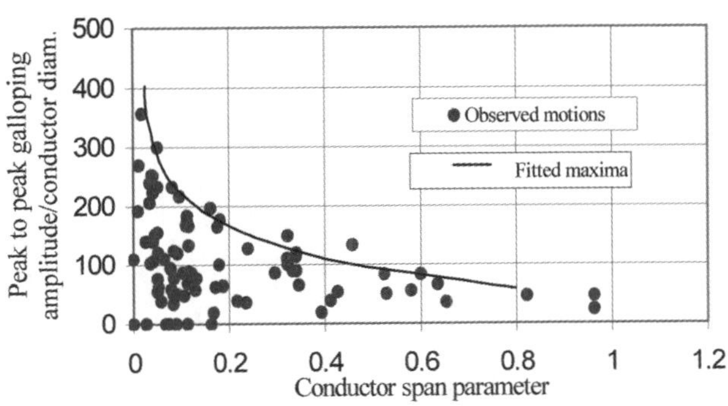

Fig. 5.38 Variation of observed maximum peak-to-peak galloping amplitude/diameter on single conductors as a function of the conductor span parameter (Lilien and Havard, 2000) (Reproduced by permission of IEEE)

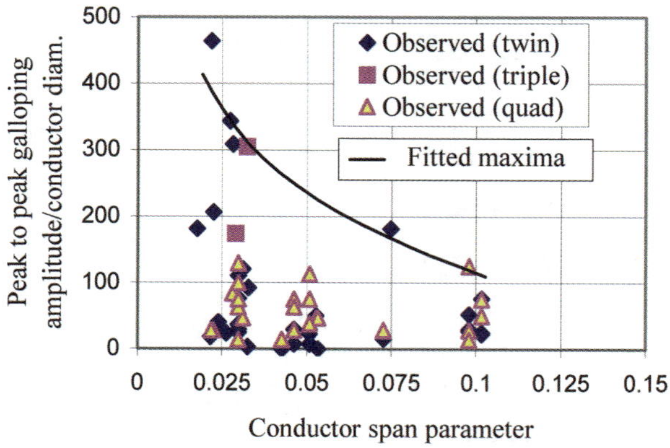

Fig. 5.39 Variation of observed maximum peak-to-peak galloping amplitude/diameter on bundle conductors as a function of the conductor span parameter (Lilien and Havard, 2000) (Reproduced by permission of IEEE)

cause the motions. The physical mechanism involved in this needs to be studied to derive a better understanding, and possibly to lead to better control methods.

The detailed evaluation of the films of galloping motions from the field data also offered an opportunity to review the existing design practices, Fig. 5.40, for incorporating clearance within tower head dimensions to avoid conductor clashing during galloping (REA 1982; Rawlins 1981). The presently used design practices inscribe an elliptical envelope around the conductors after a swing-out from vertical

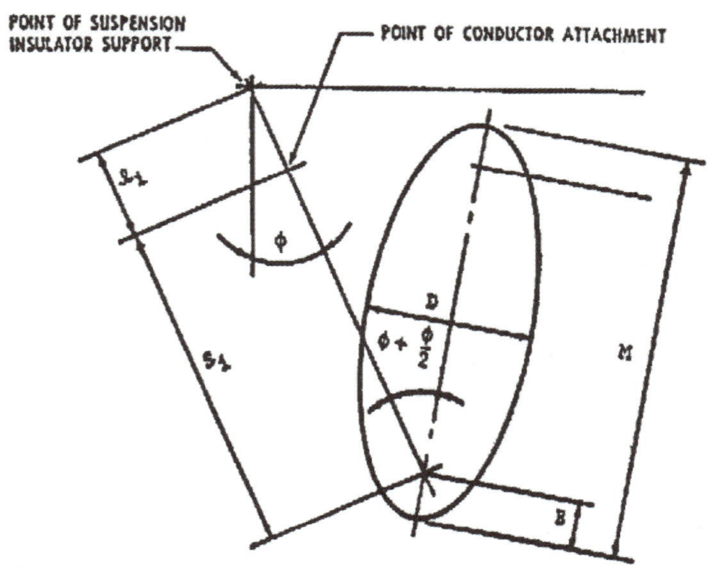

Fig. 5.40 REA Guide for Galloping Clearance ellipse Dimensions (REA 1982)

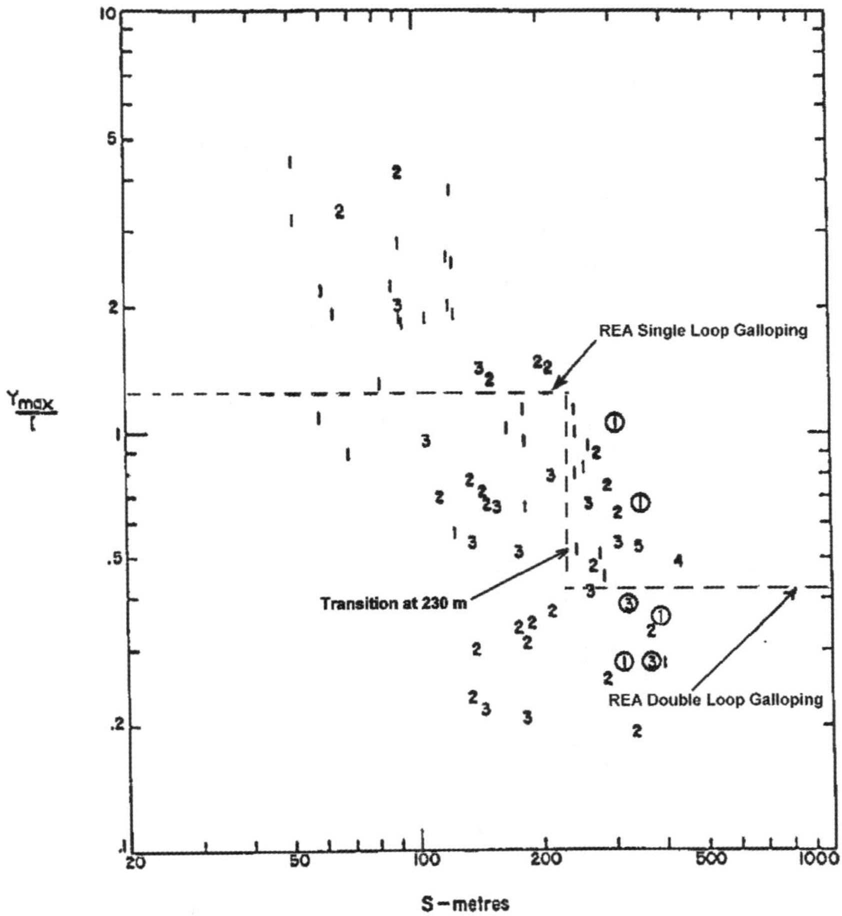

Fig. 5.41 Comparison of REA Guide recommendations on ellipse height vs span length with observed field data in the form of peak-to-peak galloping amplitude divided by sag. Data points above and to the right of the *dashed lines* are cases exceeding REA Guide recommendations.

due to wind forces. The usual ellipse profile is 2.5 times as high as wide and there is three times as much above as below the conductor. This practice has served to limit the number of conductor clashes to some extent. Comparison of actual recorded galloping motions in the form of peak-to-peak amplitude divided by sag versus span length, with the REA guide practice for ellipse height is illustrated in Fig. 5.41. That figure shows that in some cases it is non-conservative, especially for short distribution class spans, and for longer spans where two-loop, lower amplitude galloping is indicated to occur. In contrast, there have been many observations of large amplitude, single loop galloping on these longer spans.

The film analysis indicated that the profile shown in Fig. 5.42 is a reasonable alternative to the present practice. The envelope has the following features. The

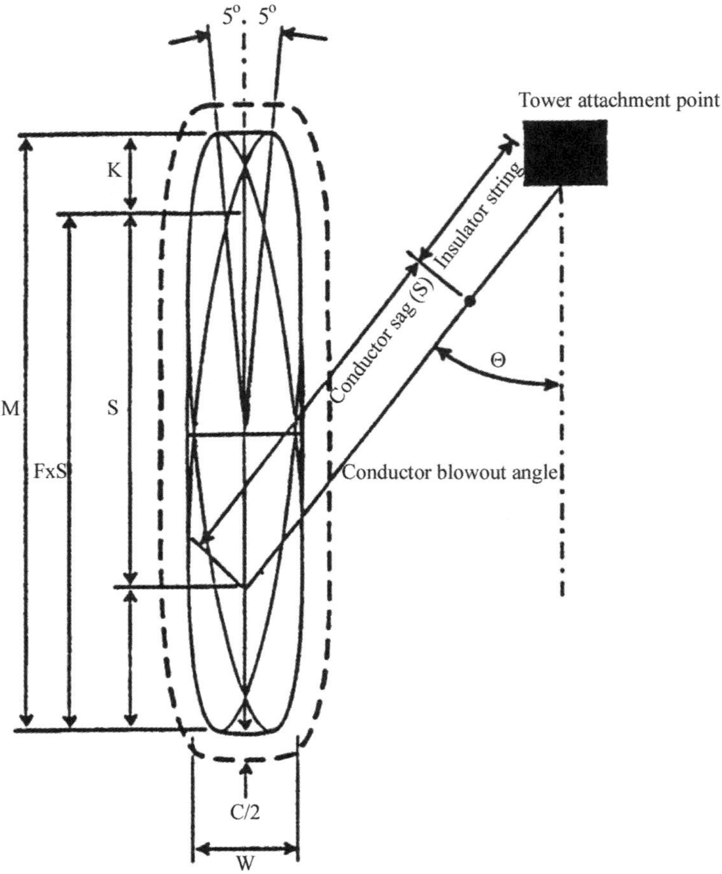

Fig. 5.42 Proposed modified galloping motion envelope based on analysis of films of observed galloping occurrences (Havard, 1996)

motions are virtually all in the vertical plane with horizontal motions always less than 10 percent of the sag. The vertical motion is asymmetrical about the stationary point of the conductor, with the average position being at the lower quarter point of the motion. The main effect of this revised clearance envelope would be reduction of the length of the cross arms of the tower. Offsetting this for some span lengths is the need to recognize larger vertical excursions following the maximum peak-to-peak amplitudes indicated in Figs. 5.37, 5.38 and 5.39.

There is also a clear difference in the galloping phenomenon, where the precipitation is due to glaze ice, as in North America, and wet snow or rime in mountains and in many European countries and Japan. In the countries where glaze is more common, galloping is seen in all overhead lines, whether of single or bundle construction. In the regions susceptible to wet snow, single conductors appear to gallop far less frequently than do bundle conductors. A study of

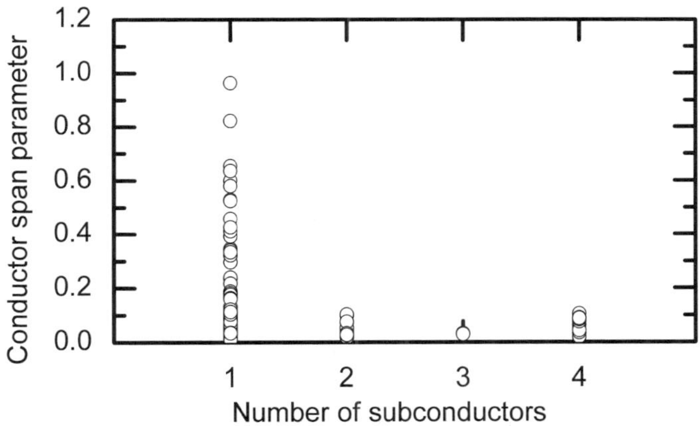

Fig. 5.43 Conductor span parameter as a function of number of subconductors. This shows a clear distinction between single and bundle conductors, and the similarity among all types of bundle conductor (Lilien and Havard, 2000) (Reproduced by permission of IEEE)

the statistics of galloping outages in The Netherlands (Wijker and Leppers 1987) showed that the country can be divided into two regions; one subject mainly to glaze ice and the other where wet snow is more common. The above difference in galloping behaviour was found to correspond well with the weather differences. This leads to galloping control approaches designed to convert bundle conductor into single conductor lines. The reasons for these differences remain unresolved.

There are also varying opinions among experts as to the relative frequency of galloping on single versus bundle conductors (Gibbon et al. 1984). This extends to applications in which bundle lines in Europe have been "despacered" as a mitigation technique against galloping. There are clearly differences in the "span parameter" derived in (Lilien and Havard 2000) above, which indicates that there could be different sensitivities, possibly related to the deposition of glaze ice versus wet snow or rime. This span parameter difference is highlighted in Fig. 5.43. This topic remains unresolved at this time.

5.7 Ice Shedding

Ice shedding may have similar consequences than galloping. It induces high dynamic loads on the lines that might be responsible for tower arm failures or even a cascade failure of several towers in a transmission line. It may provoke flashovers when the conductor rebound brings it close to an adjacent phase, ground wire or parts of the towers. Ice shedding over road crossings constitutes another hazard since big chunks of ice may fall on the vehicles below. It is not a continuous phenomenon, however, once the ice drops begin, the conductors may be excited for many cycles with a slow decrease of amplitude where the damping is mainly due to the air resistance over the conductor displacement. Ice shedding is a difficult phenomenon to observe since it happens suddenly and is not repetitive contrarily to conductor galloping.

Ice drop has been simulated mechanically on test spans to derive estimated rebound heights of the conductors (Morgan and Swift 1964; Stewart 1983). Some studies have also been made to simulate the effect of ice drop using scaled physical models and finite element analytical methods (Jamaladdine et al. 1996; Fekr and McClure 1996). The finite element approach has also been used to determine that a failure of support hardware can be attributed to damage due to ice shedding (Fekr et al. 1998).

These studies assumed that the ice on a span will drop in one piece and the iced and adjacent spans will rebound directly in response to this loss of weight. However, direct observations of the loss of ice during the end of glaze ice episodes show that it "unzips" along the length of the span rather than falling off in one piece. It appears that the observed damage and inferred motions could also be due to the more common and better documented dynamic effects associated with conductor galloping, which can be large enough and repeated often enough to cause damage through fatigue of the structural elements.

More observations of this phenomenon, including some passive monitoring, are required to determine whether the extreme motions that have been attributed to ice dropping are real, or are the by-product of galloping motions.

During the course of tests aiming at inducing ice shedding on a conductor, an impulse was given to the conductor in the middle of the span. Two types of aeolian vibration dampers were located at each end of the span: a Stockbridge damper at one end and a Hydro-Québec damper at the other end (Van Dyke et al. 2001). The interest of this test regarding ice shedding is that it induced a wave propagation along the span which may be similar to the one following ice shedding. In Fig. 5.44,

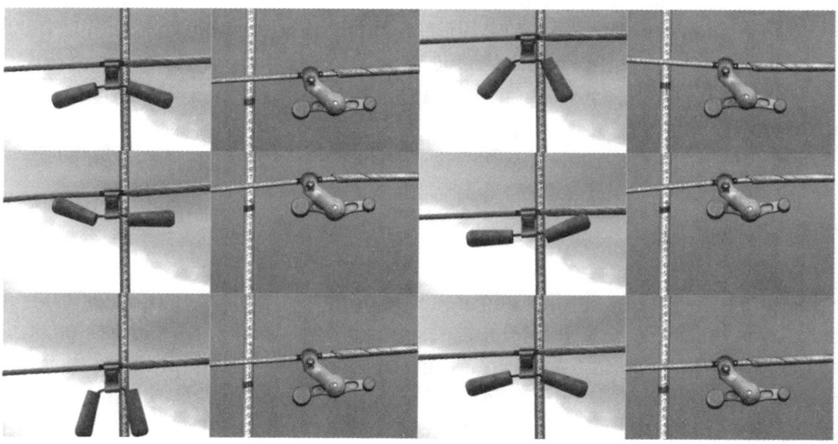

Fig. 5.44 Stockbridge Vs Hydro-Québec dampers (Van Dyke et al., 2001) behaviour after an impulse has been given to the conductor in the middle of the span (Leblond, 2004) (Reproduced by permission Hydro-Québec)

it can be seen that the Stockbridge damper masses hit the conductor thus damaging it. Moreover, the Stockbridge damper messenger wires lost their rigidity after three impulses and the masses were drooping while the Hydro-Québec damper was not damaged after 100 impulses.

Recently, three ice shedding tests were conducted on the middle span (length of 450 m) of a test line (Van Dyke 2007). To simulate the unzipping effect, the ice weight was simulated by using a conductor as a dead weight attached to the conductor under test. The dead weight conductor was held initially with pulleys at each end of the span. The strings that were used to fix it under the test conductor had a rated strength of 1 kN and were located at every 15 m. When one pulley was released downward, the strings broke one after the other along the span and the load was released from the conductor under test.

The first test was conducted on a single conductor, the second test on three conductors linked with four interphase spacers located at one third and one fourth of the span length where only the lower conductor was loaded. The last configuration was used for the third test, but the three conductors were initially loaded and the load was dropped only on the lower conductor for the test. The simulated ice load for the first two tests was 0.6 kg/m and 1.5 kg/m for the third test.

Analysis of the results showed that for the first two tests the conductor jump height above the conductor position without ice may be estimated to be equivalent to the initial deformation of the conductor due to the ice load (Fig. 5.45). For the third tests where the three conductors were loaded and only the lower one was unloaded, the jump height was about one third of the initial deformation. However, in that

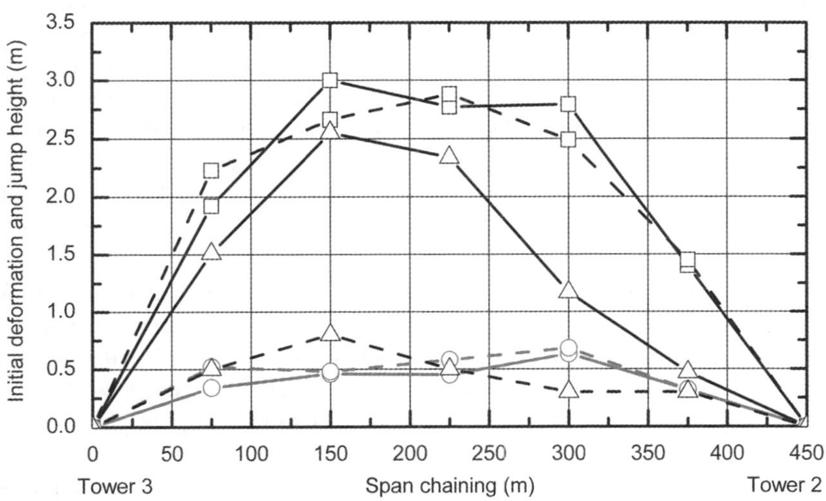

Fig. 5.45 Initial conductor deformation and maximum rebound when the ice sheds (Van Dyke, 2007) *Straight lines*: Initial deformation; *Dotted lines*: Jump height; □: Single conductor; ○: Vertical arrangement of 3 conductors with the bottom one loaded and interphase spacers; Δ: Vertical arrangement of 3 loaded conductors and interphase spacers with load drop on the bottom conductor. (Reproduced by permission Hydro-Québec)

case, the lower conductor rotated around the middle conductor following a radius equivalent to the interphase spacer's length. Thus, the interphase spacers reduced the conductor from coming together by a factor of five or more.

In conclusion, interphase spacers are an efficient way of reducing the conductors from coming together due to ice shedding, however, the lateral clearance must be sufficient to allow for the possible rotation of the lower conductor around the middle conductor. For conductors without interphase spacers, the jump height above the conductor position without ice may be estimated to be equivalent to the initial deformation of the conductor due to the ice load.

To prevent flashovers due to ice shedding, utilities usually increase the clearance between phases and between phases and ground wires including sometimes a horizontal shift of the middle phase of one to two meters on vertical arrangements. This increased clearance is also useful for galloping, as seen in the preceding section. Interphase spacers are also used but in this case they are installed in the middle span since the first span mode is mainly excited when the ice shed on the whole span. Ice removal at an early stage of the ice precipitations to avoid severe ice shedding is also considered but seldom used due to the inherent difficulties of the process.

5.8 Bundle Rolling

Power lines passing through elevated mountainous regions are subject to the accumulation of rime ice, which can accumulate into massive deposits and lead to rolling of bundle conductors. The same effect can occur with glaze ice accumulation, especially on long spans over valleys and rivers, etc. This is a form of instability that can leave the bundle in its rolled position, with damage to the spacers, and conductors, and loss of service of the line. Utility problems due to this effect have been reported in Japan and Canada (Manukata et al. 1963; Matsubayashi 1963; St-Louis et al. 1993). There is a serious lack of knowledge of the ice or wet snow accumulations and of the moment applied by wind action on this deposit.

Studies carried out at Ontario Hydro (Nigol et al. 1977) simulated this torsional unbalance using full-scale physical models and analytical approaches. The physical modeling was carried out with two- and four-conductor bundles using two span lengths, several arrangements of spacers, and six different spacer types. A total of 58 test arrangements were evaluated, as illustrated by Fig. 5.46. The tests consisted of incrementally increasing the applied moment up to the point of torsional instability. Then the moment was decreased slowly until the original orientation was restored. In each case the bundle exhibits a sudden roll at the point of instability with the two, generally longest, subspans twisting several times. Figure 5.47 shows the plot of moment versus angle of rotation of the middle of the bundle as measured and predicted from the analysis. The analytical model developed represents the behaviour of a general single span with fixed ends, in which the spacers are rigid and

Fig. 5.46 Outdoor test line at Ontario Hydro used for bundle torsional stability studies showing twisted sub-conductors in mid-subspan (*arrow*) following collapse (Nigol et al., 1977) (Reproduced by permission of IEEE)

do not slip. The applied moment and angle of rotation at the point of instability for a simplified span with equal subspan lengths are given by Eqs. (5.13) and (5.14), respectively.

$$M_{\max} = \frac{nkTa^2}{2L} \sin\left(\frac{2\theta_{\max}}{k}\right) + \frac{4ns\theta_{\max}}{L} \tag{5.13}$$

$$\theta_{\max} = \frac{k}{2} \cos^{-1}\left(\frac{-4s}{Ta^2}\right) \tag{5.14}$$

These studies provided simplified modeling of the phenomenon and a simple tool to predict the critical conditions that can cause the bundle to roll over. This model allows the number of spacers to be selected to resist the rolling instability for a preselected amount of ice and wind moment. In studies of recent rolling instabilities, the model shows that in a region of similar ice accumulation, the longest spans are at greatest risk, and the resistance to rolling can be increased by the use of a larger than normal number of spacers.

If the spacers retain their grip on the conductors, the bundle will naturally return to its normal orientation once the ice has melted off. But if the clamps slip on the conductors, the restoration of the bundle to its normal orientation can be a very difficult and costly procedure.

While there are models that can lead to numbers and locations of spacers to resist this rolling effect, the ice accretion conditions, for which special designs should be considered, need to be determined.

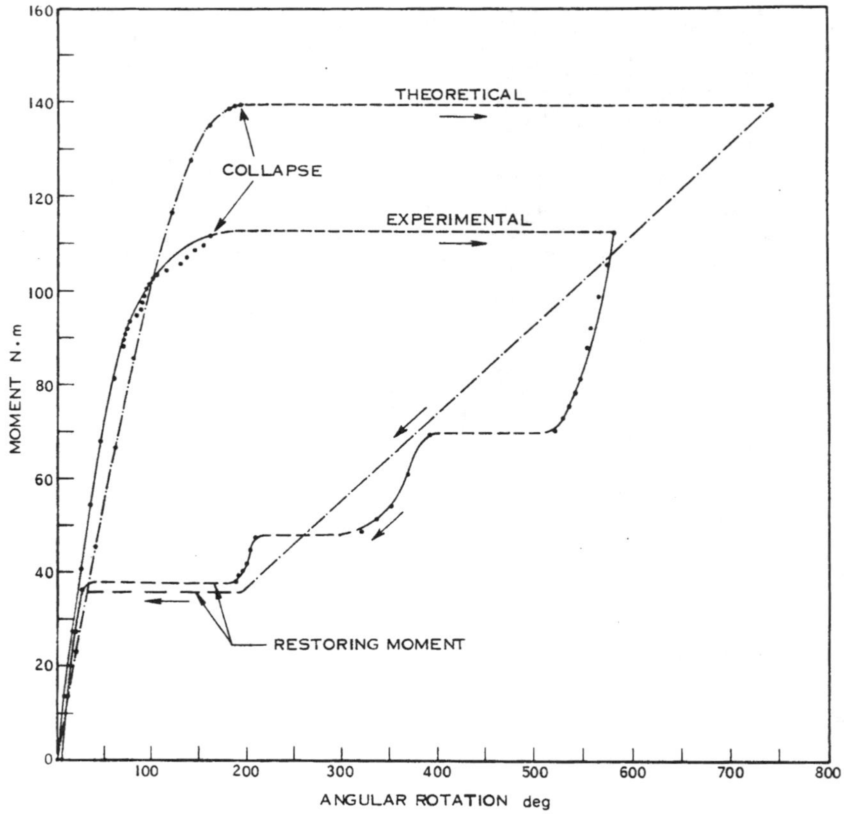

Fig. 5.47 Measured and theoretical values of applied moment versus angular rotation for a torsional stability test on a two conductor bundle (Nigol et al., 1977) (Reproduced by permission of IEEE)

5.9 Conclusion

In conclusion, especially when modified by added ice, aeolian vibrations, waked-induced oscillations, galloping, ice shedding and bundle rolling may all have a major impact on transmission lines. Those phenomena have been studied for many years from a theoretical or experimental standpoint and extensive field observations have been made, but few remedies are available for eliminating the onset of galloping or for attenuating the effects of aeolian vibrations or wake-induced oscillation on iced conductors. Moreover, the presence of ice on conductors may completely invalidate any expected tower dynamic loads or line hardware life expectancy deduced on the basis of bare conductors. However, this chapter will help the reader to identify and better understand those phenomena and it gives some indications on how to cope with it.

References

Berg A, Smart TJ, Halsan K, Papailiou KO, Schmidt J, Smart TJ, Hearnshaw D (August 1992) Results of questionnaire on interphase spacers. CIGRÉ SC22 WG11, Electra vol 143

Brika D, Laneville A (1993) Vortex-induced vibrations of a long flexible circular cylinder. Journal of fluid mechanics, vol 250: 481–508

Brokenshire RE (1979) Experimental study of the loads imposed on welded steel support structures by galloping 345 kV bundled conductors. IEEE Paper A79-551-3, IEEE PES Summer Meeting, Vancouver

Brooks NPH (1960) Experimental Investigation of the Aeroelastic Instability of Bluff Two-Dimensional Cylinders. M.A.Sc. Thesis, University of British Columbia, July

Chadha J (1974) A dynamic model investigation of conductor galloping. IEEE Paper no C74 059-2

Chadha J, Jaster W (1975) Influence of turbulence on the galloping instability of iced conductors. IEEE Trans. Power Apparatus & Systems vol PAS-94, no 5, September/October 1975: 1489–1499

Chan JK, Shah AH , Popplewell N (1992). Modelling of Conductor Galloping. Report CEA R&D. Project 321 T 672

Davis DA, Richards DJW, Scriven RA (1963) Investigation of conductor oscillation on the 275 kV crossing over the rivers Severn and Wye. Proceedings IEE, vol 110, no 1: 205–218

Den Hartog JP (1932) Transmission line vibration due to sleet. AIEE Transactions, vol 51: 1074–1076

De Tourreil C, Kuffel E (1998) Comportement des isolateurs composites rigides à socle soumis à des charges mécaniques dynamiques. Canadian Electrical Association, Report no 004D877, December

Edwards AT, Madeyski A (August 1956) A Progress Report on the Investigation of Galloping Transmission Line Conductors. AIEE Trans, vol 75, part 3: 666–686

Fekr MR, McClure G (1996) Numerical modeling of the dynamic response of ice shedding on electrical transmission lines. In: Proc 7th International Workshop on Atmospheric Icing of Structures, Chicoutimi: 49–54

Fekr MR, McClure G, Hartmann D (1998) Investigation of transmission line failure due to ice shedding effects using dynamic analysis. In: Proc 8th International Workshop on Atmospheric Icing of Structures: 11–16

Gartshore IS (1973) The effects of free stream turbulence on the drag of rectangular two-dimensional prisms. Boundary Layer Wind Tunnel Laboratory 4–73, University of Western Ontario

Gibbon RR, Juul PH, White HB, Wijker WJ (1984) Damage to overhead lines caused by conductor galloping. ELECTRA no 94: 71–76

Halsan KA, Havard DG, Fikke SM, Gerezgiher A (1998) Galloping Studies at Statnett Using Remote Monitoring. CIGRE SC22, WG11 meeting.Graz, Austria, Paper no 22–98

Hardy C, Van Dyke P (January 1995) Field Observations on Wind-Induced Conductor Motions. Journal of Fluids and Structures, vol 9, no 1: 43–60

Havard DG (1996) Fifteen years field trials of galloping controls for overhead power lines. In: Proc 7th International Workshop on Atmospheric Icing of Structures, Chicoutimi

Havard DG (1998) Analysis of galloping conductor field data. 8th International Workshop on Atmospheric Icing of Structures, Reykjavik

Havard DG (2003) Dynamic Loads On Transmission Line Structures During Galloping – Field Data and Elastic Analysis. 5th International Symposium on Cable Dynamics, Santa Margherita

Havard DG, Nigol O (January 1978) Control of torsionally induced conductor galloping with detuning pendulums. IEEE Paper no A78: 125–127

Havard DG, Pohlman JC (February 1984) Five years' field trials of detuning pendulums for galloping control. IEEE, Los Angeles, Trans. PAS: 318–327

Hillier R, Cherry RJ (1981) The effect of stream turbulence on separation bubbles. Journal of Wind Engineering and Industrial Aerodynamics, vol 8: 49–58

Hoerner SF (1965) Fluid dynamic drag. published by the author, New York, 415 p
Irvine HM, Caughay TK (1974) The linear theory of free vibrations of a suspended cable. In: Proc Roy Soc London Series A, 341: 299–315
Jamaladdine A, Beauchemin R, Rousselet J, McClure G (1996) Weight-dropping simulation of ice-shedding effects on an overhead transmission line model. In: Proc 7th International Workshop on Atmospheric Icing of Structures, Chicoutimi: 44–48
Keutgen R (1999) Galloping Phenomena. A Finite Element Approach. Ph.D. Thesis. Collection des publications de la Faculté des Sciences. Appliquées de l'Université de Liège. no. 191: 1–202
Kiessling F, Nefzger P, Nolasco JF, Kaintzyk U (2003) Overhead Power Lines – Planning, design, construction, Power Systems. Springer: 321–348
Laneville A (May 1973) Effects of turbulence on wind induced vibrations of bluff cylinders. Ph.D. Thesis, University of British Columbia, 129 p
Laneville A, Parkinson GV (1971) Effects of turbulence on galloping bluff cylinders. Presented at the 3rd International Conference on Winds Effects on Buildings and Structures, Tokyo
Laneville A, Gartshore IS, Parkinson GV (1975) An explanation of some effects of turbulence on bluff bodies. In: Proc 4th Int'l Conference on Buildings and Structures, London, Cambridge Univ. Press: 333–342
Lilien JL, Havard DG (April 2000) Galloping Data Base on Single and Bundle Conductors Prediction of Maximum Amplitudes. IEEE Trans on Power Delivery, vol 15, no 2: 670–674
Lilien JL (convenor), Van Dyke P (secretary), Asselin JM, Farzaneh M, Halsan K, Havard DG, Hearnshaw D, Laneville A, Mito M, Rawlins CB, St-Louis M, Sunkle D, Vinogradov A (2007) State of the art of conductor galloping. Cigré TFB2.11.06, Électra, technical brochure no 322, 140 p
Loudon D (1999) Vibration control of fjord crossings in Norway. In: Proc 3rd International Symposium on Cable Dynamics: 183–187
Manukata M, Yoshida Y, Ishii H (1963) Determination of spacer intervals in quadruple conductor transmission lines. Sumitomo Electric Technical Review, no 1
Matsubayashi Y (1963) Theoretical considerations of the twisting phenomenon of the bundle conductor type transmission line. Sumitomo Electric Technical Review, no 3
Matsuura Y, Suzuki Y, Arakawa K, Tanaka K (1990) Technical aspects on long-term performance of suspension insulators and its laboratory evaluation methods. Canadian Electrical Association Symposium on Insulators, Power System Planning and Operating Section, Engineering and Operating Division, Montreal: 14–25
Miner MA (1945) Cumulative damage in fatigue. Journal of Applied Mechanics, vol 12: A159–A164
Morgan VT, Swift DA (1964) Jump height of overhead line conductors after the sudden release of ice loads. Proc IEE, vol 111, no 10: 1736–1746
Morishita S, Tsujimoto K, Yasui M, Mori N, Inoue T, Shimojima K, Naito K (1984) Galloping phenomena of large bundle conductors – Experimental results of the field test lines. In: Proc Cigre, 1984 Session, Paris, Paper no 22–04
Nakamura Y (1980) Galloping of Bundled Power Line Conductors. Journ Sound & Vibration, vol 73, no 3: 363–377
Nakamura Y, Tomonari Y (1977) Galloping of rectangular prisms in a smooth and in a turbulent flow. Journal of Sound and Vibration, vol 52, issue 2: 233–241
Nakamura Y, Tomonari Y (May 1981) The aerodynamic characteristics of D-section prisms in a smooth and in a turbulent flow. Aeronautical Quarterly. vol 32: 153–168
Naudascher E, Rockwell D (1994) Flow-induced Vibrations – An Engineering Guide. A.A. Balkema Publishers, Rotterdam
Nigol O, Clarke GJ (1974) Conductor galloping and control based on torsional mechanism. IEEE Paper no C74 016–2
Nigol O, Clarke GJ, Havard DG (January 1977) Torsional stability of bundle conductors. IEEE Paper no F 77 224–9
Nigol O, Havard DG (1978) Control of torsionally induced galloping with detuning pendulums. IEEE Paper no A78 125–7

Novak M (1969) Aeorelastic galloping of prismatic bodies. ASCE Journal of the Engineering Mechanics Division, 96: 115–142

Novak M (1971) Galloping and vortex induced oscillations of structures. In: Proc 3rd International Conference on wind effects on buildings and structures, Tokyo, Paper IV–16: 11

Novak M, Tanaka H (1974) Effect of turbulence on galloping instability. ASCE Journal Engineering Mechanics Division, vol 100, no EM1: 27–47

Novak M, Davenport A, Tanaka H (1978) Vibration of towers due to galloping of iced cables. ASCE Journal of the Engineering Mechanics Division, 104: 457–473

Palmgren A (1924) Die lebensdauer von kugellagern, Zeitschrift des vereins deutscher ingenieure. vol 68: 339–341

Parkinson GV (1971) Wind-induced instability of structures. Philosophical Transactions for the Royal Society of London. Series A, Mathematical and Physical Sciences, vol 269, issue 1199: 395–409

Parkinson GV (1989). Phenomena and modelling of flow-induced vibrations of bluff bodies. Progress in Aerospace Sciences 26: 169–224

Parkinson GV, Smith JD (1964) The square prism as an aeroelastic non-linear oscillator. Quart Journ Mech and Applied Math, vol XVII, pt 2: 225–239

Pon CJ, Havard DG (October 1993) Control of distribution line galloping. Canadian Electrical Association, Montreal, R&D Project 196T367

Pon CJ, Havard DG (March 1994) Field trials of galloping control devices for bundle conductor lines. Canadian Electrical Association, Montreal, R&D Project 133T386

Pon CJ, Havard DG, Edwards AT (July 1982) Performance of interphase spacers for galloping control. Ontario Hydro Research Division Report no 82-216-K

Pon CJ, Havard DG, Currie IG, MacDonald R (1990) Aeolian vibration excitation of bundle conductors. Report on CEA R&D Project 177T510

Price SJ (1975) Wake induced flutter of power transmission conductors. J Sound Vib 38(1): 125–147

Rawlins CB (1981) Analysis of conductor galloping field observations – single conductors. IEEE Transactions on Power Apparatus and Systems, vol PAS-100, no 8

Rawlins CB (April 1988) Research on vibration of overhead ground wires. IEEE Transactions on Power Delivery, vol 3, no 2, pp 769–775

Rawlins CB (2001) Galloping Eigenmodes in a Multispan Overhead Line Section. Proceedings of Fourth International Symposium on Cable Dynamics. Montreal (Canada). May 28–30. pp 85–92

Rawlins CB, Hard AR, Ikegami R, Doocy ES (1979) Transmission Line Reference Book – Wind-Induced Conductor Motion. Palo Alto, California: Electric Power Research Institute

REA (1962, Rev 1982) Galloping Conductors. Design Report no 1, Rural Electrification Administration, Transmission Branch

Shimizu M, Shugo M, Sato J (1998) Geometric non-linear analysis of transmission line galloping. Journal of Structural Engineering, vol 44A: 951–960 (Japanese only)

Simpson A (1966) Determination of the inplane natural frequencies of multispan transmission lines by a transfer-matrix method. In: Proc IEE, vol 113, no 5: 870–878

Simpson A (1979) Fluid-dynamic stability aspects of cables, Mechanics of wave-induced forces cylinders. T.L. Shaw (ed.) Pitman, 90 p

Smith JD (1962) The square prism as an aeroelastic non-linear oscillator. M.A.Sc. Thesis, University of British Columbia

Stewart JR (1983) Ice as an influence on compact line phase spacing. In: Proc. First International Workshop on Atmospheric Icing of Structures, Hanover: 77–82

St-Louis M, Hardy C, Bellerive JP (1993) Bundled-conductor spacers: Hydro-Quebec's experience. Paper Presented to the Canadian Electrical Association, Transmission Section Meeting, Montreal

Strouhal V (1878) On Aeolian tones. Ann of Phys, 5: 216

Tunstall M (1989) Accretion of ice and aerodynamic coefficients. Study day on galloping, University of Liège

Tunstall M (Convenor), Obro M (Secretary), Couvreur M, Ervik M, Fukuda J, Havard DG, Hearnshaw D, Jürdens C, Kempner L, Lilien JL, Okumura T, Pohlman JC, Rawlins CB, Rhebergen B, Ruritz R, Schlyter C, Schmidt J, Shkaptsov VA, Smart T, St-Louis M, Sunkle D, Tavano F, Turna ÖF, Voyatzakis Y, Wolfs M (1995) Field observations of overhead line galloping – Galloping reporting forms. CIGRE SC22:WG11 Task Force on Galloping, Electra no 162

Van Dyke P (2007) Galop induit en ligne expérimentale à l'aide de profils en D sur conducteur simple avec ou sans entretoises interphases. Ph.D. thesis, Université de Sherbrooke

Van Dyke P, Laneville A (July 2004) Galloping of a single conductor covered with a D-section on a high voltage overhead test line. 5th International Colloquium on Bluff Body Aerodynamics and Applications: 377–380

Van Dyke P, Laneville A (2005) HAWS clamp performance on a high voltage overhead test line. 6th International Symposium on Cable Dynamics Proceedings, Charleston: 205–209

Van Dyke P, Paquette R, St-Louis M (2001) Design and test of a new Aeolian vibration damper. 4th International Symposium on Cable Dynamics, Montréal

Wardlaw RL, Cooper KR, Scanlan RH (1973) Observations on the problem of subspan oscillation of bundled power conductors. Intern symposium vibration problems in industry, Keswick, UK Atomic Energy Authority, Paper no 323

Wolfs M (Convenor), Sunkle D (Secretary), Diana G, Ervik M, Fujii K, Halsan K, Havard D, Hearnshaw D, Jürdens C, Kleveborn R, Lilien JL, Mikkelsen SD, Mito M, Okumura T, Papailiou K, Pohlman JC, Rawlins C, Rhebergen B, St-Louis M, Schlyter C, Schmid, J, Shkaptsov V, Smart T, Tavano F, Tunstall M, Van Dyke P, Voyatzakis Y, Wang J (2000) Review of galloping control methods. Electra no 191: 44–61

Wang J, Lilien JL (July 1998) Overhead electrical transmission line galloping. A full multi-span 3-dof model, some applications and design recommendations. IEEE Transactions on Power Delivery, vol 13, no 3: 909–916

Washizu K, Ohya A (1978) Aeroelastic instability of rectangular cylinders in a heaving mode. Journal of Sound and Vibration, 59: 195–210

Wijker WJ, Leppers PH (1987) 25 years of galloping experience in the Netherlands. CIGRE SC22-WG11TFG Report 87-0

Zdravkovich MM (1984) Classification of flow-induced oscillations of two parallel circular cylinders in various arrangement. Symposium on flow-induced vibration, vol 2, ASME: 1–18

Chapter 6
Anti-icing and De-icing Techniques for Overhead Lines

Masoud Farzaneh, Christophe Volat and André Leblond

6.1 Introduction

Combating ice deposits on overhead transmission lines has been a big challenge in cold climate regions for the last eighty years. With an expanding electric system, it has been a difficult task to prevent or remove ice from long lines with single or bundled conductors. Considerable research has been carried out and large-scale technologies have been developed to address this problem. Based on worldwide power utility experience, two different strategies regarding ice accretion on overhead lines have been adopted.

To prevent failure, power utilities try to build overhead lines that are capable of withstanding large icing events (with a low probability of occurrence). This commonly requires strengthened towers and costly lines. In the past, when these reinforced lines had to face extreme icing events, (ice sleeve overload), where collapse of transmission lines was possible, "de-icing" methods were used as the first strategy to protect them. These de-icing methods are generally put into operation when a certain amount of ice has accumulated on the conductor in order to shed the ice sleeve as soon as possible. These methods require specific ice detection systems able to follow the ice storm in local areas as the ice load accumulates on conductor line sections in order to be able to intervene in time.

The second strategy, "anti-icing", is the prevention of part of or the entire expected ice load. This strategy can be used on existing weaker lines that may not have been designed to current extreme ice loading standards or utilized on new lines to allow for a less expensive standard of construction, considering the low occurrence probability of extreme ice storms. To prevent severe ice load damage, ice accumulation on conductors must be avoided or considerably reduced. This could be achieved by using anti-icing methods to prevent or weaken ice adhesion or/and to use de-icing methods in the early stages of an icing storm to limit the size of the ice load on the conductors.

M. Farzaneh
University of Québec at Chicoutimi, 555 Boulevard de l'Université, Chicoutimi, Canada G7H 2B1
e-mail: farzaneh@uqac.ca

M. Farzaneh (ed.), *Atmospheric Icing of Power Networks*,

In order to support these two strategies aiming at combating icing damage on overhead lines, a large number of anti-icing and de-icing methods have been developed. Some of these methods have been well documented in specific reviews since the 1980s. To the best of our knowledge, the first review dedicated to these anti-icing and de-icing methods applicable to overhead power lines was presented by Polhman and Landers (1982). A few years later, another paper presented a detailed review of two methods, rolling and heating by short-circuit current, which are used on the Manitoba Hydro power network (Hesse 1988). A classification of de-icing and anti-icing methods into four categories, passive, thermal, mechanical, and miscellaneous, based on the physical principle used in the method of ice removal, was proposed later (Laforte et al. 1998). More recently, another report on power line anti-icing and de-icing techniques was presented (CEA 2002). The methods listed were classified into six categories; passive techniques, active coatings and sheathings, active methods on bare conductors, methods using external thermal energy, methods using external mechanical energy, and finally, miscellaneous methods, with less potential for application to ground-wires and energized conductors.

Another classification was proposed to illustrate the permanent or temporary character of a method, the need for line modification, and whether it is automated or manual (Volat et al. 2005). Thus, the existing potential methods presented in previous reviews can be divided into inline, limited-use, and permanent. Inline methods refer to those using Joule effect to melt the ice, using the energy from the line itself or using an external energy source, with no device or coating directly added to the energized conductor or ground wire (GW). Limited-use methods regroup all methods that are not permanently installed on the lines, but are used at specific locations, primarily for de-icing, or that can only be used once. Finally, permanent methods include all methods permanently installed on the conductors or GWs.

It appears that no permanent method, above, is presently in widespread use. Most of them, such as coatings or devices that are added to energized conductors or GWs, are in fact in development or in the conceptual stage. The development of such methods is quite complex as they have to meet specific requirements to ensure good performance and life expectancy. This consideration is developed in detail in the next chapter.

6.2 Anti-icing Techniques

Ice adhesion strength is directly affected by the physico-chemical properties of the surface, as well as the mechanical mechanisms taking place at the ice-substrate interface (Petrenko and Whitworth 1999). In order to prevent ice or snow accretion on conductors or GWs, different approaches seem to be emerging. The first of these is to weaken ice adhesion strength by acting directly at the interface between the ice and the conductor. The second approach consists in preventing the freezing of supercooled water droplets upon impact with the surface to be protected. Finally, a third approach consists in employing different combined methods in order to limit ice accretion on conductors and GWs.

6.2.1 Reducing Ice Adhesion Strength

One way to prevent ice accretion is to use a specific surface treatment coating to weaken, at different scales, the physico-chemical and mechanical interactions between ice and substrate. One solution consists in using a solid icephobic coating to protect the exposed surface. The literature review reveals that two main approaches have been proposed for developing icephobic coatings. The first approach consists in elaborating icephobic coatings from materials of lower surface energy. Poly(tetrafluoroethylene) (PTFE or Teflon®) and poly(dimethylsiloxane) (silicone or PDMS) were largely employed and tested (Mulherin and Haehnel 2003; Frankenstein and Tuthill 2002; Croutch and Hartley 1992; Hacker et al. 2000). In a recent long-term study, PTFE tapes have demonstrated sufficient wet snow accretion reduction effect on overhead transmission lines (Nakagami 2007) but no tests were conducted for atmospheric ice. Ice adhesion test results reveal that in general, silicone-based polymers performed slightly better than PTFE-based ones. Recently, an epoxy-silicone component (called Wearlon®) was developed. This commercially available low-ice-adhesion solid material provides an ice adhesion reduction factor of about 12, with reference to aluminium, compared to a factor of about 2 for Teflon® (Laforte and Beisswenger 2005). The other approach proposed is not as common, as it makes use of heterogeneous polymer coatings in order to decrease ice adhesion (Reich 1994; Murase et al. 1994; Croutch and Hartley 1992). Other studies found that better coatings than PDMS- or PTFE-based ones can be obtained by mixing polysiloxane and fluorocarbon types of materials. For example, polyperfluoalkyl(meth)-acrylate modified with a lithium compound reduces ice adhesion by a factor 25 compared to PTFE (Murase et al. 1994). Based on these interesting results obtained from heterogeneous polymer coatings, a patent for the elaboration of icephobic coatings was submitted by the Boeing Company (Byrd 2004).

Literature research so far indicates that no material can really be considered perfectly icephobic, assuming that ice can detach itself from the icephobic surface under its own weight or from the action of the wind. Recent developments in ice adhesion research, as well as in nanomaterial and surface sciences, have reinitiated interest for icephobic materials. These new developments, as well as the icephobic material theory, are described in greater detail in Section 6.7.

One way to weaken ice adhesion strength on conductors is to intercalate a thin viscous or liquid film between the ice and the substrate. If that is achieved, the physico-chemical interaction between ice and substrate is lessened, leading to ice release by gravitational force. This ice surface protection can be produced by using viscous liquids like industrial lubricants, oils, and greases (Gastonguay and Champagne 1996). Recent investigations on ice adhesion demonstrated that lithium grease and industrial lubricants can reduce ice adhesion strength on aluminium by a factor of 63 (Laforte and Beisswenger 2005). However, these temporary coatings, or viscous icephobic materials as they are named in the CEA report (CEA 2002), are not well adapted for electrical network equipment, particularly for conductors and GWs, since they require recurrent applications. This means that maintenance workers would have to travel to the problem location to apply the coating, since it would be difficult to use automatic mechanical spray systems. Reapplication of these non-permanent coatings would require accurate timing, good weather forecasts and accurate field observations to predict when a severe storm will hit. Finally, as these products are not biodegradable, they may constitute a threat to the environment. For all these reasons, such non-permanent coatings, at present, are considered to be unsuitable, only to be used as a last resort solution to protect strategic line sections when severe ice accretions are forecasted.

Some other anti-icing methods, which can be classified as exotic (no real potential) and are well indexed in the literature (CEA 2002), will not be discussed in this section. All the aforementioned anti-icing methods showing a certain potential are summarized in Table 6.1. This table lists some practical considerations regarding the different anti-icing methods. A particular interest in relation with Section 6.6 was provided concerning the applicability of the method to bare conductors or GWs. Also, details concerning the permanence of the coatings on the conductor, the possibility of direct installation on natural sites, and the length of power network that can be protected are mentioned. For this last item, considerations about the estimated cost as well as required energy were taken into account. More details about these methods relative to their cost, lifespan, energy required are well defined in (CEA 2002).

6.2.2 Prevention of Supercooled Droplet Freezing

Prevention of droplet freezing requires a different approach than that used to weaken ice adhesion strength. The idea is to prevent freezing of supercooled water droplets upon impact with the conductor surface. Depending on the method used, this approach can prevent completely the formation of ice on the conductor, or simply help to keep a thin water film between the conductor and the ice layer in order to facilitate natural ice shedding by gravity or wind action forces.

Creating a water film between ice and substrate can be simply achieved using freezing point depressant liquids which, when mixed with water, lower its freezing point. These commercially-available fluids are commonly used to protect aircraft and pavement from icing. Like viscous icephobic materials, they present the same constraints of applicability and duration as they are not permanent and require timely application on structures in order to provide lasting ice protection (CEA 2002). In this context, they may possibly be used only for specific and strategic line sections.

Preventing water droplets from freezing can be easily achieved by maintaining a positive temperature just above the freezing point during the icing event. One method to increase conductor surface temperature is to use the heat generated by the circulation of the current in conductors. Heating line conductors by Joule effect to prevent ice accretion or de-ice is recognized worldwide as the most efficient engineering approach to minimize the consequences of severe ice storms on overhead

Table 6.1 Anti-icing methods based on ice adhesion strength reduction

Method	Principle	Efficiency	Applicability to overhead lines			Duration	Installation	Protected length
			Conductor		GW			
			Single	Bundled				
Solid icephobic	Weak inter-molecular forces	Not proved yet	X	X	X	Permanent	In factory	Longer section
Viscous icephobic	Weak inter-molecular forces	Significant for atmospheric ice and wet snow	X	X	X	Need frequent applications	On site	Short section

lines (Motlis 2002). Both AC and DC currents have been used in different countries to melt ice (CEA 2002). Technologies for both types of currents are available. The methodology has been developed, and decades of operational experience with ice-melting systems have been gained around the world. Joule-effect thermal methods can be used both for anti-icing and de-icing on conductors. A dedicated review of the major Joule-effect methods is presented in Section 6.4.

As Joule-effect methods are based on the direct conversion of conductor current, some thermal methods take advantage of electromagnetic properties of energized conductors combined with physical properties of materials. One of the proposed methods uses loss in specific dielectric coating that covers the entire surface of the conductor (Petrenko and Peng 2003). All dielectrics (except vacuum) have two types of losses: one is conduction loss, representing the flow of actual charge through the dielectric, and the other is dielectric loss due to movement or rotation of the atoms or molecules in an AC electric field. By choosing an adequate dielectric coating from the ferroelectric materials family, conductor surface temperature can be maintained above the freezing point. However, anti-icing seems to be obtained at high frequency, or 60 kHz instead of the 60 Hz service voltage frequency. This last condition is problematic since working with high frequency can lead to the generation of electromagnetic interference. Also, high frequency generation requires the addition of an external source in order to superimpose the 60 kHz electric field onto the 60 Hz field.

The other methods proposed are based on the use of a ferromagnetic coating in order to sustain a positive temperature on the surface of the energized conductor (CEA 2002). Instead of absorbing energy from the electric field, ferromagnetic coatings absorb energy from the magnetic field, which is at maximum at the conductor surface, as it is for the electric field. Ferromagnetic-coating heating is based on hysteresis and induced leakage current losses generated by an AC electric field. Methods using different shapes of coating and apparatus are currently in service in Japan to prevent wet snow accretion: rings, cylindrical envelopes or spiral wires. However, as discussed in detail in (CEA 2002), these methods do not seem adequate for preventing ice accretion under severe icing conditions, especially at lower temperatures and high wind speeds, as they would require high magnetic field magnitude and consequently high current values to maintain positive surface temperature for conductors. Also, such methods induce electric losses other times of the year, which is undesirable when no icing occurs. In this context, these methods become economically less attractive than thermal methods based on the Joule effect.

Another method resides in using electrical tracers, as is the case for heating pipes in chemical plants. Electrical tracers, which are electrically insulated resistive heating wires, are wound around conductors or GWs in order to heat their surface. This is a mature technology in the chemical field but requires adaptation to power lines. The efficiency of the method has been demonstrated in laboratory.

Finally, in the case of GWs, maintaining a positive GW surface temperature in order to prevent supercooled water droplets from freezing requires the GW to be energized. In this context, the methods previously proposed and summarized in Table 6.2 can be used to prevent ice from accreting on these wires. This will require

Table 6.2 Anti-icing methods based on prevention of supercooled droplet freezing

Method	Principle	Efficiency	Applicability to overhead lines			Duration	Installation	Range
			Conductor		GW			
			Single	Bundled				
Freezing point depressant liquid	Create water film at the ice-conductor interface	[1]Significant for atmospheric ice and wet snow	x	x	x	Need frequent applications	On site	Short section
Ferroelectric coating	Maintain a positive temperature of conductor surface	[1]Significant for atmospheric ice	x	x	[2]x	Permanent	In factory	Short section
Ferromagnetic materials	Maintain a positive temperature of conductor surface	Significant for wet snow	x	x	[2]x	Permanent	On site	Longer section
Electrical tracer	Maintain a positive temperature of conductor surface	Significant for wet snow	x	x	x	Permanent	On site	Short section
Joule effect	See section 6.4							

[1] Concept not tested on natural site

[2] Require GW insulation from tower

a complete insulation from the towers, which can be a costly measure. This could be an alternative solution for critical GW sections crossing rivers or highways.

6.3 De-icing Techniques

De-icing involves specific methods to speed the shedding process after snow or ice has formed on conductors and GWs. Generally, two approaches are used: thermal methods to melt, and mechanical methods to break ice accretions. As demonstrated in (Laforte et al. 1998), mechanical methods require around 100 times less energy than thermal methods to force ice shedding. Thermal methods are currently recommended and in use by several electric utilities. This is mainly due to the fact that thermal methods, and particularly Joule-effect methods, permit the de-icing of longer line sections and require less manpower than mechanical methods. These methods are usually applied to de-icing conductors but not GWs. Nevertheless, mechanical methods are preferred for specific and fast interventions in order to de-ice short critical sections of power networks. GWs loaded with ice may stretch extensively and get too close to phase conductors, leading to flashovers causing power outages. At some locations the use of thermal methods to de-ice phase conductors can sometimes reduce the electrical clearances with the GWs as well. Such problems are usually found on a very small scale.

6.3.1 Thermal Methods

Thermal methods include all methods which cause the melting of ice in order to force its shedding (Fig. 6.1). All the aforementioned anti-icing methods preventing freezing of supercooled water droplets by increasing conductor surface temperature

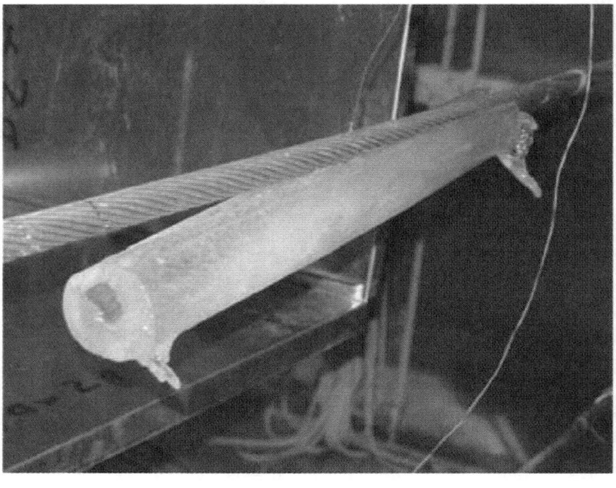

Fig. 6.1 De-icing sample with ice at near falling point (Prud'Homme et al. 2005a, reproduced by permission of P. Prud'homme, Hydro-Québec)

can also be considered de-icing methods. The only difference is that less energy is required for anti-icing than for de-icing (Prud'Homme et al. 2005a).

Along with the thermal methods presented previously, other methods can be mentioned with a potential application to GWs and conductors. Established methods based on the Joule effect are presented in greater detail in Section 6.4.

Another method is similar to those using ferroelectric materials (see Table 6.2), but in that case the ice acts as the ferroelectric coating (Sullivan et al. 2003). Thanks to superimposed high-frequency electric field of around 100 kHz, dielectric losses in ice combined with conductor skin effect lead to ice melting, which can be theoretically induced on longer power line sections, up to 50 km. This solution has only been tested in a laboratory on one-metre-long bare conductors. One disadvantage is that this could generate electromagnetic disturbance, which can interfere with telecommunications. Another is that, in order to speed ice melting, an increase in high-frequency voltage is also required, as well as the development of specific inductors to support this high voltage.

All the thermal methods presented previously can be considered as internal heat sources delivered by the conductor from the point of view of the ice layer. Some proposed methods found in the literature can be considered as external heat sources. One method consists in using radiation, specifically radio-waves, to heat the conductor/ice interface (Berry et al. 1993). Found efficient to de-ice railways, this method does not seem well adapted for power lines as its use would be limited to de-icing one span at a time and requires a specific vehicle-mounted source of radio waves. Another method uses steam as a source of heat. This method is currently used by Hydro-Québec to de-ice substation equipment like disconnect switches and post insulators (Lanoie et al. 2005). They have adapted a vehicle for specific interventions, equipped with a special insulated telescopic arm for remote steam de-icing operations in high-voltage environments. There may be some applicability for this steam de-icing vehicle to reach some critical spans of power lines, if the lines are not too high in the air.

All the thermal methods with their principal characteristics are presented in Table 6.3.

6.3.2 Mechanical Methods

Mechanical methods refer to all methods used to break ice in order to accelerate shedding. Generally, most of the mechanical methods are based on two approaches, one of which consists in directly breaking the accretion by scraping the ice, and the second uses the energy released by shock waves, vibrations, or twisting of the conductor or GW. One of the main advantages of mechanical methods is that, compared to thermal methods, they can easily be applied to GWs with little or no modifications.

The simplest scraping methods are manual scrapers, rollers or cutters attached to a rope which is pulled by maintenance workers in order to release the ice. These are generally in-house methods and used when lines are accessible (CEA 2002).

Table 6.3 Thermal de-icing methods

Method	Principle	Efficiency	Applicability to overhead lines			Duration	Installation	Protected length
			Conductor		GW			
			Single	Bundled				
High frequency	Dielectric losses in ice layer and skin effect	[1]Significant for atmospheric ice	x	x	[2]x	Permanent	On site	Short to medium section
Transportable radio wave source	Heating of conductor/ice interface	Significant for atmospheric ice	x		x	Permanent	Not required	Short section
Transportable steam source	Melting of ice	Significant for atmospheric ice	x		x	Permanent	Not required	Short section
Joule effect	See section 6.4							

[1] Concept not tested on natural site

[2] Require GW insulation from tower

Recently, automatic robots have been developed for mechanical shedding. One of these, called the Remotely Operated Vehicle (ROV), has been developed at IREQ (Montambault et al. 2000; Leblond et al. 2002). Hydro-Québec TransÉnergie was looking for a mechanical device for de-icing GWs and conductors on its transmission network. The development of a method allowing for gradual de-icing of the cables, easier on structures, was the main objective. Robust, lightweight, and compact, the device has high traction force, which allows it to perform demanding tasks. The ROV was successfully tested on live line conductors (315-kV lines). Its electronic circuitry is protected against electromagnetic interference and it has an operational range of 1 km. The de-icing tool, based on a set of steel blades, is mounted on the ROV (Fig. 6.2). This device is ready to be tested in field conditions, on ice accretions resulting from an icing event. The ROV will probably have to be installed from a helicopter or an insulated boom truck, since the icy structures prevent linemen from reaching the GWs.

Energy-release methods use conductors or GWs to transmit the mechanical energy that induces ice shedding. One way to mechanically induce ice shedding is to create a shock wave which propagates along the conductor. As ice is a very brittle material at high strain rate ($> 10^{-3}$/s), both under compressive and tensile stress (Petrovic 2003), less mechanical energy is required for breaking the ice under mechanical shocks because energy is not dissipated in plastic deformation.

The first method being used to create a shock wave is to manually hit the conductor with an insulated pole. This method is effective to de-ice one span at a time, but requires conductors to be not very high above the ground. Based on this principle, a new concept using a rope was proposed (Laforte et al. 1998). The rope is equipped with a hammer/hook head. The hammer is activated by a pneumatic piston charged with compressed air. This rope can be used from the ground by linemen or mounted on a telescopic lift. This method requires that the lines be accessible to linemen and, consequently, is not efficient to de-ice river crossings. In that case, the rope can be activated directly from the tower or a helicopter (CEA 2002).

Fig. 6.2 Prototype of the ROV de-icer (Leblond et al. 2002)

Fig. 6.3 Helicopter use of 90 kg weight with large knotted rope (Reproduced by permission of B. Anderson, BC Hydro)

In mountainous terrain with only helicopter access, BC Hydro crews have used a 90-kg weight attached to a 30-m rope with a series of large knots, which impacts, lifts and drops small sections of conductor causing gradual shedding of the snow (Fig. 6.3). A large 500-m span can typically have fifty percent of the sticky wet snow removed in less than half an hour by carrying out about 30 multiple knot pulls per half hour along the length of the span. The bottom portion of the knotted rope is positioned vertically below the conductor and the top portion of the rope is pulled up at a 30 degree angle against the conductor causing the knots to catch (impact, lift and drop) the conductor as the knots, spaced one metre apart, with one metre spacing are pulled across the conductor. Caution must be used to ensure the suspended weight does not come into contact with the ground, which could cause an outage on energized lines. This has proved much more effective than simply impacting the conductors with insulated poles from a helicopter.

More recently, Hydro-Québec TransÉnergie has proposed a new method to de-ice GWs using shock waves. This portable system, called De-icer Actuated by Cartridge (DAC), shown in Fig. 6.4, consists in using a portable cylinder-piston device equipped with blank cartridges that can be remotely fired to create shock waves (Leblond et al. 2005). The high velocity piston is activated by firing blank cartridges. The de-icing operation with the DAC is carried out entirely from the ground, which represents a major advantage compared to other mechanical methods. First, a commercially available line-thrower is used to throw a projectile which tows a line that passes over the cable to be de-iced. Next, the DAC is pulled up to the GW and held in place by a taut rope (Fig. 6.4) (Leblond et al. 2005).

At the top of the DAC, the piston rod is equipped with an open-ended clamp which is firmly held into contact with the GW to transmit the force created from the firing pressure in the combustion chamber. At the bottom, the DAC is equipped with a revolver system (Fig. 6.4) that stocks 6 blank cartridges in a barrel. A solenoid, plugged into the electronic box which is fixed to the cylinder, is used to pull the

Fig. 6.4 DAC prototype held in place by a taut rope and ready to be fired (Leblond et al. 2005)

trigger of the revolver system that can be remotely fired from the ground by means of a highly secure digitally encoded RF firing system to provide interference-free operation. Numerous tests have been carried out to assess the efficiency of the method and to optimize its physical parameters. Experimental tests on a GW 100-metre test span yielded good results for eccentric ice accretions. Multiple firings were necessary for equivalent concentric ice accretions (Leblond et al. 2005). The use of the DAC, one span at a time, may create significant unbalanced loads at the supporting structures. In order to minimize such loads, the taut rope used to hold the DAC in place (Fig. 6.4) is temporarily anchored to the ground (to vehicles like trucks or snowmobiles) at mid-span in order to keep the GW from freely moving up during the de-icing operation. This prevents significant dynamic unbalanced loads to be transmitted to the supporting structures that could otherwise lead to the slippage of the GW at suspension clamps or could damage the tower-peaks.

Electro-impulse methods have also been proposed for de-icing conductors and GWs using shock waves. One of these methods, initially developed for de-icing aircraft wings (EIDI), was tested on power lines (Egbert et al. 1989). EIDI is a de-icing method involving the short discharge of a capacitor through an electric coil to produce a strong magnetic field a few milliseconds in duration. This in turn produces a shockwave that acts on a nearby electrically conductive target. The EIDI system developed on power lines comprises an actuator containing the coil and the conductive target which is directly attached to the line conductors or GWs. The impulse current source is pole-mounted. When a current pulse is sent to the coil, a shock wave is transmitted to the conductor via the conductive target.

Experimental results from a 100-meter span showed that the EIDI method was not really effective since only about 10% of the span close to the EIDI system was de-iced. Based on these results, the same authors proposed an improved concept of EIDI actuator for power lines. The proposed actuator consists of two insulated strips of copper-ribbon wire stacked together and wrapped in a spiral around the conductor, along the entire span length. When energized under current pulse, the two wires tend to repel one another and exert a force outward from the conductor. Some successful laboratory tests were conducted on a small segment of conductor, with about 12 mm of ice thickness. No test on natural sites could be carried out because the investigators were not able to wrap the actuator tightly enough around the conductor. This led to a downward actuator motion towards the conductor and, as a result, no ice shedding. The same concept, using the EIDI principle with two wires working in repulsion under high current impulse was proposed (Laforte et al. 1998). The shock wave EIDI actuator is formed by a pair of strands on the external layer of the cable, which are isolated and connected at one extremity. The other extremity is connected to the impulse current generator. In this condition, the actuator must be tightly wrapped around the cable. Tests conducted at a natural site have been able to de-ice a 260-m cable. Even if this technology is effective, some investigations need to be carried out before moving forward to a prototype installation on GW (CEA 2002). Particular attention needs to be paid to the degradation of the insulation of the EIDI actuator by lightning.

Using the same EIDI principle, a method was developed and tested at the Institut de Recherche d'Hydro-Québec (IREQ) (Landry et al. 2000) in order to de-ice lines with twin or quad bundled conductors at rated voltages of 315 and 735 kV respectively. In this method, the EIDI actuator is formed by the bundle conductors. The high-current impulse is generated by a short-circuit current (I_{SC}) at the rated voltage of the transmission line. The subsequent action of electromagnetic forces leads the conductors to knock against each other and release the ice, as shown in Fig. 6.5. Tests were carried out on a sample overhead transmission line with twin (315 kV) and quad (735 kV) bundles installed in the switchyard at the IREQ high-power laboratory (Landry et al. 2001). In order to reduce the amplitude and duration of the short-circuit currents as much as possible, asymmetrical I_{SC} and re-closing sequences are necessary. Conductors have to be excited at a frequency close to their fundamental sub-span frequency to get a maximum dynamic motion synchronized with the re-closing sequences. Impact studies on the Hydro-Québec power system reveal that this method could likely be applied to 315-kV lines, but only in emergency conditions like severe ice storms. For 735-kV lines, the required short-circuit currents and re-closing sequences are unacceptable for network stability and, therefore, the method would not be applied.

Another set of mechanical methods aims to induce sustained vibrations in the conductors or GWs in order to shed ice. A vibrating device attached to the middle

Fig. 6.5 De-icing a twin bundle using 10 kA and an appropriate re-closing sequence (Landry et al. 2000, © 2000 IEEE)

of the span was developed by the Protura Company (Roger 2004) and called AIC for Automatic Ice Control. This AIC, presented in Fig. 6.6, is based on the same technology as the Power Line Sensor (PLS) developed by the same company and includes a current transformer for power supply, a camera, different sensors for ice detection, a control box and an HF emitter/receiver, and a commercial electromagnetic vibrator. All these apparatus are encapsulated in a rigid protective housing mounted directly on the conductor. The AIC is permanently installed in the middle of the span and is completely autonomous. Thanks to its ice detector and HF communication capabilities, de-icing sequences can be fully automated or manually activated through a signal order transmitted by the HF.

A similar method to the AIC device, called ice-shedder device, is attached to conductors and uses a motor to move an unbalanced weight, which causes the device to vibrate. The vibration frequency generated by the ice-shedder is in the same range as the natural frequency of the span. Some preliminary tests were conducted on power lines with a span of about 500 feet and a diameter of 1.2 inches (Nourai and Hayes 2003). By operating the ice-shedder device within a frequency range

Fig. 6.6 Automatic Ice Control (AIC) developed by Protura (Reproduced by permission of Mr. Roger Hansen, Protura)

of approximately 1.5–8.0 Hz, conductor displacements of about 4–13 inches were observed, with conductor accelerations within the range of 0.5–14 g. Accumulated ice was adequately shed from the conductors within these ranges, with hanging ice being the most easily shed and tubular type ice being the most difficult. As with the Protura device, this automated technique can easily be installed on existing bare conductors and GWs. Additionally, the ice-shedder device can be driven using power from the bare conductor line to which it is attached. For GWs, an external power supply must be added. In the future, this technique could be applied to bundle conductors with minor modifications.. However, the de-icing capacity of this method, and particularly the maximum radial thickness of ice and the maximum span length that can be de-iced are presently unknown. Also, large oscillations of conductors or GWs similar to galloping could cause mechanical damage to the power lines or support insulators over the long term.

Another mechanical method, recently proposed by (Laforte et al. 2005a), produces induced vibrating waves along the conductors. It consists in using a specially designed apparatus to slowly twist the wire or the conductor around its longitudinal axis. It then suddenly releases the elastic mechanical energy accumulated by torsion (Allaire and Laforte 2003). The twisting force can be applied by hand or motor, depending on the rigidity of the conductor. The efficiency of this method was demonstrated on a natural site, where a 15-metre-long GW span was entirely de-iced using a manual twisting device installed at mid-span (Laforte et al. 2005a). The main advantages of this twisting de-icing system are its simplicity and efficiency for all types of ice accretions on single conductors. This method requires very low mechanical energy. It is difficult to apply in practice since the mid-span is not easily accessible from the ground during or after an ice storm. As the deformation is below the elastic shear limit of the wire material, the twisting does not damage it by fatigue. The main disadvantages are that the automation of this de-icing process is quite complex, as it needs an electric motor mounted on a rigid attachment at the height of the conductor; a reduction gear box and a magnetic clutch coupled at the motor, a control module, and an ice detection unit. Also, it is difficult to install as some modifications to the GW or conductor pole attachments are required. Finally, this technique cannot be applied to bundle conductors.

All the aforementioned mechanical methods are presented with their principal characteristics in Table 6.4.

6.4 Joule-Effect Methods

Using Joule-effect heating to clear ice from conductors is one of the oldest methods. This method was successfully experimented by the New England Company during the 1920 storm (Olive 1925). Over the years, thermal de-icing and anti-icing methods based on the Joule effect have been developed, adopted and routinely carried out in the USA, Canada and former USSR. With an ever-growing energy demand, transmission lines became longer, leading to lower currents being carried at higher voltages. With the difficulty in getting enough current into the conductors at these higher voltages, using the Joule effect to heat conductors and melt ice has become a challenging task. Nevertheless, old Joule-effect methods have been optimized in response to the new power network configurations.

Generally, using AC does not involve adding high-cost equipment, since the required current is supplied directly to the lines. Quite often the melting current must be very high for long transmission lines. In this context, DC seems to be more advantageous, especially for longer high voltage lines with large-diameter conductors. This is because reactive losses are eliminated. A variety of conceptual electrical schemes have been considered for ice-melting technologies. Some of the main methods based on the Joule effect are as follows.

6.4.1 Conductor De-icing

6.4.1.1 Load Shifting Method

The load-shifting method, which requires no additional equipment on the lines themselves, consists in using the heating effect of increased load current to prevent icing or to remove ice from conductors. Normal operating conditions must be modified in order to force more load current through a particular circuit by transferring or shifting loads from other circuits linking the same two substations (Polhman and Landers 1982) (Prud'Homme et al. 2005a). If the load is high enough, the current in the remaining circuit will induce ice melting. This method is actually well suited to single-conductor lines as bundle conductors generally require too much current. One problem with this method is the difficulty in controlling the current flow during the de-icing period, which is mainly managed by the power load demand of customers (Prud'Homme et al. 2005b). Loss of power load could lead to de-icing failure, whereas too much power load could lead to the overheating of conductors. Moreover, the power load available must be optimum for the climatic conditions where the de-icing takes place. Efficiency of the load-shifting method requires the

Table 6.4 Mechanical de-icing methods

Method	Principle	Efficiency	Applicability to overhead lines			Transportable	Modifications to structures	Range
			Conductor		GW			
			Single	Bundled				
Manual scraping or rolling methods	Break ice	Significant for atmospheric ice	x	x	x	Portable	Not required	Span by span
ROV	Break ice	Significant for atmospheric ice	x	x	x	Portable	Not required	Span by span
Impact with a rope manually handled	Shockwave generation	Significant for atmospheric ice	x	x	x	Portable	Not required	Span by span
Helicopter knotted rope and weight	Shockwave generation by impact and drop	Significant for atmospheric ice	x	x	x	Portable	Not required	Span by span
DAC	Shockwave generation by mechanical impulse	Significant for atmospheric ice	–	–	x	Portable	Not required	Span by span
EIDI actuators	Electromagnetic impact shockwave generation	Significant for atmospheric ice	x	–	x	Permanent	Required	10% of the span length

Table 6.4 (continued)

Method	Principle	Efficiency	Applicability to overhead lines			Transportable	Modifications to structures	Range
			Conductor		GW			
			Single	Bundled				
EIDI insulated strip	Electromagnetic impact shockwave generation	[1]Significant for atmospheric ice	x	–	x	Permanent	Not required	20% of the length
EIDI pair of strands	Electromagnetic impact shockwave generation	Significant for atmospheric ice	–	–	x	Permanent	Required	Short to medium length
EIDI bundle conductors	Electromagnetic impact shockwave generation	Significant for atmospheric ice	–	x	–	Permanent	Required	Short length
AIC	Induced vibrations	Significant for atmospheric ice	x	–	x	Permanent	Not required	Short length
Ice-shedder device	Induced natural frequency vibrations	Significant for atmospheric ice	x	–	x	Permanent	Not required	Short length
Twisting device	Twisting of GW	Significant for atmospheric ice	x	–	x	Can be permanent	Can be required	Span by span

[1] Method not tested on natural site

adoption of a well-defined de-icing strategy with the help of decision-making tools, as developed by Hydro-Québec TransÉnergie (Prud'Homme et al. 2005a).

6.4.1.2 Reduced-Voltage, Short-Circuit Method

This method consists in applying a three-phase short-circuit to one end of a line and a three-phase voltage source at the other end. Many power utilities in the world have some experience with short-circuit heating. In the early 1970s, Manitoba Hydro began using 3-phase short-circuits to melt ice as an experimental procedure (Adolphe 1992). Today, they have the capability to melt ice off several thousand kilometres of lines with conductors ranging in size from 2/0 to 336.4 kcmil ACSR. Currently, 90 stations, at 33, 66, and 115 kV, are equipped for short-circuiting. Ice melting is routinely carried out by Manitoba Hydro, not only during severe, widespread ice storms, but also during less severe weather conditions, as a preventive measure against the slow build-up of ice on conductors. Current intensity is a function of the applied voltage, circuit length, and the electrical characteristics of the conductor. As an approximate rule, for applied voltages of 12, 25, or 69 kV, it is possible to get the required current intensity for circuit lengths of 12, 25, or 69 km, respectively, on Hydro-Québec's single conductor lines, within a margin of 15% (Gingras et al. 2000). This de-icing method requires that some equipment be added, such as a switch on the short-circuit side which is normally open in order to produce three-phase faults. On the source side, switches and connections are required to power the lines to be de-iced. Also, the overload of existing equipment like system protections must be increased to support the de-icing current (Prud'Homme et al. 2005b).

6.4.1.3 DC Current

Both AC and DC can be used to heat line conductors. Using AC does not involve high additional costs, since the melting current is supplied directly from the existing network. To obtain the necessary value of melting current, the melting voltage and corresponding total melting power must be sufficiently high, especially with long transmission lines. If the length of the line being heated and the required melting current and voltage are relatively small, AC can be used successfully. DC is more advantageous for long high-voltage power lines with large cross-section conductors because reactive losses are eliminated.

This technology has been developed, installed, and successfully used on a large-scale in the former USSR to melt ice on long 500-kV lines with large, bundled phase conductors (Motlis 2002). In Canada, Hydro-Québec TransÉnergie will use this method, using DC rectifiers, at the Lévis substation (Fig. 6.7) in order to de-ice 735-kV quad bundles and 315-kV twin bundles of strategic lines (Gingras et al. 2000; Déry and Gingras 2005). For these lines, a short-circuit method with an AC source would require too much power.

DC de-icing requires the formation of a closed loop using line conductors as illustrated in Fig. 6.8. As DC converter installation is relatively expensive,

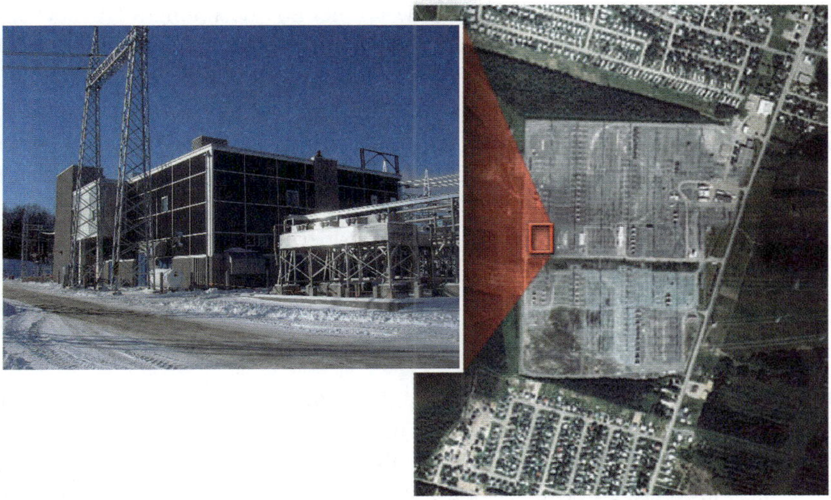

Fig. 6.7 Aerial photo of Hydro-Québec's Lévis substation DC converter (Reproduced by permission of Hydro-Québec)

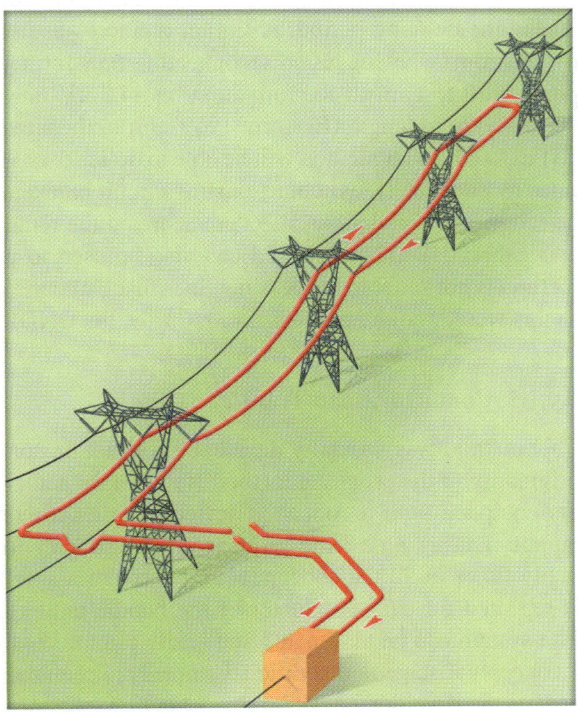

Fig. 6.8 One of the two-step sequences required to de-ice line phase conductors (Figure based on Granger et al. 2005)

Hydro-Québec TransÉnergie has solved the problem by designing a DC converter which operates as a static var compensator under normal operating conditions (Déry and Gingras 2005). The DC installation rated at 250 MW will also be used as a static var compensator on a day-to-day basis to improve the local voltage regulation and to insure its availability for de-icing purpose. In de-icing mode, the DC installation will provide up to the 7,200 amperes required to melt ice on quad-bundle conductors. Development and testing of an appropriate disconnect switch was also required to make sure it could operate safely considering the high current level with ice accumulation (Granger et al. 2005).

6.4.1.4 On-load Network De-icer

Most Joule-effect de-icing methods (except the load shifting method) require disconnecting the sections to be de-iced from the network. To overcome this problem, a new concept, on-load network de-icer (ONDI) has recently been developed by CITEQ, a joint venture between Hydro-Québec and ABB, in collaboration with Hydro-Québec TransÉnergie (Cloutier et al. 2007). The ONDI concept is based entirely on the use of phase-shifting transformers (PST). These are a special type of three-phase transformer employed for controlling power flows in transmission lines. During de-icing operations, the PST is connected in series with one of the two lines to create an AC current loop in transmission lines. By correctly adjusting the phase-shift of the PST, it is possible to increase the current flow by a factor of four in the line opposite to the PST line, while maintaining the voltage in the network during the de-icing period. A similar method was used in the USA and presented by Ekstrom, who suggested reconnecting transformer windings for a sixty-degree phase shift to combat ice formation on 34.5-kV lines. These line were limited to 40 to 46 km in length (Ekstorm 1959). From the simulation network conducted by CITEQ, the ONDI method will be able to de-ice over 900 km of 230-kV and 315-kV lines by adequately switching existing circuit breakers to configure the network for de-icing. The simulations also demonstrated the remote possibility of line de-icing using the ONDI. This method can also be used to prevent conductor icing. This method is not suitable for de-icing lines made of bundles of three or four conductors as this would require an excessive rating of the ONDI.

6.4.1.5 Contactor Load Transfer

This method was specially developed for bundle conductors (Couture 2004). The originality of the proposed method is to replace actual bundle spacers by new spacers equipped with a contactor device in order to control the current flow in the bundle. During a de-icing sequence, the contactor forces the current, which normally flows in all the subconductors, through a single subconductor. The process is repeated for each conductor of the bundle until complete de-icing is achieved. The system can be automated and easily remote controlled. This method is still at a conceptual stage and additional studies are required to estimate the real need, as well as the development and implementation costs of this method.

6.4.1.6 Pulse Electrothermal De-icer

Similar to AC and DC Joule-effect methods, the pulse electrothermal de-icer uses the current pulse to heat an external conductive coating surrounding the conductor (Petrenko and Sullivan 2005). This conductive coating can be made of a layer of stranded conductors insulated from the conductor by a dielectric layer, as illustrated in Fig. 6.9. According to the authors, the main advantage of using current pulses is that the required average power can be reduced by a factor of 100 as compared to AC or DC current. Due to the very small duration of current pulse, the thermal energy from the Joule heat of the external layer is entirely released at the ice/external layer interface, with a minimum diffusion into the dielectric layer. Also with this method, both anti-icing and de-icing can be obtained. It requires some significant modifications of the bare conductor by the addition of an insulated and a conductive layer sheath over the length of the conductor. This can provide problems during the installation, but also in the summertime where the thermal limitation of the conductor can be reached rapidly, leading to a reduction in service ampacity during warmer days. Finally, the efficiency of this method must be demonstrated as no laboratory tests on real stranded conductors have been carried out to date.

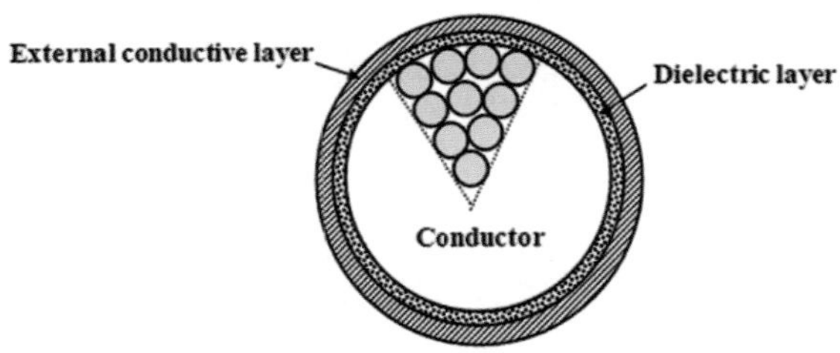

Fig. 6.9 Illustration of the principle of the pulse electrothermal de-icer applied to overhead conductors (Petrenko and Sullivan 2005, reproduced by permission of Victor F. Petrenko)

6.4.2 *Ground Wire De-icing*

GW de-icing is an important part of a global approach to avoid major transmission network breakdowns during severe ice storms. GW de-icing by Joule effect (Bourdages 2000) is possible with the "low" voltage short-circuit method. This requires a current source as well as electrical insulation of GWs at towers, as shown in Fig. 6.10. A medium-voltage AC transformer (25 kV) to which current can be supplied by the main AC circuit may be used as the current source. This method is very useful to de-ice many kilometres of GWs. The range of de-icing is limited only by the withstand voltage of insulators and arcing horns covered with ice. Also, the electrical insulation of the GWs will increase the grounding resistance (Prud'Homme et al. 2005b). In remote areas, it is possible to use an auxiliary diesel-powered generator to de-ice the GWs on strategic line sections, such as river crossings. Opinions are still divided on the final approach to use. One philosophy recommends mechanical reinforcing of GWs and tower top members to withstand the amount of ice accumulation forecast on the basis of geographical area. Another

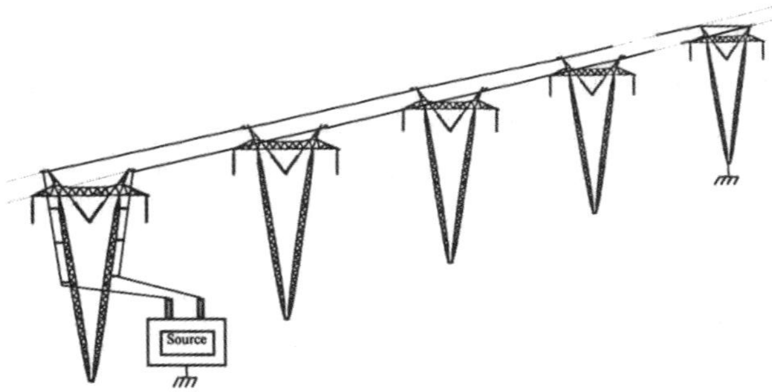

Fig. 6.10 Simultaneous de-icing of two GWs (loop configuration) (Bourdages 2000, reproduced by permission of M. Bourdages, Hydro-Québec)

philosophy recommends simply replacing damaged GWs as this operation costs little more than insulating the existing GWs. Finally, the last philosophy recommends GW insulation for Joule-effect de-icing, as associated costs with this operation are partly compensated by the elimination of yearly induction losses of about 2 kW/km for a 735-kV line at 1 kA.

All the aforementioned Joule-effect methods are presented with their principal characteristics in Table 6.5.

6.5 Methods for Limiting Ice Accretion Weight

These include methods which are not considered to be de-icing or anti-icing methods. They are usually called passive methods since they do not require any external energy supply other than from natural forces such as wind, gravity or solar radiation. These methods do not prevent ice accumulation or force ice shedding, but help to limit the problematic effects of ice loads on overhead lines. Some of these methods are already effective for wet snow but require further study for ice applications.

6.5.1 Anti-Torsion Devices

The idea of using counterweights to control galloping comes from field observations by maintenance personnel both in Japan and France. They observed that these devices could reduce the weight of snow accumulated on conductors as well as increase the speed of snow shedding (CEA 2002). Counterweights and spacers such as interphase spacers and spacer-dampers increase the torsional stiffness of conductors and GWs. This in turn limits their rotation caused by eccentric snow loads on the windward side and consequently limits the formation of cylindrical deposits. With a high eccentric snow loading, snow shedding is speeded up due to gravity and wind forces. Based on these observations of wet snow, numerical and experimental

Table 6.5 Joule-Effect de-icing methods

Method	Principle	Efficiency	Applicability to overhead lines			Additional equipment	Required a circuit disconnection	Range
			Conductor		GW			
			Single	Bundled				
Load shifting method	Force more AC load current through a particular circuit	Tested on site	x	x	–	None	YES	Medium
Reduced-voltage short-circuit method	Increase AC current by short-circuit three-phase line	Tested on site	x	–	–	Switch and increase capacity of protective equipments	YES	Medium
DC current	Use of DC current	Tested on site	x	x	–	DC converter	YES	Large
On load network de-icer	Use phase-shifting transformer to control power flow	Concept	x	[1]x	–	Phase shifting transformer	NO	Large

Table 6.5 (continued)

Method	Principle	Efficiency	Applicability to overhead lines			Additional equipment	Required a circuit disconnection	Range
			Conductor		GW			
			Single	Bundled				
Contactor load transfer	Control the current flow in the bundle	Concept	–	x	–	Specific contactors replacing actual spacers	NO	Large
Low AC voltage	Use an external AC source	Tested on site	–	–	[1]x	External AC source	NO	Medium
Pulse electrothermal de-icer	[2]Use impulse current from an external source	Concept	x	x	x	External impulse current source	YES	Small

[1] Require GW insulation from tower

[2] Require addition of an insulated layer

studies were conducted on atmospheric ice (McComber 2001; Blackburn et al. 2002). The results show that limiting the rotation of a GW under icing conditions can effectively reduce the ice load as well as shedding time under natural warming conditions. The ice load reduction is more significant for rime ice than glaze ice (Blackburn et al. 2002).

6.5.2 *Special Conductors*

Recent ice-shedding experiments conducted at a natural site on a 200-m segment of line conductors have highlighted the fact that under the same warming conditions, an ice-covered conductor presenting a smoother surface, such as trapezoidal strands, takes less time to shed ice naturally than a conductor presenting a rougher outer surface. In this study, smooth conductors with trapezoidal strands were compared to ACSS and ACSR round-strand conductors (Laforte et al. 2005b).

6.6 Practical Aspects

Most of the methods, such as specific coatings or devices that must be added to the energized conductors or GWs, are in fact in development or at the conceptual stage. The development of such methods becomes more complex as they have to meet some specific requirements to ensure good performance and life expectancy. This consideration is developed in detail in this section.

As mentioned previously, these methods have to respect some specific electrical, mechanical, and thermal constraints relative to power line operation. Also, some environmental constraints like U.V. radiation can decrease the life expectancy of certain devices. These parameters should generally be taken into account in the development of a method, as they serve to define its application field, particularly in the case of methods permanently installed on energized conductors and GWs.

6.6.1 *Electrical Constraints*

The presence of high electric and magnetic fields, as well as electrical discharges and the impact of lightning and flashovers should normally be taken into account in the development of de-icing and anti-icing methods. Also, electromagnetic disturbance, caused by the high-frequency electric fields emitted by some devices, can interfere with civil or military apparatus and must be considered in design.

Lightning induces very high impulse currents in connection with high voltages, along with large mechanically induced forces and high temperatures (IEEE Working Group 1985; Leroy and Gary 1984). Depending on the type of strike (direct or indirect), currents between 30 and 60 kA can be generated, and can sometimes reach as high as 200 kA in the worst cases (IEEE Working Group 1985). Theses high currents are accompanied by voltages higher than 1 MV. These are generally sufficient to induce flashover on or between overhead line equipment. In fact, lightning can breakdown the electrical insulation of the dielectric coating or electrical tracing of such methods as electromagnetic expulsive sheathings and vibrating devices. Consequently, lightning can directly affect active de-icing or anti-icing methods. This implies that the devices used should ideally be electrically insulated from live conductors or GWs and properly designed to withstand lightning strokes and flashovers.

6.6.2 Mechanical Constraints

Permanent methods used on live conductors and GWs are also subjected to different types of mechanical constraints.

If a coating is already on the conductor or GW before installation, it must support all the mechanical stresses caused by the pulling on and off of the reel carts, and the rolling and stretching through the stringing sheaves as well as the bending of the conductor or GW in the span (Techniques de l'ingénieur 1996a). All permanent coatings are subject to these stresses if they cannot be put on the conductor or GW after installation. This is true for future icephobic, low-adhesion, ferroelectric and ferromagnetic coatings (see Tables 6.1 and 6.2). However, some of these coatings could be deposited on the conductor or GW during the unrolling process. The coatings would still have to support the mechanical stress induced by the sheave used to adjust the sag of the conductor or GW. Once GWs and conductors are installed on towers, they undergo a bending under their own weight caused by their elasticity. For this reason, the coatings must preferably have the same or higher elasticity coefficient as their corresponding substrate.

On the other hand, all methods have to support mechanical online stresses (stretching and torsion) caused by the low-frequency vibration of large amplitude, called galloping of energized conductors or GWs, created by wind, ice shedding (Druez et al. 1994; McClure and Lapointe 2003; Hardy and Van Dyke 1995), or electrodynamic stresses induced by high-current pulses (Techniques de l'ingénieur 1996b). Under galloping, conductors or GWs oscillate to a frequency close to the fundamental of the span, low-order harmonics (0.5–3 Hz), but with amplitudes that may range from 1 m to 10 m or more, depending on span length. Galloping can also cause rigid coatings to undergo crack ignitions. Any other apparatus mechanically attached to the conductor, such as vibrating devices and ferromagnetic heating rings or envelopes, could be subjected to high acceleration forces generated by aeolian vibrations or galloping oscillations.

Under very high impulse current, parallel live conductors are subjected to electrodynamic strains resulting in high velocity mechanical shocks between them (Techniques de l'ingénieur 1996b). High-current pulses can be induced by short-circuits on active conductors. These can be from different causes such as treefall, galloping, or lightning.

Hence, the mechanical constraints inherent to the installation and the dynamic behaviour of conductors and GWs must be taken into account in the applicability of the new prevention and de-icing methods currently in development. This will contribute to decreasing the potentiality of some new concepts based on rigid ferroelectric coatings (CEA 2002). Preferably, coatings will have to be more flexible, with the same equivalent mechanical coefficient as that of the conductor or GW on which they are installed. Also, elongation of conductors during warmer summer days has to be taken into account, in that the stress must be supported by permanent coatings installed on conductors.

Methods based on the Joule effect are not affected by these considerations, as no apparatus or coatings need to be added directly on the conductors or GWs.

6.6.3 Thermal Constraints

One of the major aspects that should be considered is the thermal energy released by the high-current pulse of lightning. With the short duration of the pulse (a few tens to hundreds of μs), this is equivalent to high-frequency leakage current (from hundreds kHz to MHz) flowing mainly onto the surface of the conductor due to the skin effect. Most of the thermal Joule energy generated by the strike is dissipated at the surface of the conductor. In some cases, this thermal energy is sufficient to melt the surface aluminium conductor fibres (IEEE Working Group 1985). This could consequently melt material on the surface of the energized conductors or GWs. Any coating or apparatus installed on the surface of live conductors or GWs, or fixed to towers, can be subjected to this kind of thermal shock. This can cause permanent damage and drastically reduce their life expectancy and performance.

The second aspect deals with the thermal limitation of energized conductors. Conductors under service current can support a steady-state temperature, which depends on weather conditions, conductor characteristics, and conductor electrical current (Jakl and Jakl 2000; IEEE Std 1993). Conductors can generally support a maximum allowable temperature, above which a loss of strength, sag, line losses, or a combination of these can occur. Normally, for a given conductor type and a steady-state service current line, the temperature depends mainly on ambient temperature, as well as wind speed and direction. During winter, these conditions are not critical because conductor temperature is low enough to allow ice to accrete on it. This becomes critical during summer, when high ambient temperatures, solar radiation, and low wind conditions prevail. In this situation, temperature is only limited by current intensity.

With the presence of special coatings for ice prevention or removal, the thermal limitation of energized conductors must be considered. With these coatings, values of convection heat loss and the total heat capacity of the conductor have to be taken into consideration in the maximum-allowed-temperature calculation (IEEE Std 1993). As these coatings are permanently installed on the conductors, particular attention needs to be paid to the thermal conductivity of the coating and the different current values (steady-state and short-circuit) acceptable for the conductor and its prevention or de-icing coating. One solution to this can be solved by replacing actual conductors by high-temperature conductors like ACSS. These may be operated at temperatures in excess of 200 °C without loss of strength.

6.7 New Developments in Anti-icing Methods

While efforts to develop different active ice removal techniques have received the most attention (CEA 2002), few studies have focused on understanding the basic mechanisms of ice adhesion and the development of icephobic surfaces from the very fundamental standpoints. Recent advances in ice adhesion mechanisms, as well as recent discoveries in nanotechnologies and new active materials have led to new developments in some prevention methods.

This section presents the latest advanced concepts for the development of future icephobic coatings, as well as some active icephobic coatings which are of potential interest in the development of new anti-icing methods. Some of the methods presented are actually at a conceptual stage or are theoretically efficient, whereas others have experimentally demonstrated their efficiency on a very small scale.

6.7.1 Icephobic Materials

Atmospheric ice is generally formed from supercooled water droplets. To adhere to a surface, in the first stage of ice formation, these droplets wet the surface, thus replacing the existing surface–air interface with a surface–water interface. This wetting condition has attracted attention to low-surface-energy or hydrophobic materials as potential icephobic materials, or materials that allow for the ice to detach itself under the effect of gravity or wind. Teflon® is a good example, as it has very low surface energy, as well as the lowest ice-adhesion force among solid materials (Yoshida et al. 1991). The ice adhesion forces obtained with Teflon® demonstrated that it cannot be considered an icephobic material as it cannot completely prevent or drastically decrease atmospheric ice adhesion. Some efforts have been made to develop other solid materials with low ice adhesion. Such materials have actually not been developed for supercooled droplets. Some of them yielded good results for wet snow flakes (Colbeck 1996), for which the adhesion process is quite different.

Advances in material and surface science, such as ice adhesion research, has recently reinitiated interest for passive icephobic materials. Recent studies have established a strong correlation between hydrophobicity and the ice adhesion strength of a surface (Petrenko and Peng 2003; Somlo and Gupta 2001). These studies showed that an increase in the hydrophobicity of a surface leads to a decrease of its ice adhesion strength. The strong correlation between the work on water adhesion and the strength of ice adhesion reveals that the degree of hydrogen bonding seems to be a major factor controlling ice adhesion strength.

In this context, by increasing the hydrophobicity of the material surface, it could be possible to drastically reduce ice adhesion strength. This is an interesting research

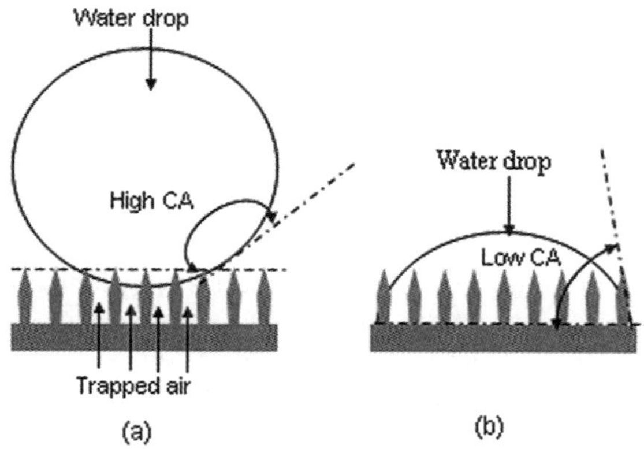

Fig. 6.11 Behaviour of a water drop on a solid textured surface. (**a**) the drop can sit on the *top* of these posts and the space between the posts is filled with air. (**b**) the water drop is impaled on the posts due to its own weight or high velocity impact

direction that has not been seriously explored in the field of icephobic materials. Over the past few years, new technologies have been developed in order to create high hydrophobicity with high water contact angle (CA), and more recently very high hydrophobicity with CA greater than 150° (Petrenko and Peng 2003; Somlo and Gupta 2001). These specific surfaces have attracted much interest because of potential practical applications as self-cleaning or anti-soiling materials (Feng 2002).

Conventionally, highly hydrophobic surfaces have been produced mainly in two ways. One is to cover a rough surface with a low-surface-energy material (CA > 90°), and the other is to roughen the surface of hydrophobic materials (Feng 2002; Kiuru 2004; Shiu et al. 2004). Relative to the studies done on highly hydrophobic materials, the contact angle or hydrophobic properties of a surface increase with an increase of surface roughness. In fact, rough surfaces allow for the creation of a texture in which air is likely to remain trapped when it comes into contact with water. The drop is then partially sitting on air and it behaves much like a fakir on a bed of nails: it sits comfortably on top of the posts, as illustrated by Fig. 6.11a. Likewise, it should become very difficult for supercooled water droplets, from freezing rain for example, to wet the textured surface totally. Intermolecular interactions at the ice/solid interface could be considerably reduced as most of the substrate is formed by air. As a consequence, a drastic reduction of the ice adhesion strength could be obtained.

This ideal icephobic material should be able to function in the case of a single water drop smoothly deposited on the surface. If other water drops add weight to the initial drop, or if the water drop strikes the surface with a certain velocity, it can be impaled on the posts, as shown in Fig. 6.11b. In this situation, the solid surface loses its hydrophobicity. Then, supercooled water drops can easily wet the entire surface, at the same time increasing the anchorage or interlocking mechanical effect at intermolecular interactions. The consequence could be a drastic increase in ice-adhesion strength.

In order to avoid this problem, another parameter must be taken into account in the characterization of textured surfaces. While the water CA has been commonly used as a criterion to evaluate the hydrophobicity of a solid surface, this alone is insufficient to assess the sliding properties of a water droplet on the surface (Feng 2002). This sliding property can be quantified by the sliding angle (α). This is the critical angle at which a water drop begins to slide down an inclined plane. It reflects the de-wetting property of a surface (Feng 2002; Shiu et al. 2004). Low α means a high rolling capacity of water drops on the surface of a horizontal plane. With this new property of textured surface, supercooled water drops could be easily evacuated from the solid surface before freezing, thus preventing ice accumulation.

Up to now, many methods have been developed to produce highly or super-hydrophobic surfaces. Interest has now turned towards methods that can be applied on a large scale to materials which can be found on aerial power lines. These methods deal with ultra-thin film deposition based on self-assembled monolayers (SAMs) and diamond-like carbon (DLC).

SAMs are attractive as a potential icephobic material for various reasons. First, recent theoretical and experimental studies have demonstrated the potentiality of SAMs as a highly hydrophobic surface (Petrenko and Peng 2003; Somlo and Gupta 2001; Kulinich and Farzaneh 2003). In particular, theoretical studies have demonstrated the possibility of combining high CA and low sliding angle with SAMs (Kulinich and Farzaneh 2004; Ji et al. 2002). From a practical point of view, SAMs can be chemically deposited at relatively low cost by simply dipping the substrate in specific solutions which adhere well to the substrate. And finally, SAMs formed on aluminium substrates have been reported to reduce ice-adhesion strength (Petrenko and Peng 2003). The results obtained for ice-adhesion reduction were performed with non-exotic SAMs, in order to make the surface hydrophobic. No particular attention was paid to optimize the assembly of SAMs in order to create a very highly hydrophobic surface and to study its effects on ice-adhesion strength. It may be expected that further improvement in this direction can be achieved, along with better understanding the mechanisms of SAM formation and ice adhesion to them. It may also be expected that the proper choice of SAM molecules with optimal chain length and chemical composition could yield some positive results. Likewise, the mechanical and environmental behaviour of SAMs should be of great interest as they are organic substances and remain fragile.

DLC is generally advantaged by its high mechanical hardness, chemical inertness (Ji et al. 2002), and better life expectancy, as compared to SAMs. DLC is not used alone because it is not a hydrophobic material ($CA > 90°$). It can be used to create a rough surface and an adherence layer on which is deposed the hydrophobic material. Recent studies have demonstrated the potentiality of ultra-thin films based on DLC associated with fluorocarbon (CF_x) (Kulinich and Farzaneh 2004; Ji et al. 2002). Obtained by common PECVD technology, these coatings exhibit high hydrophobicity with high adhesion to aluminium and porcelain, and good mechanical properties. No ice-adhesion tests were done with these films. More recent studies showed that adding suitable hydrophobic polymers (PTFE or PDMS) during DLC deposition can considerably enhance DLC hydrophobic properties and produce a novel diamond-like-carbon-polymer-hybrid (DLC-p-h) (Kiuru 2004). This novel structure has the great advantage of exhibiting a high CA and a very low sliding angle (around 0.15 °), which can lead to a new type of hybrid DLC coating.

6.7.2 Potential Active Icephobic Coatings

As developments in nanotechnology seem to open up a new direction for research in passive icephobic coatings, recent discoveries in ice adhesion have led to new developments in active icephobic coatings. Active icephobic coatings need energy to be effective, which comes from an electrical source. Like passive icephobic ones, active coatings must prevent or reduce considerably ice-adhesion strength on surfaces by breaking the chemical bonding involved in ice adhesion and/or intercalating air or other gases between the solid surface and the ice. Mechanical methods can also be used to weaken ice-adhesion strength. For this concept, an interfacial stress between ice and substrate must be generated with sufficient amplitude in order to produce adhesion failure of the ice layer. This can be achieved using conversion of electrical energy to mechanical vibrations by active materials.

The first method identified in the literature is the application of a DC voltage on ice-adhesion strength (Petrenko 2000). Results of tests showed that two different mechanisms dominate in the reduction of ice-adhesion strength: reduction of the electrostatic interactions, and electrolysis of ice. Ice electrolysis seems to be dominant in the case of an ice/solid metal interface, as proposed in (Petrenko 2000). The ice electrolysis method acts as a passive icephobic material; electrolysis gases are intercalated between the ice and the solid surface and behave like air trapped in a textured surface. Gas also accumulates in the form of bubbles, which contribute to produce interfacial cracks. The other advantage of this method is that it requires little electrical energy. Figure 6.12 illustrates the principle of this method. DC voltage is applied between the grid-electrode and a conductive surface. The grid-electrode is insulated from the conductive surface (Fig. 6.12a), which is the surface to be protected and must be conductive, as it acts as the second electrode of the circuit. When ice forms on the surface, it bridges the circuit, and DC voltage is applied to the interface between the conductive surface and the ice. The grid-electrode can have different configurations, as shown in Figs. 6.12b and 6.12c. As the grid-electrode and its insulating layer can be as thick as 1 mm, this method can be considered to be a very lightweight active coating.

This method is well suited to conductive surfaces and for application to insulating surfaces which should be made conductive. Also, this method requires a minimum of ice conductivity to ensure electrical current conduction. This completes the electrical circuit formed by the electrodes and the ice layer. This is not really a problem, as atmospheric ice is generally conductive due to the fact that it carries various contaminants (Farzaneh and Melo 1990). If the ice is not conductive enough, grid-electrode spacing must be reduced considerably, or the applied voltage increased in the same manner.

The configuration shown in Fig. 6.13 was proposed for energized conductors (Petrenko 2000). It seems more convenient to use it on GWs, as the conductive surface could be the wire itself. With this configuration, an axial grid-electrode surrounding the wire must be used. This seems to be difficult to set up and install as the grid-electrode must be electrically insulated from the wire. More studies and experiments will need to be carried out, under different icing conditions

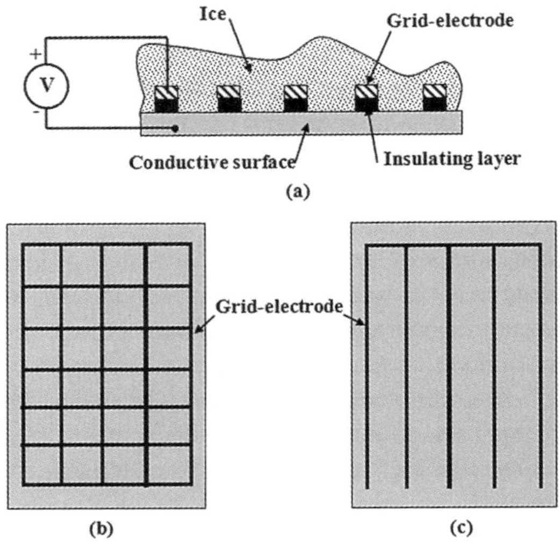

Fig. 6.12 Principle of generation of ice electrolysis by application of a DC voltage (US Patent 6027075, reproduced by permission of Victor F. Petrenko)

and using different electrode sizes and materials, in order to optimize method efficiency.

One method was proposed recently to avoid ice adhesion on surfaces (Wang 2002). This method, which can be considered as an active icephobic material, is based on the assumption of ice surface electrical properties caused by the presence of a liquid-like layer. Present at temperatures below −20 °C and with a thickness of a few nanometers, the liquid-like layer can provide an ordered orientation of the randomness of the water molecules. This results in a high density of electrical charges at the ice surface interface, either positive or negative, dependently of the exposed surface charge. If a charge is generated on the surface coming in contact with the ice, it is possible to modify the adhesion between the two surfaces. As like charges of same polarity repel, DC voltage from an external source that matches that of the charge occurring in the liquid-like layer is applied to the surface. This reduces the adhesion between the ice and the surface.

This method aims to break intermolecular forces between the ice layer and the exposed surface. The principle is quite different from the electrolysis method

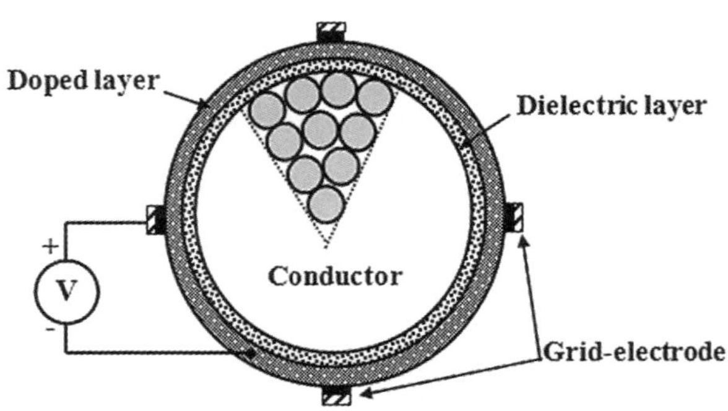

Fig. 6.13 Ice electrolysis method for bare conductor application (International patent 2005083862, reproduced by permission Victor F. Petrenko)

described previously. Here, the exposed surface, which must be conductive, is coated with a dielectric layer. The DC voltage of negative or positive polarity is applied to the conductive surface. By polarization, the dielectric surface exposed to ice accumulation acquires a charge of the same polarity as the conductive surface. If this charge is of the same polarity as that of the ice, then the substrate becomes an icephobic surface as it and the ice repel each other.

The advantage of this method is that no grid-electrode insulated from exposed conductive surface is needed, since the DC voltage is only applied to the conductive surface. The amount of electrical energy required is small and the dielectric surface can be a very thin coating, like paint, applied on the conductive surface. Finally, the ice needs not necessarily be electrically conductive.

This method, much like ice electrolysis, is better adapted to GWs than bare conductors, as the wire can act as the conductive surface. The disadvantage of this method is that the dielectric coating must cover the entire conductive surface to be protected, thus the entire GW surface. As mentioned in (Wang 2002), this method needs an ice detection system and inverse the polarity of the dielectric coating by changing the polarity of the conductive surface. The efficiency of the method on a cylindrical surface has not yet been experimentally demonstrated. As mentioned previously, the addition of an active coating on a wire must satisfy some specific constraints, which could affect the applicability of such a method to GWs.

Another method consists in employing active material as electromechanical coating directly deposited on the surface of bare conductors or GWs. For this, piezoelectric films as well as electro-active polymers can be used as thin active coatings. Piezoelectric films were proposed by (Broussoux et al. 1992) for de-icing optical and radio-optical windows. Piezoelectric polymers like PVDF seem in principle well adapted for this purpose as they are very thin, between 9 and 110 μm, flexible, and can support high mechanical and electrical stresses compared to thin ceramic piezoelectric films, which are particularly brittle. Commercially available piezoelectric polymers are expensive and their efficiency regarding overhead conductors or GWs needs to be demonstrated. Electro-active polymers and particularly the dielectric elastomers generally used as electromechanical actuators to produce artificial muscles can also provide thin-film active coatings (Bar-Cohen 2001; Pelrine et al. 2000). Dielectric polymers consist of a polymer material sandwiched between two compliant electrodes. These conductive electrodes make it possible to apply a high electric field to the thin polymer film in order to stretch the polymer under Maxwell stress. Depending of the polymer employed, high strains of up to 200 % and stresses up to 7 MPa under HV can be obtained (Bar-Cohen 2001). Such stress generated at the ice/substrate interface is theoretically sufficient to debond the ice layer (Yoshida et al. 1991; Javan-Mashmool et al. 2005). These dielectric polymers are currently only available in very small samples that are not convenient for experimental testing. Similar to piezoelectric polymers, it seems difficult to use these electromechanical coatings on energized conductors as an insulator layer between conductor. A coating should be added in order to prevent disruptive discharges within the coating. Also, some experimental investigations on conductors need to be carried out in order to verify their efficiency on cylindrical surfaces.

6.8 Conclusions

Many different anti-icing and de-icing methods can be applied to power-network conductors and GWs. Most of the current methods currently in use or projected for use have been discussed, except for some methods of little interest for power networks.

Strategies adopted by power utilities have tended towards the use of Joule-effect thermal methods for the large-scale protection of power-network overhead lines. Having been in use in many countries for over several decades, most Joule-effect methods have reached maturity. The efficiency of these methods relies on adequate choice and use within the framework of well-established emergency procedures, taking into account both timely availability of power on the network and the prevailing atmospheric conditions along the line sections. Synchronicity between the various operations and maintenance services is therefore a must in order to properly use the windows of opportunity for energy-efficient de-icing sequences, and thus keep de-icing time to a minimum.

For the protection of shorter, more strategic line sections, mechanical methods seem to be the most practical. This overview has highlighted a few methods of interest, some of which have been successfully utilized on actual lines. These are mechanical methods that are not permanently installed on the lines and thus require human intervention. This can cause problems for de-icing certain hard-to-reach sections of lines such as river or highway crossings. The use of mechanical methods to de-ice short segments of overhead lines can sometimes damage the line, depending on the severity of the accretion. In effect, the portable methods can only de-ice a single section at a time, which can unbalance the distribution of weight between towers, leading to possible damage to the attachments or towers themselves. Permanent mechanical methods, such as vibration-inducing systems, could possibly provide a solution to problems inherent to manual methods. This depends on conclusive experimental results from natural test sites and the ability to withstand the constraints related to installation on high-voltage conductors.

Considerable progress has been made in the field of icephobic coatings over the past ten years. It seems that the development of an icephobic coating, which would once and for all settle the atmospheric icing issue, can no longer be considered the utopian idea it once was. This is because of the recent progress made in the fields of nanotechnology and active materials. There is still a long way to go however, before an active, passive or combined icephobic coating can be used to protect overhead cables from ice overloads. Optimistically, one would hope that the discovery of an icephobic coating, on which ice cannot accumulate or be sheared off by its own weight, can be achieved before the end of the decade. Once the concept has been demonstrated and validated, an important step before actual application of the coating on power networks can be made, is that the material will have to be developed to withstand all the aforementioned mechanical and thermal constraints that are inherent to the installation and functioning of power networks. It is also important to mention that the development of an icephobic coating must include an application and/or surface treatment procedure for conductors and GWs. This would need to be done in a large-scale manufacturing environment, if it is not possible to "spray" them directly onto existing cables. The development of icephobic coatings specific to power networks is a monumental challenge and a very-long-term project, leaving plenty of room for further development of traditional methods and optimization of current approaches. To this end, the overview of the current anti-icing and de-icing methods shows that a growing number of companies seem to have targeting well-defined, proven methods, and have integrated them into well established de-icing plans. The 1998 ice storm that hit a large part of Northeastern North America undoubtedly initiated this movement.

Finally, having looked at all the available methods, it is interesting to note that a fair amount of work remains to be done regarding the protection of GWs. Notwithstanding mechanical methods, few methods are currently operational for application to long GW spans. It would seem that one solution is replacing damaged GWs. Another solution would be to remove the GWs in high-risk icing areas. This decision could be made by comparing, for the same line section, icing events and associated damage to GWs with the power outages associated to lightning. In high-risk icing areas, the use of lightning arresters might also be of interest in replacing the GWs to protect overhead lines against lightning.

Acknowledgments The authors would like to thank Mr. Barry Anderson of BC Hydro for useful and constructive comments on this chapter.

References

Adolphe WD (1992) Ice melting of sub-transmission lines. Manitoba Hydro, CEA paper, Toronto: 1–27

Allaire M, Laforte JL (2003) Method and apparatus for breaking ice accretions on an aerial wire. U.S Patent 6518497

Bar-Cohen Y (2001) Electroactive Polymers as Artificial Muscles – Reality and Challenges. Proceedings of the 42nd AIAA (SDM), Seattle, paper 2001–1492: 1–10

Berry WB, Sachs JL, Kleinman RL (1993) Radio Frequency (RF) Third Rail De-icing – A Comparison with Heated Rail. IEEE/ASME Joint Railroad Conference: 41–45

Blackburn C, Laforte JL Gagnon D, St-Louis M (2002) Comparative Study of the Quantity of Rime and Glaze Accreted on Free-Rotating and Fixed Overhead Ground Cables: Effect of Anti-Torsion Device. In: Proc 10th International Workshop on Atmospheric Icing of Structures, paper 9–2: 1–6

Bourdages M (2000) Ground Wire De-Icing by Joule Effect. In: Proc International OPGW Symposium, Montréal

Broussoux D, Ceccaldi MC, Leclerc P (1992) Device for the removal of the ice formed on the surface of a wall, notably an optical or radio-electrical window. US Patent 5172024

Byrd NR (2004) Polysiloxane (amide-ureide) anti-ice coating in Patent. Patent, Editor, The Boeing Company, Chicago

CEA (2002) De-Icing Techniques Before, During, and Following Ice Storms

Cloutier, R, Bergeron A, Brochu J (2007) On-Load Network De-Icer Specification for a Large Transmission Network. IEEE Trans on Power Delivery 22(3): 1947–1955

Colbeck SC (1996) Capillarity bonding of wet surfaces. Surface and Coating Technology 81: 209–214

Couture P (2004) Switching Modules for the Extraction/Injection of Power (Without Ground or Phase Reference) From a Bundled HV Line. IEEE Trans. on Power Delivery 19(3): 1259–1266

Croutch VK, Hartley RA, (1992) Adhesion of ice to coatings and the performance of ice release coatings. Journal of Coatings Technology 64: 41–53

Déry A, Gingras JP (2005) Hydro-Québec–TransÉnergie's Lévis Substation De-Icing Project. Proc. of 11th International Workshop on Atmospheric Icing of Structures, Montréal: 1–6

Druez J, Louchez S, McComber P (1994) Ice shedding from wires. Cold Regions Science and Technology 23: 377–388

Egbert RI, Scharag RL, Bernhart WD, Zumwalt GW, Kendrew TJ (1989) An Investigation of Power Line De-Icing by Electro-Impulse Methods. IEEE Trans P&D 4(3): 1855–1861

Ekstorm IR (1959) More angles of phase shift added to previously known ice-melting methods, AIEE Paper 58–1156: 1449–1452

Farzaneh M, Melo OT (1990) Properties and effect of freezing rain and winter fog on outline insulators. Cold Regions Science and Technology 19: 33–46

Feng L (2002) Super-Hydrophobic Surface: From Natural to Artificial. Adv. Mater. 14(24): 1857–1860

Frankenstein S, Tuthill, AM (2002) Ice adhesion to locks and dams: Past work; future directions. Journal of Cold Regions Engineering 16(2): 83–96

Gastonguay L, Champagne GY (1996) Problems Related to Ice Formation on the Guywires of Hydro-Québec Telecommunications Towers: A Critical Review of Literature. In: Proc 7th International Workshop on Atmospheric Icing of Structures, Chicoutimi: 1–6

Gingras JP, Breault S, Brodeur R, Crevier C, Déry A, Vallée A (2000) Stratégie de renforcement du réseau d'Hydro-Québec à la suite du verglas exceptionnel de janvier 1998: une démarche pour sécuriser davantage l'alimentation électrique, CIGRÉ Session, Paris, Paper 37–101

Granger M, Dutil A, Nantel A, Laurin S (2005) Performance Aspects of Lévis Substation DC De-Icing Project. Proc. of 11th International Workshop on Atmospheric Icing of Structures, Montréal, Canada: 1–6

Hacker D, White B, Ferrill R (2000) Ice release coatings for helicopter rotor blades. 78th Annual Meeting of the International Coatings Exposition 2000, Chicago

Hardy C, Van Dyke p (1995) Field observations on wind-induced conductor motion. J of Fluids and Structures 9: 43–60

Hesse KH (1988) The management of devasting ice storms. In: Proc 3rd International Workshops of Atmospheric Icing of Structure, Paris: 330–334

IEEE Standard for Calculating the Current-Temperature Relationship of Bare Overhead Conductors, IEEE Std 738–1993

IEEE Working Group on lightning performances of transmission line (1985) A simplified method for estimated lightning performance of transmission line. IEEE Trans PAS 104: 919–927

Jakl F, Jakl A (2000) Effect on Elevated Temperatures on Mechanical Properties of Overhead Conductors Under Steady State and Short-Circuit Conditions. IEEE Trans P&D 15(1): 242–246

Javan-Mashmool, Volat C, Farzaneh M (2005) A New Method for Measuring Ice Adhesion Strength at an Ice/Substrate Interface. Hydrological Processes 20: 645–655

Ji H A, Côté D, Koshel B, Terreault G, Abel P, Ducharme G, Ross S, Gagné M (2002) Hydrophobic fluorinated carbon coatings on silicate glaze and aluminum., J Thin Solid Films 405: 104–108

Kiuru M (2004) Experimental Studies on Diamond-Like Carbon and Novel Diamond-Like Carbon – Polymer – Hybrid Coatings. Academic Dissertation, Faculty of Science of the University of Helsinki: 1–36

Kulinich SA, Farzaneh M (2003) Hydrophobic properties of surfaces coated with fluoroalkylsiloxane and alkylsiloxane monolayers. Surf Sci 573: 379–390

Kulinich SA, Farzaneh M (2004) Alkylsilane self-assembled monolayers: Modeling their wetting characteristic. Appl Surf Sci 230: 232–240

Laforte C, Beisswenger A (2005) Icephobic Material Centrifuge Adhesion Test. In: Proc 11th International Workshop on Atmospheric Icing of Structures, Montréal: 1–6

Laforte JL, Allaire MA, Laflamme JL (1998) State-of-the-Art on Power Line De-Icing. Atmospheric Research 46: 143–158

Laforte JL, Allaire MA, Laforte C (2005a) Demonstration of the Feasibility of a New Mechanical Method of Cable De-Icing. Proc. of 11th International Workshop on Atmospheric Icing of Structures, Montréal, Canada, pp 1–6

Laforte JL, Allaire MA, Gagnon D (2005b) Ice Shedding of 200m-Long Artificially Iced Overhead Cables at an Outdoor Test Site. In: Proc 11th International Workshop on Atmospheric Icing of Structures, Montréal: 1–6

Landry M, Beauchemin R, Venne A (2000) De-icing EHV overhead transmission lines using electromagnetic forces generated by moderate short-circuit currents. 2000 IEEE ESMO Conference, Montreal: 94–100

Landry M, Beauchemin R, Venne A (2001) De-icing EHV Overhead Transmission Lines by Short-circuit Currents. IEEE Canadian Review – Spring: 10–14

Lanoie R, Bouchard D, Lessard M (2005) Using Steam to De-Ice Energized Substation Disconnect Switch. In: Proc 11th International Workshop on Atmospheric Icing of Structures, Montréal: 1–6

Leblond A, Montambault S, St-Louis M, Beauchemin R, Laforte JL, Allaire MA (2002) Development of New Ground Wire De-Icing Methods. In: Proc 10th International Workshop on Atmospheric Icing of Structure, Brno, paper 9.6: 1–6

Leblond A, Lamarche B, Bouchard D, Panaroni B, Hamel M (2005) Development of a Portable De-Icing Device for Overhead Ground Wires. In: Proc 11th International Workshop on Atmospheric Icing of Structures, Montréal: 1–6

Leroy G, Gary C (1984) Les propriétés diélectriques de l'air et les très hautes tensions. Collection de la direction des Études et Recherches d'Électricité de France, ed Eyrolles

McClure G, Lapointe M (2003) Modeling the structural dynamic response of overhead transmission lines. Computers and Structures 81: 825–834

McComber P (2001) A non-circular accretion shape freezing rain model for transmission line icing. In: Proc 9th International Workshop on Atmospheric Icing of Structures, Chester: 1–6

Montambault S, Côté J, St-Louis M (2000) Preliminary Results on the Development of a Teleoperated Compact Trolley for Live-Line Working. 2000 IEEE ESMO Conference, Montréal: 21–27

Motlis Y (2002) Melting Ice on Overhead-Line Conductors by Electrical Current. CIGRE SC22/WG12 document, draft no. 4, revised for the WG12 Meeting, Paris

Mulherin ND, Haehnel RB (2003) Progress in Evaluating Surface Coatings for Icing Control at Corps Hydraulic Structures. Ice Engineering, Technical Note 03–4

Murase H., Nanishi K, Kogure H, Fujibayashi T, Tamura K, Haruta (1994) Interactions between heterogeneous surfaces of polymers and water. Journal of Applied Polymer Science 54(13): 2051–2062.

Nakagami M (2007) Verification on the Effects of PTFE Tapes on Reducing Snow Accretion on Overhead Power Transmission Lines. In: Proc 12th International Workshop on Atmospheric Icing of Structures, Yokohama: 1–6.

Nourai A, Hayes RM (2003) Power line ice-shedder. U.S Patent 6660934

Olive CR (1925) Sleet and ice troubles on transmission lines in New England. AIEE paper, May

Pelrine R, Kornbluh R, Joseph J, Heydt R, Pei Q (2000) High-field deformation of elastomeric dielectrics for actuators. Materials Science and Engineering 11: 89–100

Petrenko VF, Whitworth RW (1999) Physics of ice. Oxford University Press, Oxford

Petrenko VF, Sullivan CR (2003) Methods and systems for removing ice from surfaces. U.S. Patent 6653698

Petrenko VF, Sullivan CR (2005) Pulse electrothermal deicier for power cable. International Patent 2005083862

Petrenko VF, Peng S (2003) Reduction of ice adhesion to metal by using self-assembling monolayers (SAMs). Can J Phys 81: 387–393

Petrenko VF (2000) Systems and methods for modifying ice adhesion strength. U.S Patent 6027075

Petrovic JJ (2003) Mechanical Properties of ice and snow. J of Materials Science 38: 1–6

Polhman JC, Landers P (1982) Present state-of-the-art of transmission line icing. IEEE Trans. PAS 101(8): 2443–2450

Prud'Homme P, Roux MO, Guilbault P, Séguin PA, Hounkpatin E (2005a) Determination of Current Required to De-Ice Transmission Line Conductors. In: Proc 11th International Workshop on Atmospheric Icing of Structures, Montréal: 1–6

Prud'Homme P, Dutil A, Laurin S, Lebeau S, Benny J, Cloutier R (2005b) Hydro-Québec TransÉnergie Line Conductor De-Icing Techniques. In: Proc 11th International Workshop on Atmospheric Icing of Structures, Montréal: 1–6

Reich A (1994) Interface influences upon ice adhesion to airfoil materials. AIAA, 32nd Aerospace Sciences Meeting and Exhibit, Reno, USA, paper 1994–714

Roger H (2004) Means and Method for Removing Extraneous Matter Like Ice/Snow an Overhead Line. Patent US2004065458

Roger H (2004) Monitoring System and Device for an Electric Power Line Network. Patent US2004038891

Shiu J et al. (2004) Fabrication of Tunable Superhydrophobic Surfaces by Nanosphere Lithography. Chem of Mat 16(4): 561–564

Somlo B, Gupta V (2001) A hydrophobic self-assembled monolayer with improved adhesion to aluminum for de-icing application. J Mech Mater 33: 471–480

Sullivan, CR, Petrenko, VF, McCurdy, JD (2003) Using Dielectric Losses to De-Ice Power Transmission Lines with 100 kHz High-Voltage Excitation. Conf on IEEE Industry Applications Magazine, 9(5): 49–54

Techniques de l'ingénieur (1996a) Lignes aériennes: échauffement et efforts électrodynamiques. Traité Génie électrique, section D 4439: 1–9

Techniques de l'ingénieur (1996b) Mise en place des câbles. Traité Génie électrique, section D 4429: 11–14

Volat C, Farzaneh M, Leblond A (2005) De-icing/Anti-icing Techniques for Power Lines: Current Methods and Future Direction. In: Proc 11th International Workshop on Atmospheric Icing of Structures, Montréal: 1–11

Wang ST (2002) Method and apparatus for autonomous de-icing. US Patent 640209

Yoshida M, Ohichi T, Konno K, Gocho M (1991) Adhesion of Ice to Various Materials. Cold Regions Technology Conference, Sapporo

Chapter 7
Effects of Ice and Snow on the Electrical Performance of Power Network Insulators

Masoud Farzaneh and William A. Chisholm

7.1 Introduction

Ice and snow accretion has been identified as a significant risk factor in the satisfactory reliability of overhead power lines and outdoor substations. Large-scale ice storms leading to extensive mechanical damage are one aspect, already covered in previous chapters.

The sight of twisted towers or conductors weighted down to the ground by ice or snow accretion and high wind speed makes a strong impression on the public and the press. In most cases like these, the public and press are sympathetic to some of the delays and costs involved in restoring the mechanical failures – provided these delays do not extend to more than a day or two. However, there are other important problems related to ice and snow accumulation on line and station insulators that do not excite such sympathy. Road salt, used for transport safety, degrades the electrical strength of nearby insulators further by acting as a strong local pollution source.

This chapter describes the electrical flashover problems along iced or snow-covered insulators, which are also important in power system operations. More precisely, the roles of several environmental and meteorological factors on the electrical characteristics of the ice layer are discussed. Numerical models that quantify the risk of electrical faults on different types of ice and snow are then elaborated. The physical basis and predictions lead to good agreement with the reported test and application experience. The chapter finishes up with some practical options and recommendations that can lead to more reliable electrical insulator performance in winter conditions.

M. Farzaneh
University of Québec at Chicoutimi, 555 Boulevard de l'Université, Chicoutimi, Canada G7H 2B1
e-mail: farzaneh@uqac.ca

M. Farzaneh (ed.), *Atmospheric Icing of Power Networks*,

7.2 Insulator Functions, Dimensions and Materials

7.2.1 Insulator Functions

Outdoor power network insulators are the devices that support high-voltage conductors mechanically, and separate them from other conductors or from ground electrically. Requirements for the mechanical strength of insulators are determined by loading conditions, as discussed in previous chapters. Insulators should withstand normal operating electrical stresses with low probability of flashover failure. They should also withstand short-duration surges that occur under lightning or switching conditions. If they do not, insulator arcing will cause flashover, which is the complete bypass of electrical insulation by an arc plasma path, of the air gap between the structure and conductor.

7.2.2 Insulator Dimensions

In the accepted technical terms (IEEE Std 100 2000), the *leakage distance* is bridged by the snow or ice accretion, leaving a snow or ice path across the *dry-arc distance*. These and other insulator dimensions are defined formally as follows:

Leakage distance (creepage distance) is the sum of the shortest distances along the insulating surfaces measured between the metal parts of the insulator. Leakage distance is typically determined by the maximum contamination levels expected in the area of application. For porcelain posts, a number of ribs or "petticoats" are cut from a solid cylinder of clay before the insulator is glazed and fired. For strings of porcelain or glass insulator disc units, the shed spacing and diameter are fixed. Leakage distance is increased instead by incorporating corrugations or ribs on the under-surface or by adding insulator discs to the chain.

Dry-arc distance is the sum of the shortest distances in air between the metal parts of the insulator. The dry-arc distance of an insulator determines the BIL (basic impulse insulation level) of the insulator and meets safety clearances stated in national electric safety codes (NESC 2002; IEC 60071 2006).

Shed spacing is the distance between convolutions in the insulator surface. For chains of disc insulators, it is the distance from the bottom of an insulator to the top surface of the one beneath. It takes more ice or snow accumulation to build a continuous ice or snow path across dry-arc distance for insulators with larger shed spacing.

Shed diameter is usually of interest in conditions of heavy ice or snow accretion. A crescent accumulates on the windward side, spanning the full shed diameter of the insulator.

Shaft or Cap diameter is of interest mainly for moderate ice accretion on horizontal insulators.

7.2.3 Insulator Materials

Since the start of the 1900s, insulators have been generally made of electrical porcelain or electrical glass materials. The native materials are clay, feldspar and quartz for porcelain, and silica, soda ash, Dolomite, limestone, feldspar and sodium sulfate for glass (Looms 1988). Porcelain is covered with a glaze that provides better mechanical strength and higher electrical surface resistivity after firing. Glass disc insulators remain competitive with porcelain discs for overhead lines, but porcelain posts and bushings are more common in stations.

Since the 1980s, the electrical industry has moved towards the use of non-ceramic insulators. Materials such as ethylene-propylene diene monomer (EPDM) and polydimethylsiloxane (PDMS) silicone rubbers offer much higher surface resistivity than ceramic or glass, when wetted and even under conditions of high relative humidity. The best materials also bead water, improving wet performance. The polymer insulators are more cost-effective and, to an increasing extent, more reliable than ceramic insulators, mainly because there are fewer components per meter of dry-arc distance.

Several specialized books (Looms 1988; Gorur 1999) provide more details on the evolution of high-voltage insulator technology and on the processes used to manufacture them today. These construction details are important for mechanical performance in cold conditions, but less relevant for this short chapter on electrical aspects. Insulators, whether ceramic or non-ceramic, generally serve as an inert substrate on which the layer of electrically important ice or snow accumulates. In winter conditions, it is the electrical performance of this ice or snow layer that needs to be quantified.

7.3 Ice and Snow Accretion on Insulators

7.3.1 Visual Observations of Ice and Snow Accretion

Typically, ice and snow accumulate on the side of the insulator that faces into the wind, known as the "windward" side. The extent of accumulation usually forms a crescent shape that extends across the full insulator diameter.

Most insulators have convoluted surfaces, to extend the *leakage distance* across the surface compared to the *dry-arc distance* across the metal ends. The main reason for having a convoluted shape, with sheds along the insulator length, is to break up the column of water which might otherwise run down as a continuous vertical channel lengthwise along the insulator. The sheds work as multiple umbrellas to keep some parts of the surface dry. Properly designed sheds increase the wet flashover strength of insulators in heavy rain conditions.

For vertical insulator posts, the shed spacing between convolutions is typically 20 to 50 mm, with a shallow depth of less than 100 mm. Figure 7.1 shows that snow

Fig. 7.1 Accretion of snow and ice on 735-kV ceramic station post insulators (Reproduced by permission of J-F. Drapeau, Hydro-Québec/IREQ)

or ice can fill most shed spaces completely. In contrast, accumulation on insulators in the horizontal orientation tends to form on the top of the insulators.

Under conditions of heavy icing, some of the area between multiple chains of horizontal disc insulators may also fill in as shown in Fig. 7.2.

Fig. 7.2 Accumulation of snow on horizontal EHV transmission line disc insulators (Reproduced by permission of J-F. Drapeau, Hydro-Québec/IREQ)

7.3.2 Characterization of Ice and Snow Severity

Ice and snow occurrence and severity are climate features that are observed by meteorologists in many countries. Chapter 1 provides a survey of the modern observation methods. The results of these observations are used to manage the overall risk of transport accidents (air and ground) in adverse winter weather. On that basis, specific forecasts are prepared to initiate warning of the need for remedial action such as de-icing or road salting.

Major electrical utilities employ meteorological staff to forecast ice accretion on overhead lines and stations. This is done partly to deploy resources in anticipation of mechanical damage caused by the combined effect of high wind on the additional surface area of a heavy ice load. In the previously mentioned 1998 storm in Quebec, the ice accretion weight was approximately 1,500 kg between adjacent wood poles. In addition, since the 1980s, utilities are also reconfiguring to guard against the possible occurrence of electrical flashover problems on critical high-voltage systems.

Occurrence of snow and freezing rain is tabulated hourly in weather stations using the World Meteorological Organization (WMO) protocols (1997). Automated weather stations sometimes use heated tipping bucket rain gauges to gather precipitation rate and report snow as water equivalent depth. Manual measurements of precipitation use graduated cylinders, supplemented by actual snow or ice depth on the ground or on a reference plane. Specialized *icing-rate meter (IRM)* instruments were described in Chapter 2. The IRM report the ice accretion directly, usually using the change in vibration frequency of a thin probe.

Occurrences of freezing rain and the long-term average of days with snow cover in winter are fairly sensitive measures of global warming trends. Some climate models suggest that future conditions will have greater climate variability, with more frequent, intense and extensive winter storms. These predictions, along with the high variation in climate values in previous decades, do complicate the design process for areas where atmospheric icing is presently moderate.

Utilities have developed empirical rules to relate standard measures of ice or snow accretion on the ground to the thickness and weight of ice or snow that accumulates on horizontal conductors and structures. This is covered in detail in Chapters 3 and 4. In the absence of wind, a crescent of ice will accumulate on the conductor. The thickness of this crescent can reach the same maximum thickness as the deposit on the ground in some conditions. Under moderate wind, however, the deposit rate will vary, and the eccentric ice load can twist conductors as they have low torsional stiffness in long spans. In the severe 1998 ice storm in Quebec, 100 mm of ice accumulation on the ground in three successive storms led to a maximum of 75 mm equivalent radial accumulation on overhead conductors.

The methods to measure natural ice accretion on insulators are not as well developed as those for accretion on conductors. It is not practical to remove ice samples from the energized insulators. Instead, measurement of accumulation is usually established with a horizontal monitoring cylinder of 25-mm diameter, rotating at 1 rpm, mounted near the insulators. The radial thickness of ice on this monitoring cylinder will be up to one-half that of the accumulation on the ground, depending on the icing conditions. Tests in icing laboratories (Farzaneh et al. 2003) then develop the relation between ice accumulation on the reference cylinder and ice accumulation on each specific insulator.

The electrical resistance of an ice or snow layer deposit depends on its thickness and length. It is also affected by the conductivity of the deposit, σ_{Ice}. The resistance of a rectangular slab of dry ice with dry-arc length L, width W and thickness t is given by:

$$R = \frac{L}{\sigma_{Ice} \cdot W \cdot t} \tag{7.1}$$

Normally, field observations of precipitation conductivity are based on either liquid samples (rain) or melted samples of snow or ice. The values are usually corrected to a reference temperature of 20 °C, since the conductivity varies at a rate of about 1.6 to 2.2% per centigrade degree (CRC 2007).

The conductivity of ice is much lower than that of melted-ice water at 20 °C. The ice conductivity increases by an order of magnitude in the narrow range from −1 to 0 °C. Also, the electrical resistance of an ice slab is strongly influenced by the presence of a water film on its surface, for two reasons:

- The melted-ice water conductivity, σ_{water}, is typically 100–300 times higher than that of dry ice.
- The well-known process of purification by crystallization occurs, rejecting ionic impurities to the ice surface and increasing its surface conductivity by another order of magnitude.

The total ice layer resistance R can be expressed as the parallel combination of two terms as follows:

$$R = \frac{L}{W}\left[\frac{1}{\sigma_{ice}t_{ice} + \sigma_{water}t_{film}}\right] \tag{7.2}$$

The relative magnitude of the two conductivity terms depends mainly on the temperature of the ice layer.

As insulator geometry modifies the ice layer away from the simple rectangular shape described in Eq. 7.2, a geometric correction factor is introduced. This follows the traditional practice of integrating the surface area to obtain a *form factor* (IEC 60815 1986; Looms 1988). The influence of other factors, such as the arc-root voltage drop and radius, are also described in Section 7.

7.3.3 Electrical Consequences of Ice and Snow Accretion

Generally, the line-to-ground electrical faults occur at nearly the same time on several different parallel insulators when these are exposed to the same conditions. This tends to focus the problems at large transformer stations with hundreds of insulators in parallel rather than on lines where each tower has only three or six insulator strings in parallel. Also, the leakage currents that flow across insulators can damage wood insulation on distribution systems, leading in some cases to spectacular arcing and widespread pole fires throughout cities.

There are no ideal insulators that are completely non-conductive. Even if such an insulator existed, outdoors it would be exposed to environmental and meteorological conditions that would degrade the insulation. For example, inert mineral matter, carbon or metal oxides, soluble salts and water will form on the insulator surfaces at various times. Corona discharge will also generate oxides of nitrogen that add to the pollution level.

If a polluted insulator becomes fully wetted, the convoluted shed shapes still have some benefit by increasing the leakage distance. The electrical stress across the surface is given by the normal power system operating voltage, divided by this leakage distance. At average voltage stresses above 25 kV_{rms} per meter of leakage distance, for high levels of surface pollution, discharges will initiate, propagate and extinguish in a process called “dry-banding”. The arcs have low current, less than 1 A. This is still sufficient to heat the arc plasma to 4,000 K, a temperature that ensures that the arc is electrically conductive and shorts out portions of the leakage distance. If an arc grows sufficient length along the insulator, the remaining air gap is too weak to withstand the line voltage. The resulting flashover causes a severe line-to-ground fault with currents that would be at least 10 kA. Power systems are designed to detect these faults and clear them by operating automatic circuit breakers that interrupt the current for a period of three to thirty cycles (0.05 to 0.5 s for 60 Hz, 0.06 to 0.6 s for 50 Hz). Normally, circuits with the highest voltages are most critical and their circuit breakers close in most quickly after an interruption.

The most common reason to provide extra leakage distance on electrical insulators is to reduce the electrical stress to accommodate the anticipated local pollution levels. The shape of the insulator can be adjusted in order to increase the ratio of leakage distance to dry-arc distance. Practical insulators have ratios of 2:1 to 4:1. Insulators with higher ratios also have larger shed diameters. An increase in diameter will reduce the per-unit-length surface resistance of the insulator and this offsets some of the advantage of a higher path length. Also, for high ratios of leakage distance to dry-arc distance, the deep skirts or ribs are bypassed more easily by arcing. Moreover, bringing the sheds closer together reduces their efficiency under water cascade conditions.

Under icing conditions, the accumulation of atmospheric ice on the insulators represents a rather severe form of conductive deposit, as it is much more stable than a layer of water. Under any of these conditions, leakage currents can be established that cause heating of the contamination layer, formation of electrochemical products and electrical discharges. This situation sometimes leads to flashover. For icing conditions, both dry-arc and leakage distance dimensions have some additional selection criteria to establish reliable electrical service.

At present, revisions to international standard IEC 60815 (1986) are under discussion to establish values of unified specific creepage distance (mm/kV_{rms}) that are respected for all commercial outdoor insulation materials at all temperatures. These conditions are described in the left column of Table 7.1. In cold regions, the highest levels of insulator pollution accumulate in the winter, where there is no natural rain to wash away soluble deposits. A fraction of this pre-contamination will dissolve into ice or snow layers, increasing electrical conductivity. Also, as shown in Fig. 7.1, the accumulation of snow or ice can reduce the leakage distance. For design purposes, it is thus common to consider only the dry-arc distance, bridged by contaminated ice or snow, as the most critical dimension and situation. The right column of Table 7.1 provides additional details.

Table 7.1 Comparison of electrical flashover processes under pollution and icing conditions

Normal contamination problems	Icing problems
Deposit of contaminants (few hours to a few months)	Deposit of ice or snow (about an hour)
Deposit of moisture	Formation of water film on ice surface or inside snow
Formation of dry bands on insulator surface	Formation of air gaps in ice or snow
Arc initiation over dry bands	Arc formation across air gaps
Propagation of arc and its rapid growth	Propagation of arc and its rapid growth

IEC standard 60815 (1986) also provides guidelines for applying porcelain and glass insulator shape and size, shed diameter and minimum distance between sheds. Traditional transmission line insulators have settled on 146-mm disc separation and 254-mm diameter for chains of individual discs. Leakage distance on these units is controlled by the depth of convolutions on the bottom surface. Near the sea, or in areas of heavy industry, it is common to use IEC recommendations of 55 mm/kV, while levels of 28 mm/kV are recommended for areas of low pollution.

In contrast, IEC standard 60815 does not provide any guidance for cold conditions, even though ice and snow insulator flashover problems are observed in many countries, as shown in the next section.

7.3.4 Case Studies of Electrical Flashovers on Ice and Snow

Records of power system operations after severe icing conditions normally focus on how many electrical faults occurred, and on whether the power system remained stable as a consequence. In a detailed industry survey (Yoshida and Naito 2005), it was found that 35 utilities in 18 countries had ice and snow electrical flashover problems. In particular, a total of 83 events (69 on lines, and 14 in stations) were reported by 17 utilities on transmission lines from 400 to 735 kV in relatively clean conditions.

Some of the events reported in this survey have enough documented detail to answer the basic questions raised in Table 7.2. In all cases, multiple flashovers leading to a high risk of system disturbance were reported.

These reported problems tend to have certain common features:

- Flashovers occur after some period of rapid accumulation of pollution on insulator surfaces.
- Flashovers occur near the ice or snow melting temperature, which may be a few degrees above or below ambient, depending on solar input, electrical input, dew-point temperature, wind velocity and other thermodynamic factors.
- Flashovers occur in response to relatively moderate frost or ice accretion, or heavy snow accretion.
- Flashovers occur mainly in restricted areas within 3 km of pollution sources.

Table 7.2 Examples of conditions that cause winter flashovers

Location, date	Ice and contamination type	Melting phase	Surface pre-contamination	Altitude	System voltage
UK, 1935–36	Hoar frost	Yes	Heavy	Low	132 kV
UK, 1962–63	Hoar frost	Yes	Heavy	Low	275 kV, 400 kV
Switzerland, 1966–67	Heavy, wet snow 0.8 g/cm density	Yes	No	High	400 kV
Ontario, Canada, 1986	Ice	Yes	Moderate	Low	500 kV only
Norway, 1987	Contaminated snow; Thin rime ice and fog	Mostly Yes	Heavy	Medium	300 kV
Quebec, Canada, 1988	Contaminated snow	Yes	Low	Low	735 kV
Yugoslavia, 1989	Contaminated ice/fog; Snow	Yes	Heavy	600 m	400 kV
Ontario, Canada, 1989	Ice	Yes	Heavy	Low	230 kV
Norway, 1993	Thin rime ice and fog	Yes	Yes	Medium	300, 420 kV
Idaho, USA 1994	None stated	Yes	Yes	1,400 m	345 kV
Ontario, Canada, 1994	Ice	Yes	Heavy	Low	115, 230 kV
Quebec, Canada, 1995	Snow	No, −10°C	Low	400 m	735 kV
New England, USA, 1995	Rain, bird droppings in February	n/a	Yes	Low	345 kV
Texas, USA, 1996	High winds and snow	n/a	n/a	1,100 m	345 kV
Ontario, Canada, 1997	Contaminated ice (road salt)	Yes	None	Low	230 kV semi-conductive glaze insulators
Czech Republic 2002	Contaminated ice (cooling tower)	n/a	Heavy	Low	110, 220, 400 kV
Japan, 2004	Contaminated snow (sea salt)	Yes	No	Sea level	33 kV
Ontario, Canada, 2004	Contaminated snow	−2 °C	Heavy	Low	115 kV
China, 2000–07	Heavy ice, moderate conductivity	yes	Not a factor	High	HVDC

Forrest 1936; Meier and Niggli 1968; Vuckovic and Zdravkivic 1990; CIGRE TF 33 04 09 1999; CIGRE TF 33 04 09 2000; Jiang et al. 2005; Farzaneh et al. 2005a; Farzaneh et al. 2007a

- Flashovers occur under normal operating voltage, rather than during switching operations or in response to lightning, which occurs in winter mainly during the frontal storms that also cause ice and snow accretion on insulators.
- Multiple flashovers may occur in response to each event.

7.4 Ice Flashover Processes and Mechanisms

7.4.1 Initiation of Streamers on an Ice Surface

Extensive studies by many researchers have developed a comprehensive picture of the evolution of an arc on ice surfaces, at the relevant time scales ranging from 10 ns to 10 s. The evolution from shortest to longest time interval is illustrated on distance scales of 35 to 3,500 mm.

In the range of 100 ns to 1 μs, partial discharges develop. Figure 7.3 shows the geometry used to study the initial stages of ice flashover at −12 °C and −2 °C under lightning impulse voltage with a photomultiplier tube (PMT) and high-speed image processing. A typical sequence of corona streamer development across a 3.5-cm ice surface at a frame interval of 5 ns is shown in Fig. 7.4 (Ndiaye et al. 2007).

For a constant form (peak to average field) factor of 0.1, the streamer inception field and propagation velocity for air and ice surfaces with different conductivities for two temperatures were compared. Table 7.3 shows there were significant differences in streamer inception electric fields for all three parameters.

For a constant field of 100 kV/cm on the high-voltage electrode in Fig. 7.3, Ndiaye et al. (2007) reported that streamer propagation velocity at −2 °C increased from 0.5 m/μs in air to 1.2 m/μs across an ice surface, made from water with 80-μS/cm conductivity. The streamer velocity fell to 0.3 m/μs in air and 0.9 m/μs on ice at −12 °C. For larger 100-cm rod-to-plane air gaps at 20 °C under lightning impulse conditions, Pigini (1989) established an average streamer velocity of 2–4 m/μs. Pigini also predicted streamer time using only the voltage overstress, independent of gap size, and confirmed that the streamer velocity increases with electrode separation distance.

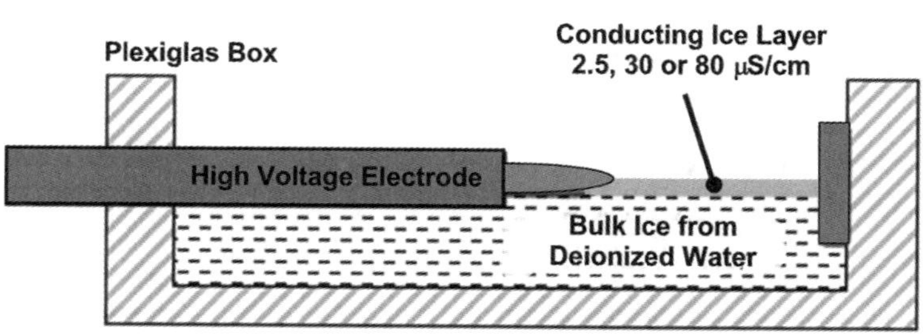

Fig. 7.3 Experimental setup for study of streamer propagation on the ice surface

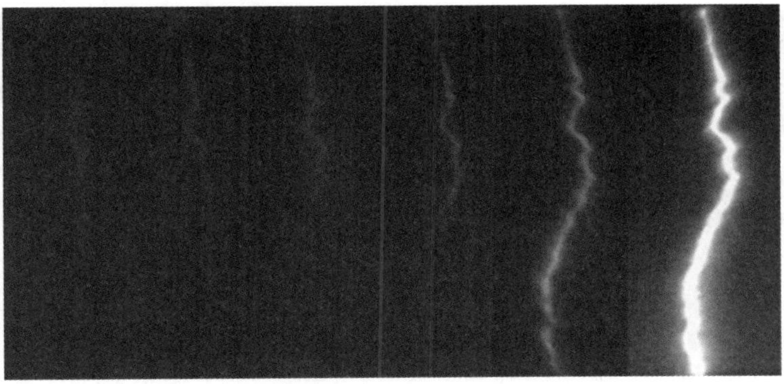

Fig. 7.4 Observation of streamer propagation at 5-ns frame rate (Reproduced by permission of IEEE)

Table 7.3 Measured positive-impulse inception electric fields on 3.5-cm ice surface

Dielectric	Inception field at $-12\,^{\circ}C$	Inception field at $-\mathbf{2}\,^{\circ}\mathbf{C}$
Air	130 kV/cm	120 kV/cm
Ice layer from 30 μS/cm water	90 kV/cm	70 kV/cm
Ice layer from 80 μS/cm water	80 kV/cm	50 kV/cm

7.4.2 Arc Initiation and Propagation on an Ice Surface

In the range of 1 to 10 ms, an arc can develop in the zone bridged by streamers if the voltage is high enough. Under ac conditions somewhat below the minimum flashover voltage, a repetitive sequence of arc development and extinction will occur, as illustrated in Fig. 7.5. A sequence of photos, 1 ms apart, shows the arc development from an initiating 10-mm gap cut into the triangle apex at the top of each photo. The ice sample is triangular, 280-mm high and 200-mm wide at the base (Zhang and Farzaneh 2000).

The 60-Hz ac voltage reaches its peak at about 4.2 ms, and the arc development persists after the voltage peak for another 3 ms. The time lag after voltage peak and the persistence of the arc at 9 ms, after the ac voltage crosses zero, give some guidance about the hysteresis present in the arc development process on ice surfaces.

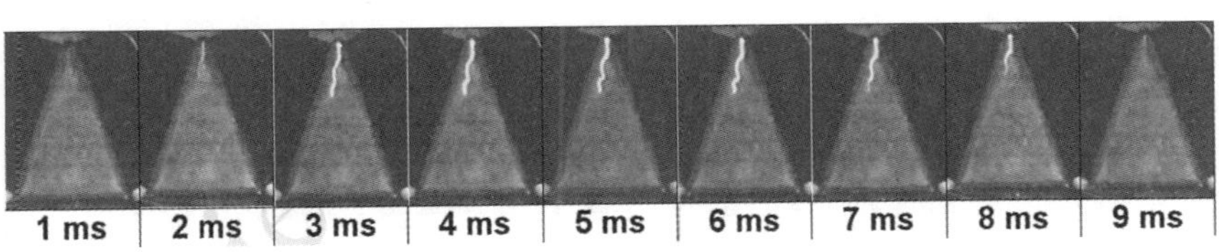

Fig. 7.5 Variation of arc length on 280-mm high by 200-mm wide ice sample under 60-Hz ac voltage (Reproduced by permission of IEEE)

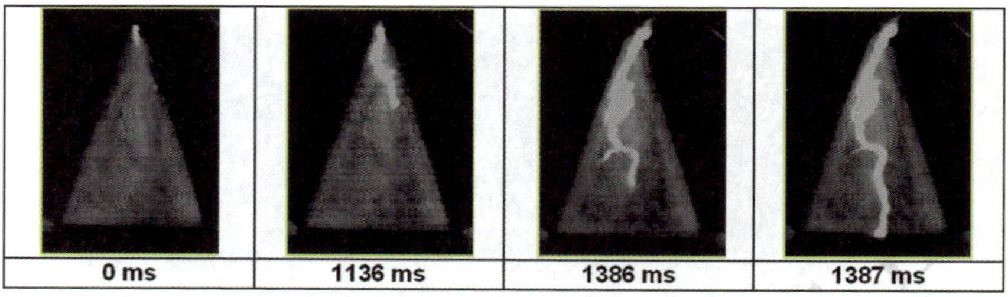

Fig. 7.6 Arc development along the triangular ice surface (Reproduced by permission of IEEE)

In the time range of 10 ms to 2 s, at or above the minimum flashover voltage, the arc can propagate along the ice surface, and terminate with a final jump to the opposite electrode. The speed achieved by the arc when it fully bridges the ice surface has also been studied with high-speed camera images. Four slides in Fig. 7.6, taken at 0, 1,136, 1,386 and 1387 ms show that the arc speed is 0.073 m/s for the first third of the sample, increases to 0.33 m/s for the second third, and bridges the final third of the ice surface in 1 ms, at a speed of about 83 m/s (Farzaneh and Zhang 2007).

The speed of the arc propagation in the final stage of development on the ice surface, at about 83 m/s, in Table 7.4 is much faster than the initial stages at 0.073 m/s, but still very much slower than the speed of streamer propagation (1 m/μs).

For large insulators, the arc propagation time scale extends beyond 2 s. For industrial insulators of 1,400 and 3,500 mm dry-arc distance, Fig. 7.7 shows how two or three gaps form during ice accretion and melting periods. Also, the arc on the ice surface is stable enough that photographs in daylight are feasible at normal shutter exposure times.

A mathematical model for the flashover performance of large insulators with multiple air gaps is developed and presented in Section 7.6.

Figure 7.8 shows the terminology applied to the description of arcs on ice surfaces. The "white arc" is distinct from those observed in most other arcing and flashover processes, including contamination and cold fog. The *arc-root radius* at the point of contact between the arc and ice surface establishes the contact resistance between the arc and the ice layer. The arc root can be visualized as a disc of radius $r = 3$ mm for typical arc currents of 0.25 A.

Figure 7.8 (a) illustrates how ice usually forms on the windward side of the insulator string. If the ice surface is dry, the electrical performance of the insulators is not significantly reduced. However, the presence of a highly conductive water film on the surface, created usually during a melting period, will cause most of the applied voltage to appear across air gaps. The air gaps tend to form in the areas of greatest electric field divergence, which, for a suspension insulator string, is near

Table 7.4 Interpretation of arc development speed in Fig. 7.6

Δt	1,136 ms	250 ms	1 ms
Δx	83 mm	83 mm	83 mm
Speed	0.073 m/s	0.33 m/s	83 m/s

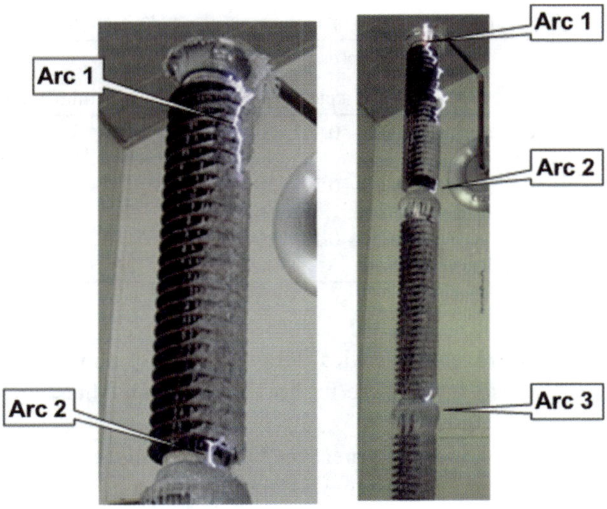

Fig. 7.7 Typical arcs at air gaps on electric utility insulators in icing flashover test. *Left*: 1.4 m dry-arc distance and two arcs. *Right*: 3.5 m dry-arc distance and three arcs (Reproduced by permission of IEEE)

the high-voltage conductor hung from the base of the bottom insulator in the chain. The high electric stress leads to partial discharge and arcing, shown at the base of the insulator in Fig. 7.8 (a).

Figure 7.8 (b) shows how the local arc in Fig. 7.8 (a) changes to a white arc if the applied voltage is high enough. The white arc can then extend along the ice surface.

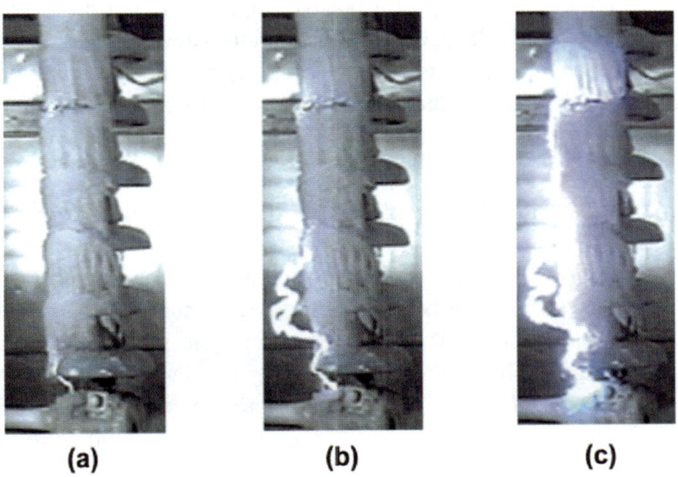

Fig. 7.8 Development of arc on string of six, 146-mm cap-and-pin suspension insulator discs for 115-kV transmission line. (**a**) Formation of local arcs (**b**) White arc formation and extension (**c**) Flashover arc (Reproduced by permission of IEEE)

Table 7.5 Summary of arc propagation speeds on 250-mm ice layer

		Arc propagation velocity, m/s		
		First stage	Second stage	Maximum value
Positive arc	Surface	0.05 – 0.3	20–50	100
	Bulk	None	3–7	50
Negative arc	Surface	0.05 – 0.3	35–60	100
	Bulk	None	10–20	50
AC arc	Surface	0.04 – 0.15	16–30	440
	Bulk	None	2–7	260

When the white arc reaches a critical length, typically about two-thirds of the insulator dry-arc distance, a complete flashover arc such as that shown in Fig. 7.8 (c) will occur.

In the photographs of arc development on 250-mm triangles of ice, shown above in Fig. 7.5, the arc propagation velocity was measured using a high-speed camera at 1 ms/s frame rate. The arc propagation velocity was not constant. Instead, it developed slowly at first, and bridged the final third of the sample very quickly. A similar sequence evolves in Fig. 7.8.

The arc propagation speed was also found to vary with voltage polarity and with the arc path, either along the surface or in the bulk of the ice. Arcs taking the internal route in dry conditions do not have a slow first stage of propagation, but their second-stage speed is slow compared to surface arcs. Table 7.5 provides a summary of observations for these experiments (Farzaneh and Zhang 2000).

The arc propagation velocity on ice surfaces is somewhat slower than those observed on polluted surfaces (Ghosh and Chatterjee 1996; Li et al. 1990). The arc propagation velocity on ice is very much slower than the speed of leader development, for example 0.1 to 2 m/μs for positive lightning impulse voltages applied to rod-to-rod gaps (Pigini 1989). Clearly, the streamer development process on ice surfaces noted in Section 7.3.1 has common features, but the final jump of the ac or dc arc on ice differs from the leader progression process in impulse air-gap flashover.

7.4.3 Voltage-Current Characteristics of Arcs on an Ice Surface

Studies of the flashover of water columns, such as those of Rumeli (1976), gave some initial guidance to researchers about the need for an accurate model of the voltage-current relationship of the arc on each specific surface or medium. The voltage-current relationship for the arc is normally fitted with a power-law relation of the form $V_{arc} = A\,x\,I_{arc}{}^{-n}$, where A and n are arc constants and x is the arc length. Table 7.6 shows some values of arc constants from the literature for various surfaces and media.

It is clear from Table 7.6 that the arc constants for the ice layer are considerably different from those measured experimentally in steam, water or wet pollution layers.

Table 7.6 Coefficients of arc voltage gradient – arc current relation $V_{arc} = A\,x\,I_{arc}{}^{-n}$

Researcher	Rumeli 1976	Peyregne et al. 1982	Rumeli 1976	Ghosh and Chatterjee 1995	Ghosh and Chatterjee 1995	Ghosh and Chatterjee 1995	Farzaneh et al. 1998
Medium	Steam	Steam	Water vapour	$FeCl_3$ electrolyte	$CaCl_2$ electrolyte	NaCl electrolyte	Ice surface
A	518	530	63	270	461	360	205
n	0.273	0.24	0.76	0.66	0.42	0.59	0.561

7.4.4 Ice Conductivity Variation near **0 °C**

The electrical conductivity of ice varies considerably just below the freezing point. Figure 7.9 shows results taken in a coaxial electrode system by Vlaar (1991), using air-free ice with varying conductivities and ion contents. The ac current through the ice was mainly capacitive, 90° out of phase with the applied voltage. The highest conductivity reached by the ice samples at 0 °C was a median 208 times lower than the conductivity of the solution at 25 °C. The ice conductivity dropped by another factor of seven from 0 °C to −15 °C, with most of this change occurring in the narrow temperature range of 0 to −2 °C.

Vlaar (1991) also noted that the standard practice of adjusting the electrical conductivity upwards by adding pollution to de-ionized water important. The tap water sample in Fig. 7.9 had a conductivity of 345 μS/cm at 25 °C, from its dominant ion content of chlorine and carbonate. Ice made from tap water had ten times the conductivity of ice made from controlled samples with the same electrical conductivity.

7.5 Cold-Fog Flashover Process and Mechanisms

Flashovers at normal service voltage have been observed at the melting point in the absence of ice bridging in high-contamination areas, such as stations near high-traffic road salting. The factors that lead to cold-fog flashovers include (Chisholm et al. 1996):

- Build-up of a pre-contamination layer on insulator surfaces over a period of days or weeks in freezing conditions without rain prior to the event.
- Chilling of insulator to below −2 °C.
- Natural fog with 10 μm volume mean diameter and 3 m/s wind speed, with an associated dew point temperature depression of less than 2 centigrade degrees (C°).
- A slow increase in ambient temperature and dew point towards 0 °C.
- Flashovers of several nearby insulators, one at a time, under normal service voltage stress, rather than on switch-on or during switching transients.

The following features differentiate cold-fog from clean-fog flashover of polluted insulators:

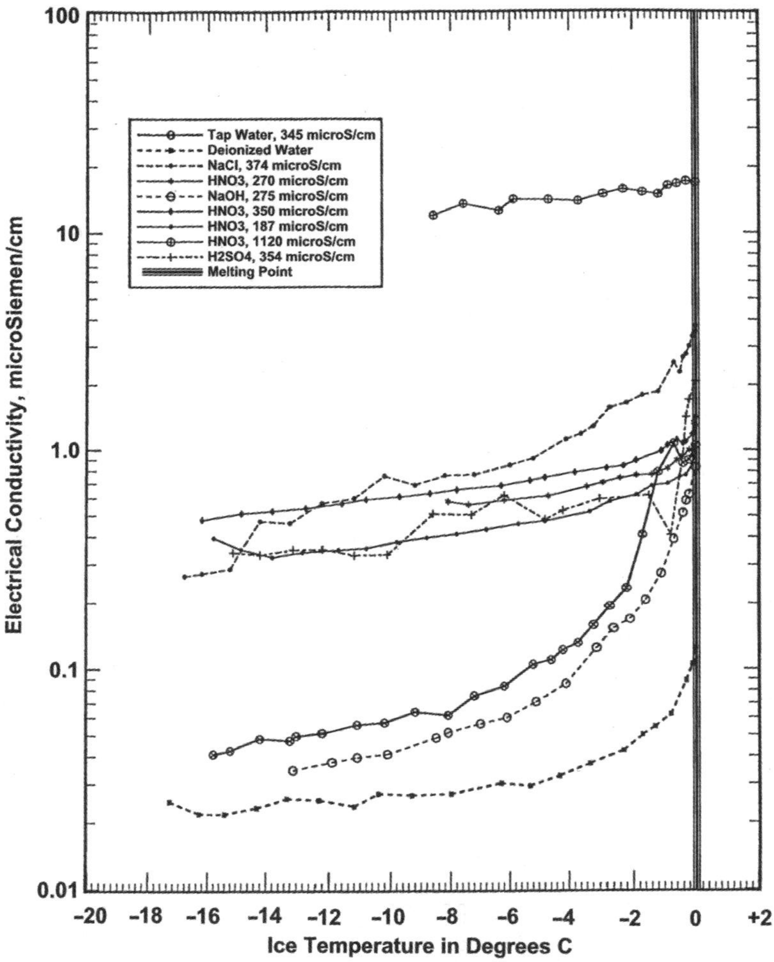

Fig. 7.9 Electrical conductivity of clean and contaminated ice below 0 °C

- The pollution accumulation period in dry conditions (with no rain) occurs mainly when temperatures are below freezing for the cold-fog conditions.
- Complete wetting of the polluted insulator surfaces occurs from accumulation of frost that melts at the initiation of arcing.
- The resistance of the wetted pollution layer reduces with increasing temperature, and this reduction will tend to increase the arc current. For clean-fog conditions, this positive feedback process is limited by quasi-stationary dry band formation as illustrated on long-rod porcelain insulators in Figure 7.10. There is stronger positive feedback in cold fog flashover, since the change in series resistance below and above freezing is more significant. Instead, the aphorism "the one that arcs is *not* the one that sparks" describes cold-fog comparison test results well.

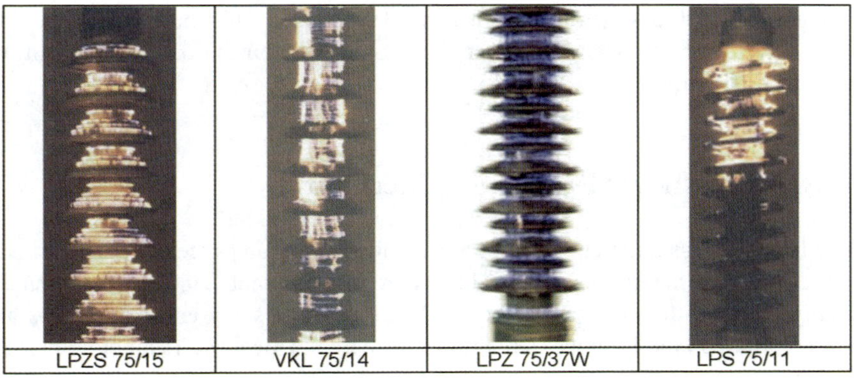

Fig. 7.10 Photograph of arcing on long-rod porcelain insulators in clean-fog conditions (Reproduced by permission of Krystian L. Chrzan)

Time exposure photographs of arcing in clean fog conditions in Fig. 7.10, with the same apparatus as used in Fig. 7.7, show multiple, weak sources of arcing and dry-banding. In the case of cold fog, only the flashover process can be photographed, and this only in darkness using time exposure.

Generally, the results of cold-fog flashover voltages are similar to those observed in clean-fog tests at the same contamination level, but the standard deviation of test results is 4% for cold fog rather than 7–12% for clean fog. For strings of porcelain insulators, Fig. 7.11 shows that the observed specific flashover strength, expressed

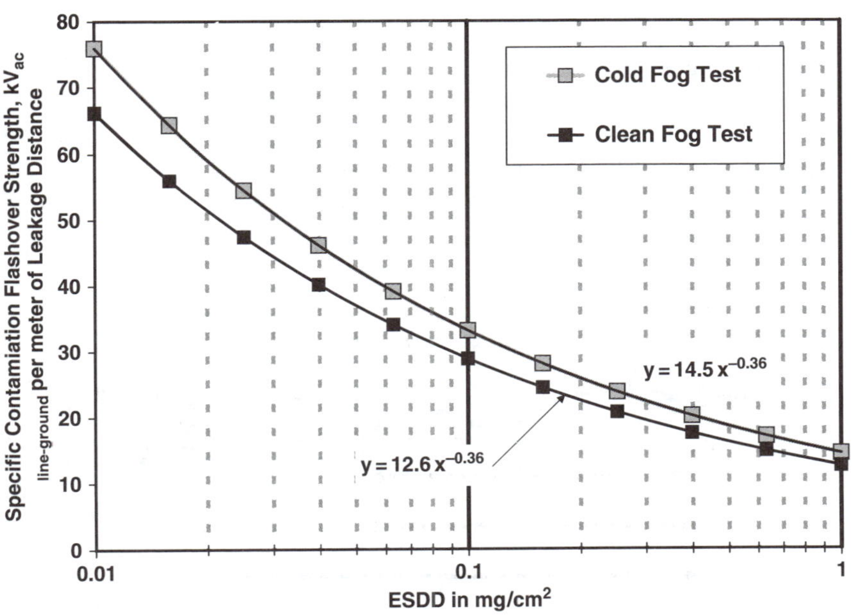

Fig. 7.11 Comparison of clean-fog and cold-fog electrical flashover strengths of leakage distance as a function of contamination level, expressed as equivalent salt deposit density (ESDD)

in kV per metre of leakage distance, is about 15% higher. This means that the cold-fog flashover problem is mainly a manifestation of the conventional contamination flashover problem.

7.6 Snow Flashover Process and Mechanisms

Snow flashover has not received the same detailed attention as ice or contaminated-surface flashover mainly because problems are less frequent. Thick layers of snow actually increase the dry-arc distance. When the snow is dry and cold, it has almost no effect on the flashover strength of insulators. However, Yasui et al. (1988) and Hemmatjou et al. (2007) have demonstrated that wet snow density, conductivity, and layer dimensions can reduce the flashover voltage of horizontal insulators.

7.6.1 Arc Development and Propagation Inside Snow

Arc development and propagation inside snow layers bridging the dry-arc distance of insulators differs in a fundamental way from those on ice surfaces. The arc extends from a metal electrode into the snow, and then propagates internally for much of the snow path. Figure 7.12 shows an overall picture of the situation and also lays out the dimensions that will be used below in modelling the arc characteristics in air and snow.
Where:

x_1 is the length of the air gap (cm)
x_2 is the length of the arc inside the snow layer (cm)
x is the total arc length (cm)
L is the dry-arc distance of the insulator (cm)
$R_s(x)$ is the residual resistance to ground of the non-bridged snow layer (Ω)

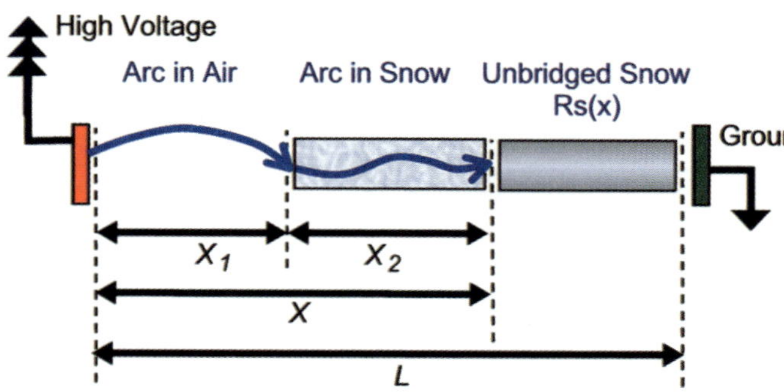

Fig. 7.12 Equivalent electrical circuit of an arc propagating inside wet snow

The electrical equation for the arcing process in snow under AC voltage is the expressed as follows:

$$V_m = A_a x_1 I_m{}^{-n_a} + A_s x_2 I_m{}^{-n_s} + I_m R_s(x) \tag{7.3}$$

where:

V_m is the peak value of the applied voltage (V)
I_m is the peak value of leakage current inside wet snow (A)
A_a and n_a are the arc constants in the air gap of length x_1 (cm)
A_s and n_s are the arc constants inside the snow, with an internal arc of length x_2 (cm)
L is the dry-arc distance of the insulator from high-voltage terminal directly to ground when the insulator is fully bridged by snow (cm)
x_1 is the length of the air gap (cm)
x_2 is the length of the arc inside the snow layer (cm)
x is total arc length (cm)
$R_s(x)$ is the residual resistance to ground of the part of the snow layer that is not bridged by arcs (Ω)

The arc constants A_a, n_a, A_s and n_s have been found to vary in depending on snow characteristics and ambient conditions.

7.6.2 Voltage-Current Characteristics of Arcs in Snow

The electrical characteristics of snow have been studied in general with simplified physical models. Hemmatjou and colleagues (2007) have carried out the most detailed studies to date. They used a glass cylinder 30 cm in height and 11.4 cm in diameter, filled with snow in a cold chamber at −12 °C, as shown in Fig. 7.13. The potentials relative to ground were measured at $D1$, the high-voltage electrode, and at electrodes inside the snow ($D2$) and at the bottom of the snow sample ($D3$). The potential differences across the air gap and inside the snow layer at a given position inside the snow were then derived. The length of the arc was adjusted to vary between 0.5 and 4 cm for the air gap and 1 to 7 cm inside the snow.

From Fig. 7.14, the leakage current is initially very low after each zero-crossing of the applied voltage, $D1$. The potentials inside the snow ($D2$) and at the end of the snow layer ($D3$) track the applied voltage until an arc forms in the air gap. At this point in time, the current increases and potentials in the snow layer drop. After each ac voltage peak in this example, the arc current extinguishes and the snow potentials recover.

Near the voltage peak of each ac cycle in Fig. 7.14, the arc voltage $D3$ reaches a plateau for each arc current. Overall, this relation defines a voltage gradient of the arc approximated by the following equation:

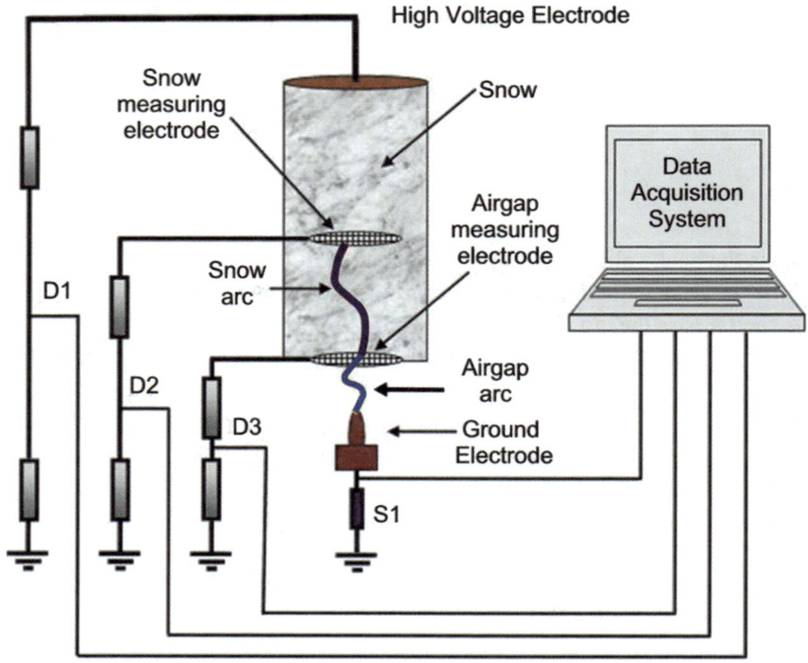

Fig. 7.13 Measurement method of potentials along snow for arc voltage-current characteristics (Reproduced by permission of IEEE)

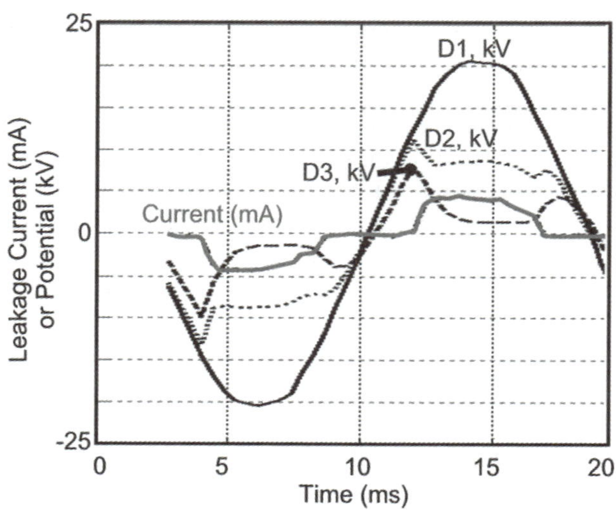

Fig. 7.14 Typical waveforms for current and potentials $D1$, $D2$ and $D3$ at different electrodes for arc in air gap (Reproduced by permission of IEEE)

$$E_{arc} = A_a I_m^{n_a} \approx 100 \cdot I_m^{-0.66} \tag{7.4}$$

where:

E_{arc} is the peak voltage gradient averaged across the air gap x_1 (V/cm)

I_m is the peak of the current (A)

A_a and n_a are arc parameters for the air gap

The values of parameters A_a and n_a for snow are different from those determined for ice surfaces and for the other media, as shown in Table 7.6. The arc constants A_a and n_a for snow are relatively close to those reported for dc arcs in air, typically ($A = 63$, $n = 0.76$) (Rumeli 1976). Some variation in values of A_a and n_a for snow was also found with arc length (Hemmatjou et al. 2007). In the range of $x_1 = 0.5$ to 3 cm, A_a varied from 108 to 84, and n_a varied from 0.64 to 0.75.

When an arc forms inside the snow, the situation changes considerably. Figure 7.15 shows that the current is now a sine wave; it has a peak magnitude of 23 mA, completely in phase with the applied voltage, $D1$. The arc potential $D3$ is low, following the E-I relation in Eq. 7.4. The potential $D2$ at the electrode inside the snow initially tracks the applied voltage, and then falls to a plateau.

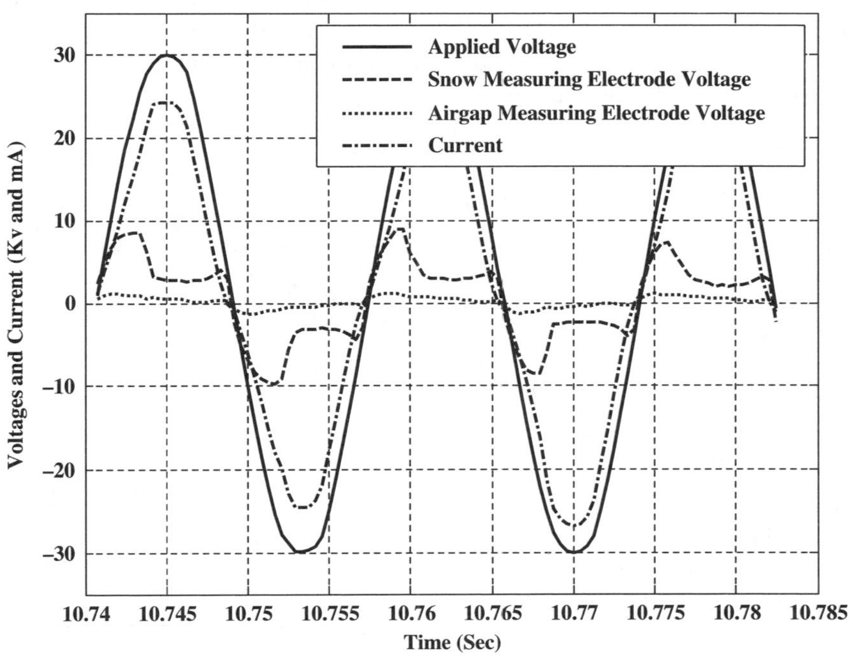

Fig. 7.15 Waveforms for current and potentials $D1$, $D2$ and $D3$ at different electrodes for arc inside snow layer (Reproduced by permission of IEEE)

In the snow layer, analysis of waveforms like those in Fig. 7.15 lead to an overall expression for the voltage gradient of the arc in snow, E_{arc} in V/cm across the distance x_2, of:

$$E_{arc} = A_s I_m^{n_s} \approx 826 \cdot I_m^{-0.36} \tag{7.5}$$

where:

E_{arc} is the peak voltage gradient averaged across distance x_2 inside the snow (V/cm)

I_m is the peak of the current (A)
A_s and n_s are arc parameters for the air gap

In the range of $x_2 = 1$ to 4 cm, A_s varied from 1,065 to 667, and n_s varied from 0.26 to 0.52.

The speed of arc propagation inside snow has not been the subject of detailed study. Figure 7.15 suggests that, at a potential $D1$ of 30 kV peak, the arc takes 1 ms to propagate a length of 2 cm inside the snow layer, corresponding to 20 m/s. This is in the same order as the speed of the arc on the ice surface in the second stage of its development, as shown in Table 7.5. With the negligible thermal mass of a thin hoar-frost layer associated with cold-fog conditions, initiation of arcing will either proceed very quickly to flashover or it will degenerate into random arcing activity, similar to that shown for clean-fog conditions in Fig. 7.10. Once this dry-banding starts in cold-fog conditions, it generally eliminates the risk of flashover because the arcing activity is not sufficiently strong to propagate.

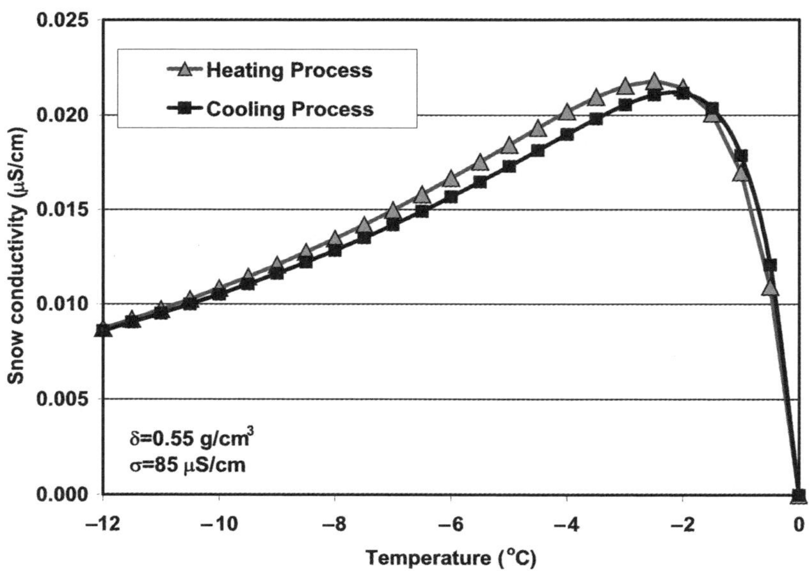

Fig. 7.16 Snow conductivity variation with temperature using two evaluation methods

7.6.3 Snow Conductivity Variation Near 0 °C

Hemmatjou et al. (2007) studied the electrical conductivity of snow and did not find such significant variation in the range of −2 to −12 °C. Instead, Fig. 7.16 shows that snow temperatures above −2 °C decreased the snow conductivity, thereby increasing the snow layer resistance and making flashover less likely.

7.7 Mathematical Modelling of Flashovers on Insulators Covered with Ice or Snow

Flashover on insulators bridged by ice or snow is a complex problem as it is not just an electrical phenomenon, but also involves thermal and electrochemical factors.

The formation of air gaps along the ice layer depends on the voltage distribution along the insulator string during the ice accumulation process. The voltage distribution is affected by the type and length of the insulator. Generally, insulator strings with less than 1 m of dry-arc distance (550 kV BIL) form a single air gap under line voltage during icing tests. For longer insulators used in HV and EHV networks, the voltage distribution is non-uniform, particularly after the appearance of icicles. Under these conditions, more than one air gap may be formed. The number, lengths and positions of the air gaps vary for different insulator types and for different lengths of the same insulator disc. With their large 100-mm shed-to-shed separation, a series of air gaps on long suspension insulator strings may form. For multi-section station post-type insulators, the space between the sheds is relatively small (>50 mm) and that between the shed and metallic flanges is relatively large. Therefore, the shed spacing can easily be bridged by icicles and the air gaps on EHV station posts usually form between the shed and a metallic flange.

Thermal and electrochemical factors also have a strong influence on the residual resistance of the ice or snow layer. Heat input from the passage of arcing current through ice will increase the electrical conductivity. Ozone from sustained arcing on ice will produce nitric acid that will increase conductivity further. These factors will reduce the resistance of the ice layer, leading to higher arc currents in a positive feedback process.

Meteorological and environmental conditions control the rate of change of temperature, the solar input and evaporation rate during the melting phase on ice, snow or cold-fog conditions. The sequence then becomes a race – will the temperature of the ice or snow layer reach a critical level before ions drip away or a number of sections of ice fall off, leaving air gaps that add electrical strength? The thickness of the water film on an ice surface, its important contribution to conductivity and the presence of air gaps also play major roles in the process of arc development and flashover on ice-covered insulators, as described mathematically in Section 7.6.4.

7.7.1 Empirical Model Using "Icing Stress Product"

The resistance of the ice, snow or cold-fog layer is an important factor in establishing whether an arc, once initiated, will reignite and propagate. If this resistance is low, the arc current is higher, and it will take a lower voltage to reignite the arc, giving more time for extension. Prior to detailed development of models using the Obenaus circuit concept (Rizk 1981), we will describe an empirical method used to consolidate experimental results from several laboratories and insulator types into a simple design process, based on a dominant term in the residual resistance of the ice or snow layer.

For a homogeneous deposit, the mass of ice per unit length, m, can be multiplied by the conductivity of the water melted from the ice accretion, $\sigma_{Melt\ Water}$, to lump together all effects of ice and pre-contamination conductivity into a single value, called an ***Icing Stress Product (ISP)***:

$$ISP = \sigma_{Melt\ Water} \cdot m = \sigma_{Melt\ Water} \frac{v}{\rho} \tag{7.6}$$

Each type of ice has its own relative density, ρ. Snow accretion, for example, has a lower density ($\rho = 0.3$) than glaze ice ($\rho = 0.9$), and needs three times the volume per unit length, v, to achieve the same mass per unit length. The use of mass in the ISP of $\sigma_{Melt\ Water} \times m$ per unit length has proved to be a good empirical predictor of flashover performance by an IEEE Station Insulator Task Force (Farzaneh et al. 2005a). The sampling process is shown in Fig. 7.17.

The following relations have been observed between flashover stress V_{50}, in kV per meter of dry-arc distance, and ISP in g/cm·μS/cm:

$$V_{50} = 396(ISP)^{-0.19} \tag{7.7}$$

The weight per unit length of accretion on an insulator is related to the radial accretion on a reference cylinder. For 254-mm diameter ceramic disc insulator strings, this relation is expressed in Table 7.7 (Farzaneh and Kiernicki 1997).

A linear relation of 3.2 g/cm ice weight per mm of accretion fits the observations up to a level of 25 mm. This is also near the upper limit of the validity of the ISP

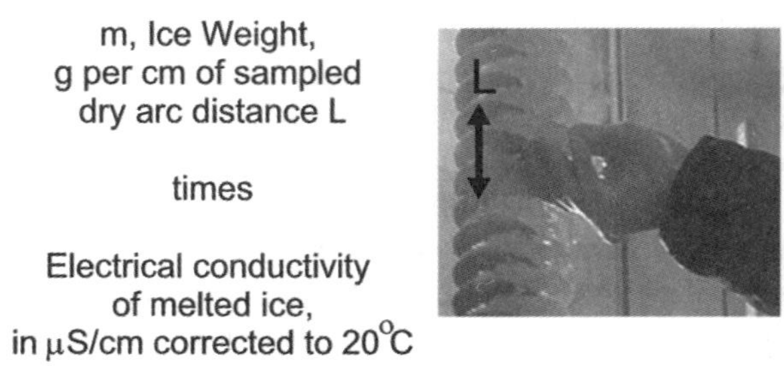

Fig. 7.17 Sampling of icing stress product on non-ceramic insulator

Table 7.7 Relation between ice accretions on reference cylinder and ice accretion on insulator string

Ice accretion level (mm)	Ice weight, g/cm dry arc	Ice accretion on 146-mm × 254-mm insulator string
5	16	Ice caps
10	32	Icicles start to form
15	24	Partial bridging
20	64	80% bridging
25	80	Maximum bridging

approach on heavily iced insulators. Local transmission line ice loading maps give the recommended levels of ice accretion for the mechanical design of lines. These values are also appropriate to use for the ice accumulation estimates on insulators. In most parts of the USA and Canada, the maximum level of ice accretion considered in mechanical design is 25 mm, representing the upper limit of application of the ISP approach from Table 7.7.

As a starting point, the IEEE Task Force (Farzaneh et al. 2005a) recommends that the 5th percentile value of precipitation conductivity (P = 0.05) should be used for establishing icing stress. If the line has high security requirements or is exceptionally long, it would be prudent to select a lower value of P, such as 0.02 or 0.01. With median precipitation conductivity $\sigma_a = 25.4\,\mu S/cm$, equation 7.23 below would be inverted to give:

$$\sigma = \sigma_a \left(\frac{1-P}{P}\right)^{1/2.3} = 25.4\left(\frac{1-0.05}{0.05}\right)^{1/2.3} = 91\,\mu S/cm \qquad (7.8)$$

Substituting P = 0.05 into Eq. 7.8 gives a design value of $\sigma = 91\,\mu S/cm$.

Using a typical ice accretion level of 25 mm on a reference cylinder (80 g/cm dry-arc distance) and the fifth-percentile value of 91 µS/cm, an ISP of 7,280 g/cm · µS/cm is obtained. The 50% flashover stress of an individual insulator string would be 73 kV rms, line to ground per meter of dry-arc distance.

The ISP concept can also be adapted for the case of cold fog on the insulator leakage distance. In this case, the 50-µm thickness of the melted ice layer does not play an important role on ISP magnitude. The dominant term in the residual resistance of the thin ice layer is related to the surface pre-contamination level. As layer thickness increases, electrical conductivity decreases. Table 7.8 shows how the ISP in this case varies with some typical levels of pollution severity.

Equation 7.9 shows the relation of flashover gradient along the leakage distance of the insulator to the ISP. The exponent in Eq. 7.9 varies slightly from $n = -0.36$ to $n = -0.37$ as a result of the correction for conductivity as a function of salt concentration in thin layers of water.

$$V_{50} = 1196(ISP)^{-0.37} \qquad (7.9)$$

Table 7.8 Typical pollution severity, ISP and leakage-distance flashover stress on line post insulators in cold-fog conditions

	ESDD ($\mu g/cm^2$)	Surface conductivity (μS at 20 °C)	Equivalent icing stress product ($g/cm_{leakage} \times \mu S/cm$)	Flashover gradient along leakage distance (kV/m leakage)
Very light	< 6	< 10	<1,000	> 93
Light	6–25	10–40	1,000–4,000	93–55
Medium	25–100	40–160	4,000–15,000	55–34
Heavy	100–400	160–600	15,000–60,000	34–20
Very Heavy	> 400	> 600	> 60, 000	< 20

Hemmatjou et al. (2007) and Yasui et al. (1988) provide flashover tests on snow-covered insulators with sufficient detail to establish the ISP defined by the *snow weight* (g/cm of insulator dry-arc distance) multiplied by the *snow conductivity* (melted and corrected to 20 °C in units of μS/cm). Their flashover voltage stress results are consolidated in Fig. 7.18.

The moderate accumulation of snow with low pollution on a non-ceramic insulator 15 cm in diameter (Hemmatjou et al. 2007) gave an ISP lower than 100 g/cm × μS/cm. This snow deposit actually increased the insulator strength above the typical 380 kV rms per meter of dry-arc distance shown in Fig. 7.18. For these cases, arcing at flashover followed the surface of the snow, which extended the flashover path by a factor of 2.2, compared to the dry-arc distance without the snow. Yasui et al. (1988)

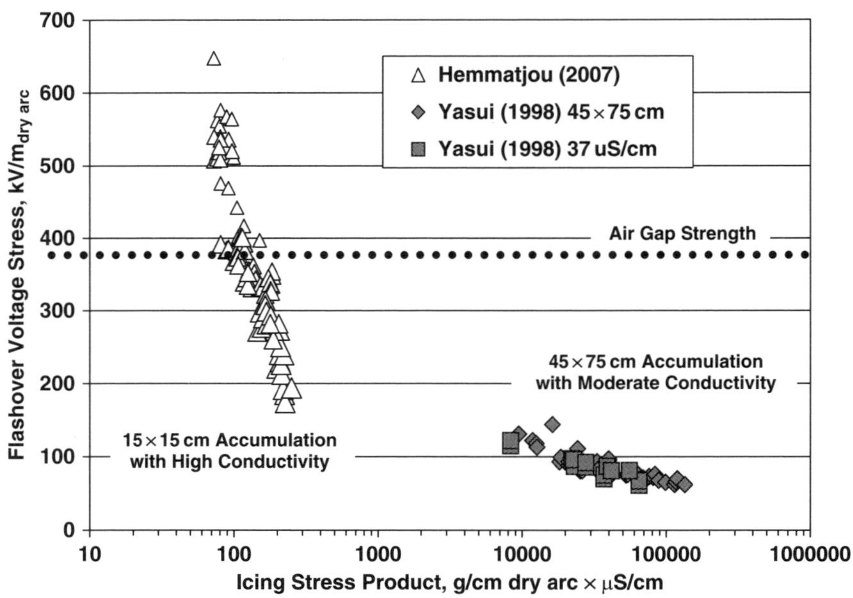

Fig. 7.18 Flashover stress versus icing stress product for moderate and heavy layers of snow

tested twin strings of 25-cm diameter disc insulators that could support thick snow layers of 45×75 cm. Their findings showed a significant loss of electrical strength, down to as low as 80 kV per meter of dry-arc distance. As a comparison:

- HV power systems ($\leq$ 250 kV) are operated at 70–75 kV per meter of dry-arc distance.
- EHV power systems ($\geq$ 315 kV) are operated at gradients of 80–90 kV per meter of dry arc.

The high gradient per meter of dry-arc distance is a recurring feature in the systems affected by icing problems. Critical EHV systems have more icing problems than HV networks. Most other power system reliability problems, such as animal and lightning outages, are far less severe on EHV lines because the basic insulation level (BIL) and corresponding dry-arc distance are larger.

The ISP can be adapted for insulators covered with snow, as previously illustrated in Fig. 7.18 of Section 7.3. However, test data are limited, and the nature of the flashover changes from external to internal, making it necessary to use independent empirical fits for each type of deposit. For the heavy, wet snow deposit of Yasui et al. (1988):

$$V_{50} = 1303(ISP)^{-0.26} \tag{7.10}$$

7.7.2 Mathematical Model Using the Obenaus Concept

The flashover on an insulator covered with ice or snow starts with the initiation of an arc and its propagation along a layer of uniform resistance per unit length. This is the same situation that can lead to flashover of a simple, polluted insulator such as a cylinder or rectangle. The Obenaus (1935) model was developed to describe flashover under DC conditions. Rizk (1981) added suitable criteria to adapt the model for ac flashover. While it has only moderate success when applied to complicated insulators under dry-band arcing conditions shown in Fig. 7.10, the Obenaus model with reignition conditions has proved to be accurate and adaptable to a wide range of ice and snow flashover situations.

The flashover on an ice surface is considered as an arc in series with a residual resistance consisting of an ice layer which is not bridged by the arc, as already shown in Fig. 7.8. With a single arc, the circuit model of Fig. 7.19 and the corresponding circuit equation for this model is as follows:

$$V_m = A\,x\,I_m^{-n} + I_m R(x) \tag{7.11}$$

where:

V_m is the peak value of applied voltage (V)
x is the local arc length (cm)
I_m is the peak value of leakage current (A)

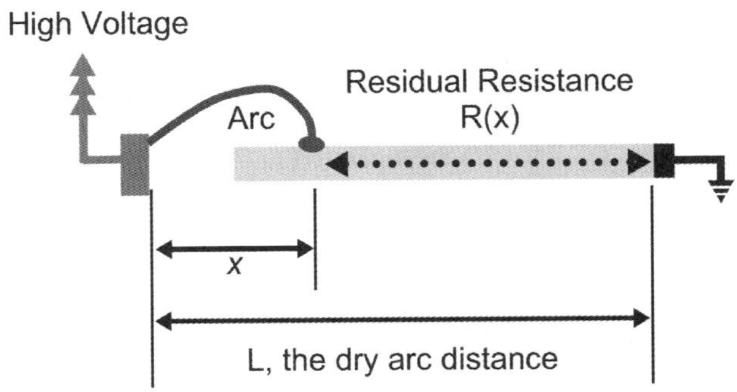

Fig. 7.19 Obenaus model for flashover on ice or snow covered insulators

$R(x)$ is the residual resistance of ice layer (Ω)
A and n are the arc constants.

7.7.3 Arc Re-ignition Conditions for Ice

Under ac voltage, as the current passes through zero twice in each cycle, the local arc extinguishes and reignites twice. Therefore, not only Eq. 7.11, but also the arc re-ignition conditions (Rizk 1981) expressed by Eq. 7.12 and Eq. 7.13, must be satisfied.

$$V_m \geq \frac{kx}{I_m^b} \tag{7.12}$$

The critical condition in which the arc retains just enough thermal energy to reignite on the next half-cycle of power frequency voltage is:

$$V_m = \frac{k_i x}{I_m^{b_i}} \tag{7.13}$$

where k_i and b_i are the arc re-ignition constants for ice. A number of studies (Farzaneh et al. 1997) (Farzaneh-Dehkordi et al. 2004) have established these coefficients for ice layers at 60 Hz:

$k_i = 1,120$ for an arc propagating upward, with natural arc buoyancy
$k_i = 1,300$ for an arc propagating downward, against natural arc buoyancy
$b_i = 0.53$
V_m is the peak value of applied voltage (V)
I_m is the peak value of current (A)
x is the arc length (cm).

For a typical 10-cm gap between insulator disc surfaces, an arc current of $I_m = 0.2$ A would need a peak ac voltage of $V_m = 26$ kV to reignite for an arc

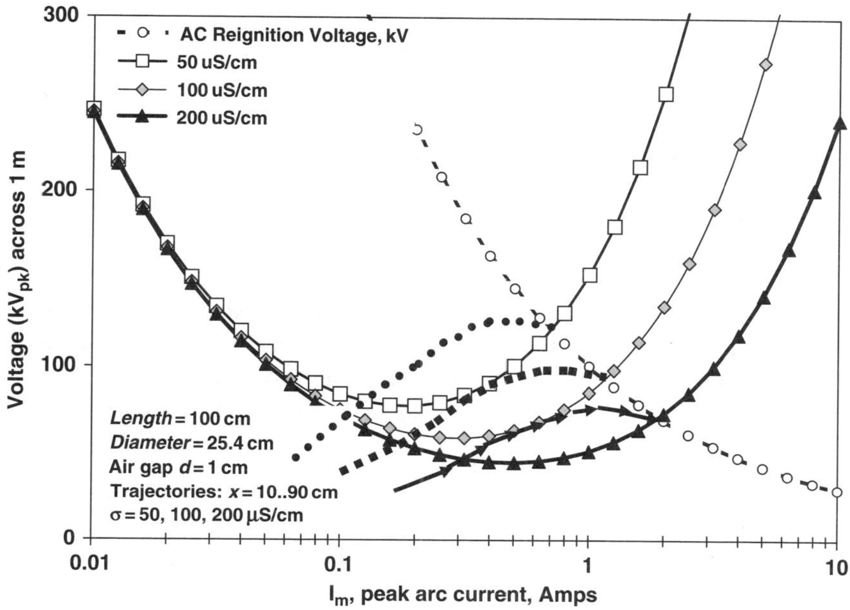

Fig. 7.20 Relations among arc, re-ignition voltages and arc current for ice flashover

starting at the base of an insulator chain and propagating upwards. This is the normal direction of propagation, as the high-voltage conductor is hung from the bottom of the chain and the top of the chain is grounded to a tower.

Arcs are buoyant and tend to move upwards with the thermal energy they are radiating. An arc at the base of an insulator string is easier to maintain and propagate than an arc at the top of the insulator string. This leads to different values of the arc re-ignition constant k for an arc propagating from the top to the bottom of the insulator than for one from the bottom to the top.

The value of k is a little lower for the arcs that are propagating upwards. In the numerical example above, with a 10-cm gap and $I_m = 0.2\,\text{A}$, it would take $V_m = 30\,\text{kV}$ to force the arc downward.

The balance between increasing arc voltage and decreasing re-ignition voltage determines the flashover level. As resistance $R(x)$ decreases, for example with increasing melted-ice conductivity, I_m increases and the re-ignition voltage V_m decreases, as shown in the numerical example found in Fig. 7.20. The balance point shifts down from $120\,\text{kV}_{\text{pk}}/\text{m}$ to $75\,\text{kV}_{\text{pk}}/\text{m}$, thus changing conductivity by a factor of four.

7.7.4 Arc Re-ignition Conditions for Snow

The arc re-ignition constraint for flashover of ice layers also applies to ac flashover of snow-covered insulators. For an arc inside wet snow, the ac re-ignition condition was found to be (Hemmatjou et al. 2007):

$$V_m \geq \frac{k_s x}{I_m^{b_s}} \tag{7.14}$$

where:

k_s and b_s are the arc re-ignition constants for snow
I_m is the peak voltage in amps
V_m is the peak voltage in volts
x is the arc distance in cm.

The values of $k_s = 6.37$ and $b_s = 0.486$ were established by experiment. In the critical case where the arc just reignites, the equality condition in Eq. 7.14 applies. This can then be inverted to obtain the arc current needed for re-ignition:

$$I_m = \left(\frac{k_s x}{V_m}\right)^{\frac{1}{b_s}} \tag{7.15}$$

This current can be substituted in the electrical circuit model of Fig. 7.12 to yield:

$$V_m = A_a x_1 \left(\frac{k_s x}{V_m}\right)^{\frac{-n_a}{b_s}} + A_s x_2 \left(\frac{k_s x}{V_m}\right)^{\frac{-n_s}{b_s}} + R(x) \left(\frac{k_s x}{V_m}\right)^{\frac{1}{b_s}} \tag{7.16}$$

Equation 7.16 shows a relationship between the peak value of alternating applied voltage (V_m) and the length of arc (x). If the constants A_a, n_a, A_s, n_s, k_s and b_s are known, then V_m is uniquely determined by arc lengths. Following this terminology, A_a, and n_a are the arc constants for the air gap, A_s and n_s are the arc constants for the arc in the snow, and k_s and b_s are the re-ignition constants of the entire arc.

7.7.5 Resistance of Several Ice and Snow Layers in Series

The ice layer formed on an insulator can be considered as a half cylinder (Farzaneh et al. 2004). The residual resistance $R(x)$ can be calculated using Eq. 7.7:

$$R(x) = \frac{1}{2\pi \gamma_e} \left[\frac{4(L-x)}{D+2d} + \ln\left(\frac{D+2d}{4r}\right)\right] \tag{7.17}$$

where:

γ_e is the surface conductivity of the ice layer in μS
L and D are the length and diameter of the insulator string in cm
d is the thickness of the ice layer in cm
r is the arc root radius.

The surface conductivity and arc-root radius were established by Farzaneh et al. (1997) to be:

$$\gamma_e = 0.0675\sigma + 2.45\,\mu S$$
$$r = \sqrt{\frac{I_m}{0.875\,\pi}} \tag{7.18}$$

where:

σ is the conductivity of the applied water measured at 20 °C (μS/cm)
I_m is the peak current (A)
r is the arc root radius (cm).

For insulators with less than 1 m of dry-arc distance, the equation for the single air gap will normally apply. As the length of the insulator increases, as shown in Fig. 7.7, two or more air gaps will usually appear during icing under energized conditions. Farzaneh and Zhang (2007) noted that, for three different insulator types with dry-arc distances of 2.7 to 3 m, normally there were three arcs: one at the top, one at the middle part, and one at the bottom of the insulator string or stack of posts. Each air gap adds to the residual resistance by increasing the number of arc roots in series.

For an ice-covered insulator with N air gaps, there will be $2N$ arc roots. The contribution of arc roots on metal fittings to $R(x)$ is ignored. Those arc roots that form on ice surfaces cause a non-uniform current distribution along the ice surface which in turn influences the resistance of the residual ice layer. Farzaneh and Zhang (2007) describe a multi-arc model to account for the number of arcs with a single root on the ice surface (N'), and the number of arcs with two roots (N''). The total number of arcs is $N = N' + N''$. Thus, the residual resistance for N arcs can be expressed by Eq. 7.19:

$$R(x) = \frac{1}{2\pi\gamma_e}\left[\frac{4(L-x)}{D+2d} + (N' + 2N'')\ln\left(\frac{D+2d}{4r}\right)\right] \tag{7.19}$$

7.7.6 *Flashover Voltage of Insulators with Several Ice Layers in Series*

The multi-arc model for insulator flashover consists of Eqs 7.11, 7.13 and 7.19 for insulators partially covered with ice.

This model can be used to predict the critical flashover voltage of insulators covered with ice, for any insulator type and voltage level if the number and type of local arcs are known. For example, the model predicts that an ice layer with a central air gap will have a balance point of 124 kV, compared to 94 kV for the single-arc model, for the dimensions given in Fig. 7.21. This difference of 30 kV is much larger than would be predicted from a fixed value of 0.95 kV per arc root on water surfaces from Wilkins (1969).

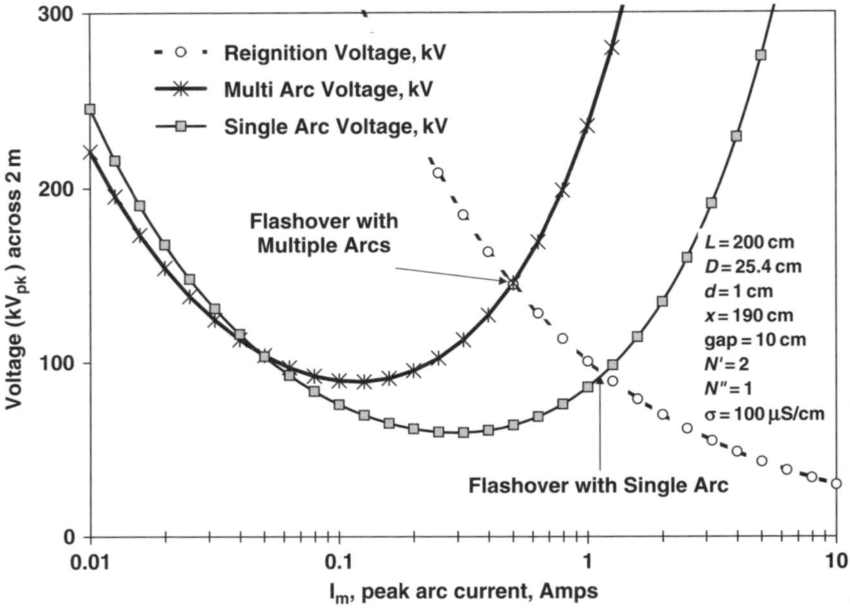

Fig. 7.21 Effect of multiple arcs on balance point between arc and re-ignition voltage

The multi-arc model of Eqs. 7.11, 7.13 and 7.19 predicts the critical flashover voltage of different insulator types and configurations used in networks with a service voltage up to 735 kV. It is important to adjust for the increasing number of air gaps with voltage level. Table 7.9 (Farzaneh and Zhang 2007) lists the recommended number of arcs to be used with the multi-arc model for predicting the critical flashover voltage of post or line insulators under icing conditions.

The multi-arc model should also be used to evaluate the improvement in strength when additional air gaps are deliberately introduced using insulator accessories such as booster sheds and creepage extenders. It is also possible to apply the three-arc model to insulators with more than 6-m dry-arc distance used in UHV systems above 800 kV. However, the validity of this approach remains to be experimentally validated.

Table 7.9 Recommended number of arcs for predicting the critical flashover voltage of the ice-covered insulators used at different voltage levels

Voltage level (phase-to-phase)	Typical dry-arc distance	Number of arcs in series for multi-arc model
$V \leq 110\,kV$	< 1 m	1
$110\,kV < V < 315\,kV$	1–2.5 m	2
$V \geq 315\,kV$	> 2.5 m	3

7.7.7 Flashover of Snow-Covered Insulators

The icing flashover model can be adapted for mathematical modelling of ac flashover of snow-covered insulators. The model is modified in Fig. 7.12 to include the basic distance x_1 for the arc to propagate in air, added in series with a new distance x_2, where the arc is observed to propagate inside the snow layer.

The residual resistance of a volume of snow, rather than a surface layer of ice, is:

$$R(x) = \int_{x_2}^{L} \frac{1}{\sigma_e(y)} \frac{1}{\pi R_y^2} dy \tag{7.20}$$

where:

$\sigma_e(y)$ is the volume conductivity along a cylindrical snow sample (μS/cm)
R_y is the radius of the cylindrical snow sample
x_2 and L are the dimensions of the snow sample defined in Fig. 7.12

Also, the average conductivity of the snow layer σ_e differs from Eq. 7.18 for ice. Instead, for snow, Hemattjou et al. (2007) found:

$$\sigma_e = 0.0057\sigma\delta + 1.2893\delta - 0.00135\sigma - 0.2346 \tag{7.21}$$

where:

σ_e is the equivalent volume conductivity of the snow sample (μS/cm)
σ is the conductivity of the melted snow at 20 °C (μS/cm)
δ is the density of the snow (g/cm^3)

The model for insulator flashover consists of Eq. 7.16, Eq. 7.20 and Eq. 7.21 for insulators partially covered with snow.

7.8 Recommended Test Methods

With a large number of unknown factors such as sunshine, wind, salt exposure rate and temperature, it has been common for the electrical utilities to construct exposure stations near the sea coast or known pollution sources. Examples include South Padre Island (Texas, USA), Moss Landing (California, USA), Brighton (UK), Noto, Takeyama and Nagasaka (Japan), Koeberg (South Africa), Anneberg (Sweden), Glogow (Poland) and many others. These have proved to be excellent for establishing cumulative damage rates on existing and new insulation systems, because the pollution exposure rates are relatively constant.

One characteristic of most test sites is that the bottom-surface contamination levels are considerably higher than top-surface levels. This reflects the washing action of rain or sea spray.

The winter conditions that lead to flashovers generally involve a long period of accumulation in cold conditions, leading to higher levels of contamination on upward-facing surfaces.

The winter conditions that lead to flashovers also include accumulation of freezing rain, wet snow or hoar frost (cold fog). These are relatively rare conditions, occurring less than 15% of the time with considerable variation in accumulation from event to event. Most insulator test stations are also located in temperate areas near the sea. It has thus proved to be impractical to establish insulator response to repeatable icing conditions from most insulator exposure sites, with Anneberg (Sweden) being an exception.

7.8.1 Standard Electrical Tests of Insulators

Standard electrical test methods (IEEE Standard 4 1995; IEC 60507 1991) for insulators describe three types of contamination tests, along with the technical requirements such as source impedance, pollution levels and other factors that ensure repeatability. These standard tests evolved from the exposure experience of various utilities. The tests include:

- Heavy rain tests, where water with controlled 100-μS/cm conductivity and rain rate is applied for a period of several minutes, then the insulator is energized.
- Salt-fog tests, where salt aerosol with conductivity ranging from 4,300 to 200,000 μS/cm is sprayed onto energized insulators.
- Clean-fog tests, where pre-contaminated insulators with salt-deposit densities of 25 to 400 $\mu g/cm^2$ are energized and then wetted with steam or water aerosol.

There is some merit in adopting existing test levels and equipment where these are appropriate. For example, many laboratories are set up to deliver 100 μS/cm rain water. It makes sense to adopt this same conductivity in standard icing tests as well, especially since this represents a reasonable upper bound of natural values of precipitation conductivity. Also, power supply requirements call for adequate voltage regulation when arcing occurs, and the arc currents in pollution and icing tests are similar.

7.8.2 Icing Test Methods

The ice conductivity variation near the melting point, as shown in Fig. 7.9, argues for the inclusion of a melting phase in electrical tests, to ensure that the ice reaches its maximum conductivity and minimum electrical strength. Tests carried out without a melting phase do not excite the line-voltage flashover mechanism at 0 °C with moderate ice accretion, a common characteristic of most important flashover events in natural conditions. Instead, tests using the Ice Progressive Stress (IPS) approach (Gutman et al. 2002) develop flashovers with heavy deposits of ice that seldom occur in nature. This means that the role of IPS tests is to rank insulator performance rather than to establish design criteria such as maximum withstand voltage (V_{WS}) and critical flashover (V_{50}) level.

In the ice deposit test method, the recommended experimental conditions are adjusted to form glaze ice with icicles. This type of ice is associated with the highest probability of flashover on energized insulators. Adequate control of air temperature, wind velocity, and the vertical and horizontal components of precipitation intensity should be established and maintained. The ice deposit parameters recommended by the IEEE Task Force on Icing Test Methods (Farzaneh et al. 2003) are summarized in Table 7.10.

The average thickness of ice in laboratory testing of insulators should be measured in the exposure zone near the energized insulators using both rotating and fixed cylinders, 25 to 30 mm in diameter and 600 mm in length. The long axes of the cylinders are horizontal, and positioned to receive the same general wetting as the insulators under test. The thickness of ice on monitoring cylinders for each test series should be based on the icing maps for the desired 10- to 500-year return period, or the more sophisticated meterological and statistical treatments in Chapters 1 and 2 .

The ice test normally calls for a preparation stage at the end of the ice accretion stage, while the temperature is still below 0 °C, and before the test voltage is reapplied for flashover voltage evaluation. The procedure may be adjusted according to two approaches; "icing regime" or "melting regime".

- The icing regime test evaluates flashover performance tests shortly after ice accretion is completed, while a water film is still present on the ice surface. In such a case, the preparation stage is short, about 2 to 3 minutes.
- The melting-regime sequence calls for an "ice hardening" delay of 15 minutes with the insulator voltage turned off, but with the same wind speed and air temperature as during the icing period. This delay ensures complete hardening of the ice and equalization of insulator and ice temperatures. After this delay, the melting sequence is started. Service voltage is re-applied, and the air temperature is increased progressively from sub-zero to melting level. At a "critical moment" of the highest flashover probability, the voltage is raised rapidly to the estimated

Table 7.10 Summary of the recommended ice deposit parameters

Ice deposit parameter	Recommended value
Type	Glaze ice with icicles
Thickness	5–30 mm on rotating cylinder
Freezing water flux	$60 \pm 20\,\mathrm{l/h/m^2}$
Water conductivity	100 μS/cm @ 20°C
Air temperature	−5 °C to −15 °C
Wind speed (if used)	3–5 m/s
Precipitation direction	45° ± 10°

flashover test level and then kept constant. The insulator will either flashover or withstand during a 15-minute stage at the test voltage. The critical moment of test voltage application is established by a combination of visual observations for the presence of a glossy layer of surface water on the ice, and measurements of leakage current peak and root mean square (rms) magnitudes.

Standard procedures for establishing the maximum withstand (V_{WS}) or critical flashover (V_{50}) level of an insulator call for an up-and-down method (IEC 60507 1991), where a 5% adjustment to the test level is made depending on whether the previous test result gave a withstand or flashover. Normally, at least five tests, each on a fresh ice layer, are needed to establish V_{WS}. Ten separate tests are needed to establish V_{50} on iced insulators because the layer is modified by melting and ice shedding during high discharge activity.

7.8.3 Cold-Fog Test Method

A specialized test method was developed to reproduce line voltage flashovers across insulator leakage distance, observed at the melting point in the absence of heavy ice accretion or full bridging. This method (Chisholm et al. 1996) consists of the following initial conditions:

- Pre-contamination using flow coating of kaolin/salt mixtures or dry-spray methods.
- Chilling of air temperature and insulator to –4 °C.
- Application of service-voltage stress.
- Circulation of fog with 10-μm volume mean diameter and 300-μS/cm conductivity.
- Wind speed of 3 m/s, with turbulence intensity of less than 10%, similar to natural conditions.

A chilled-mirror hygrometer should be used for accurate measurement of the ice-point temperature in the cold room. With constant service voltage and air flow, when the dew point is within 2 centigrade degrees (C°) of ambient temperature and a dense fog is visible, the test-chamber temperature is increased from −4 °C to −2 °C in no less than one hour. The temperature rise of the test chamber is then increased at the rate of 0.6 C°/h, from −2 °C to +1 °C, in 5 hours. The dew point should remain within 2 C° of ambient temperature during this sequence.

The flashover voltage in cold-fog tests usually declines considerably as the dew point increases towards 0 °C, and then turns up again when the insulator temperature rises above 0 °C, or when dry-banding and arcing processes occur. Since fog accretion occurs rapidly, it is possible to evaluate the flashover voltage repeatedly by increasing the test voltage in steps of 5%, with one-minute hold intervals, until an upper limit is reached or flashover occurs. The normal service voltage is re-applied and kept constant between these intervals to allow for ice to re-form and reach thermal equilibrium.

At the end of the sequence, declining flashover values are interpolated to establish V_{50} for 30 min at a dew point of 0 °C. The results from cold-fog testing are similar to well-controlled clean-fog testing at the same contamination levels (Chisholm 2007). Some of this control is related to the stabilizing action of the thin ice layer on the pollution deposit. In this sense, the artificial frost acts mainly as an insoluble (non-conducting) deposit that ensures the layer is fully wetted at the moment that it starts to melt.

7.8.4 Snow Test Method

Systematic studies of flashover on snow-covered insulators with a wide range of snow conductivities have been reported by Yasui et al. (1988) and recently by Hemmatjou et al. (2007). The electrical conductivity of snow was noted to reach a maximum at about −2 °C, as shown in Fig. 7.14. This suggests that the minimum electrical strength of a snow-covered insulator will occur slightly below the freezing point.

Two methods have been proposed for depositing snow on insulators in the laboratory:

1. Insulators are covered with soft rime produced from 15-μm diameter supercooled water droplets at −12 °C (Farzaneh and Laforte 1998). Then, the deposit is heated by raising the air temperature, in order to increase the liquid water content. Applied water conductivity is adjusted by adding sodium chloride to the de-ionized water feeding the nozzles. No voltage needs to be applied during soft-rime accretion. This approach, still under study, is the only one recommended for simulating snow on vertical insulators.
2. Insulators are covered with natural snow. If necessary, the snow conductivity is increased by adding salt (Yasui et al. 1988) and mixing with a snow blower (Hemmatjou et al. 2007). The snow is then dumped into a form (Wieck et al. 2007), and used to cover the insulator at sub-zero ambient temperature.

Once the test insulators are properly covered with snow, the test-chamber temperature is progressively raised from the snow deposit temperature to around 0 °C. The volume density, conductivity, height, and water content of snow are carefully measured.

Leakage current should be monitored closely as the applied voltage is increased at the end of each test. The current magnitude in the snow layer can vary exponentially with applied voltage. Sinusoidal currents in snow were found to be in phase with voltage, resistive rather than capacitive, as noted for ice samples. The phase relation can be monitored efficiently by observing the Lissajous figures of voltage versus current.

The IEC 60507 (1991) procedures to establish V_{WS} and V_{50}, adapted for ice layers, should also be followed to establish these values for snow tests.

7.9 Insulation Coordination for Ice and Snow Conditions

Insulation coordination is the process of selecting the appropriate type of station or line insulator, and its dry-arc and leakage distances, for the levels of icing, snow, pollution and other environmental stresses expected in the area of application. Insulation coordination is normally performed in the electric power network design stage. A large number of factors are considered in this process. Some factors related to other severe weather conditions, such as lightning and others, are related to network conditions that establish levels of switching overvoltage. At extra-high system voltage levels at or above 345 kV, issues of contamination generally place the most severe constraints on insulator leakage distance. Ice and snow are considered to be special forms of contamination, with specific influences on suitable insulator dry-arc distance.

Utilities sometimes calculate the outage rate of overhead lines or stations due to pollution level, based to a large extent on the following input parameters:

- Pollution flashover stress for different insulators derived from laboratory tests or field tests, as discussed in Sections 7.7 and 7.9.1.
- Number of flashover threats per year, as discussed in Section 7.9.2 for icing events.
- ESDD and NSDD variations based on single, average or statistically reliable measurements, covered in Sections 7.9.3 and 7.9.4.
- Variation in the natural snow or ice conductivity, in Section 7.9.5.
- Number of insulators exposed to the same flashover threat, in Section 7.9.6.

In many cases, application advice in guides such as (IEC 60815 1996) allows confirmation that a suitable insulator has been selected. However, these general guides do not consider winter conditions and specific IEEE work (Farzaneh 2005a, 2005b, 2007a, 2007b) treats these in more detail.

The calculation of outage rates from winter contamination and icing conditions follows a similar process to normal evaluations above freezing, with different parameters. The distributed resistance of the ice layer accumulated on an insulator is highly variable parameter that depends on many factors, including:

- Duration of periods without rain that end in ice or snow accretion events.
- Distance from pollution sources.
- Height above ground.
- Wind speed, direction and variability.

7.9.1 Icing Flashover Stress

Icing flashover stress, expressed in kV/m for different insulators in winter conditions, can be estimated from some characteristics of the ice or snow deposit, such as thickness, type, liquid water content, electrical conductivity, and air temperature as well. These characteristics are used to estimate the resistance per unit length along the deposit, rather than along the insulator surface.

The simplest models, such as an ISP first identified in (CIGRE, TF 33.04.09 1999), use only the resistance per unit length along the ice or snow deposit. The ISP model of test data tends to be accurate but is relatively specific for each type of deposit.

Models that are more general take into account insulator dimensions, configuration, surface material and other factors. Theoretical models for flashover, described by Farzaneh and Zhang (2007), can be generalized for several different types of insulators, such as posts or strings of discs, and for several different forms of ice deposit, such as ice, rime or snow.

7.9.2 Extent of Storm Area and Severity of Ice Deposit

The normal freezing rain activity in an area depends on the local climate. Near the sea coast, freezing rain and freezing fog exposure is very much a function of the distance from shore. In some areas, local topography has a similar effect. In these areas, freezing rain accumulation occurs at a slow rate for a long duration. Episodes lead to 200-mm accumulation of rime icing in the St. Lawrence Valley in Canada.

In other areas, such as Ontario, Canada, and Virginia and Tennessee, USA, freezing rain is normally part of a frontal storm. The most severe ice accretion occurs in storms that have transitions from sleet to freezing rain to snow. The "width" of a frontal storm in these regions ranges from tens to hundreds of kilometres. The thickness of the front where freezing rain accumulates is usually less than 10 km. The accumulation of ice will thus be strongly affected by the speed at which the front passes, which will range from 5 to 50 km/h.

Chapters 1 and 2 provide many more details on the regional variation in the ice and snow accumulation processes.

To select appropriate insulators, the local design level of ice accretion should be expressed as a return period of icing events. Then, if warranted by reliability evaluations, the insulation coordination process of station and line insulators will start to diverge from the procedures established for insulators exposed to high pollution levels (IEC 60815 1996). For moderate to heavy accretions, a significant proportion of the leakage distance is no longer effective because it is bypassed by icicles, ice or snow layers. This means that additional dry-arc distance, rather than leakage distance, is needed. Dry-arc distance increases the Basic Impulse Level (BIL) of a power system. This improves reliability for other weather-related problems (such as lightning), but also increases the size and cost of towers and stations.

The level of ice accretion that causes partial or complete bypassing of leakage distance depends on the type of insulator. As an example, for a string of 146-mm by 254-mm ceramic disc insulators, Table 7.11 describes the state of ice bridging

Table 7.11 Relation between ice accretion of reference cylinder and accretion on the standard ceramic disk insulator string and uniform-shed station post

Ice accretion level (mm)	Ice accretion on 300-mm uniform shed station post (g/cm dry arc)	Ice accretion on 146 × 254-mm insulator string
5	Ice caps (20)	Ice caps (16)
10	32 Fully Bridged (40)	Icicles start to form (32)
15	48 Fully bridged (60)	Partial bridging (48)
20	64 Fully bridged (80)	80% bridging (64)
25	80 Maximum bridging (100)	Maximum bridging (80)

as a function of ice accretion measured on a rotating 25-mm reference cylinder (Farzaneh and Kiernicki 1997) and a uniform-profile station post (Farzaneh et al. 2005a).

Each type of insulator string or post will have a different accumulation weight and ice bridging state at the same ice accretion level, as detailed in the next section.

7.9.3 Soluble Components of Insulator Pre-contamination

Proximity to expressways is one of the factors that lead to high levels of ESDD in the winter. The measured ESDD on top and bottom surfaces of insulators are usually different.

- The top-surface levels rise quickly with time at the deposit rate $D(x)$ and small amounts of winter rain (> 0.4 mm) may wash most of the top-surface pollution away.
- The bottom-surface levels rise slowly with time and moderate rain does not clean the pollution off. Instead, 10 mm of rain may achieve a 50% reduction in ESDD of the bottom surface.

The electrical performance of the insulator depends more on the top-surface ESDD in heavy icing with full bridging of the dry-arc distance. The overall value of ESDD, including top and bottom insulator surfaces, affects the flashover level in moderate and light icing conditions with partial ice bridging or in cold-fog conditions.

Recent work in Sweden (Lundmark and Olofsson 2007) evaluated the spread of road salt from expressways into sensitive aquifers and other areas. Measurements were taken upwind from a six-lane expressway, with 90,000 vehicles per day and a 2003–2004 season salt deposit rate of 8 tonnes (8 Mg) per km. Figure 7.22 from that study shows the total chloride deposit rate on the ground as a function of distance from the edge of their southbound lanes, located downwind from the measurement sites.

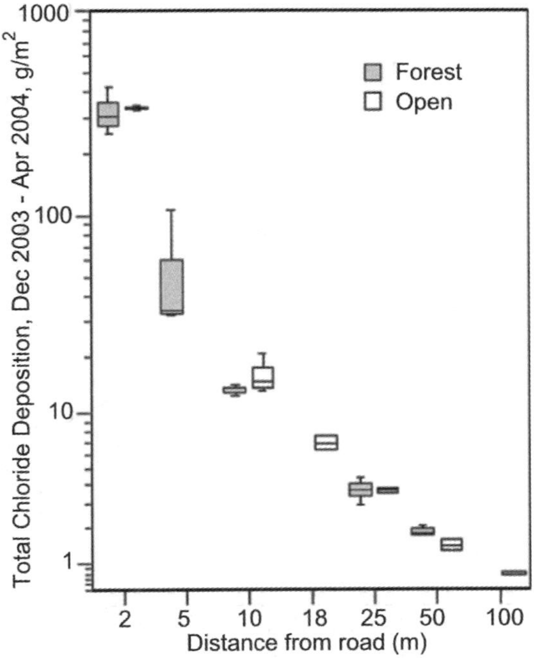

Fig. 7.22 Measured winter season (December-April) chloride deposit rate versus distance from edge of 6-lane 110 km/h expressway in Sweden (Reproduced by permission of Springer)

The total daily deposition $D(x)$ of chloride at a distance from the edge of the road x was fitted with a pair of exponential functions, one for splash and another for the spray transport mechanism, along with a constant background level. The fitted function was:

$$D(x) = a_{Splash} \cdot e^{-x/2m} + a_{Spray} \cdot e^{-x/20m} + a_{Background} \qquad (7.22)$$

where :

$D(x)$ is the chloride deposit density in g/m^2 per day
$a_{Splash} = 8\,\text{g/m}^2\text{/day}$
$a_{Spray} = 0.12\,\text{g/m}^2\text{/day}$
$a_{Background} = 0.003\,\text{g/m}^2\text{/day}$

The numerical values for Eq. 7.22 can be converted to ESDD escalation rate in μg/cm^2/day by changing the units to $1\,\text{g/m}^2 \cdot 10^6\,\mu\text{g/g} \cdot 1\,\text{m}^2/10^4\,\text{cm}^2$ and by multiplying the chloride weight by a factor (23 + 35.5)/35.5 to correct for the stoichiometric ratio of NaCl to Cl^-. This gives a value of $a_{Splash} = 1320$, $a_{Spray} = 20$ and $a_{Background} = 0.5\,\mu\text{g/cm}^2$ per day upwind from a six-lane expressway. The density values can be scaled by the number of traffic lanes. The background level may be influenced by long-range transport of pollutants (Fikke et al. 1993).

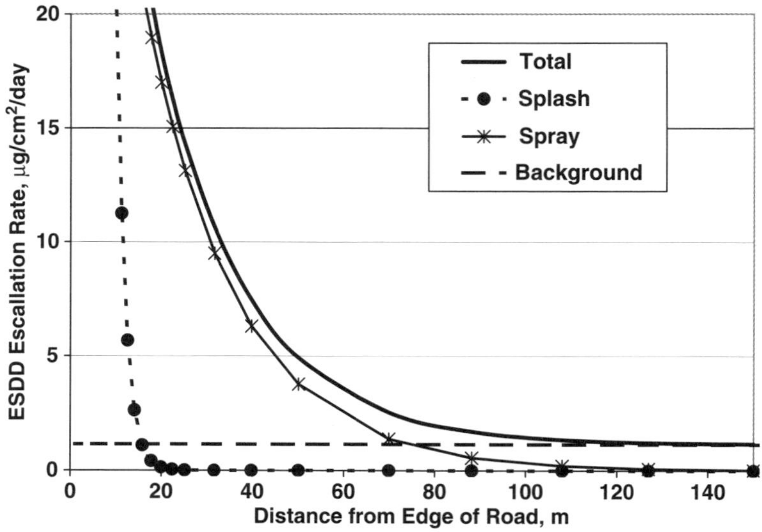

Fig. 7.23 Components of road salt accumulation

Figure 7.23 shows that splash can be ignored for stations that are more than 20 m from the edge of an expressway. A distance constant of 33 m for measurements of dustfall next to an Ontario expressway is similar to the 20-m value found for the "spray" term of salt deposition in Sweden.

Figure 7.24 shows the levels measured at four locations along a 40-km transmission corridor in Canada during a winter season. The locations of these stations relative to local expressways are shown in Table 7.12.

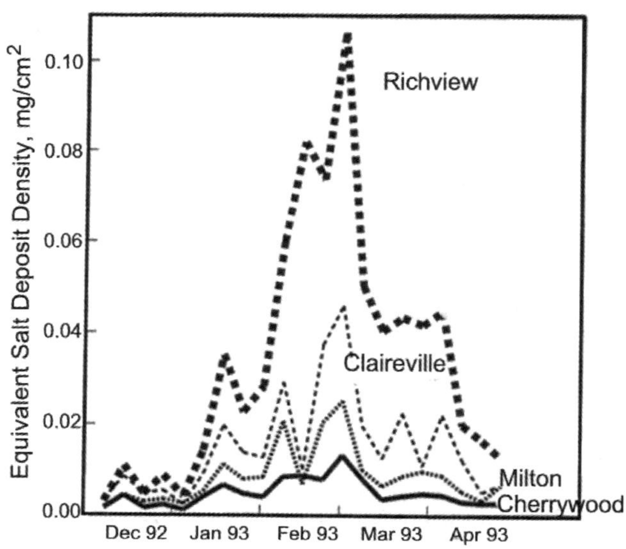

Fig. 7.24 Observed ESDD at four locations within 40 km (Reproduced by permission of IEEE)

Table 7.12 Calculated and observed (Fig. 7.24) rates of ESDD increase

Site	Distance from edge of hwy	# of lanes	ESDD from splash	ESDD from spray	ESDD from back-ground	Predicted Δ ESDD	Observed increase, μg/cm ^{2}/ day
					μg/cm^2/day		
Cherrywood	> 1000 m	4	0.00	0.00	0.33	0.33	0.5
Milton	600	6	0.00	0.00	0.49	0.49	1
Claireville	206	4	0.00	0.00	0.33	0.33	1–1.5
Richview 4	127	14	0.00	0.08	1.15	1.23	1.4
Richview 2	108	14	0.00	0.21	1.15	1.36	2.6
Richview T	37	4	0.00	2.07	0.33	2.40	6.6

The levels of salt accumulation on insulators were observed to be higher than those computed from the Lundmark and Olofsson model (2007). This is interesting because the assumption that all salt deposit attaches to the insulator can be challenged, and this would give a bias in the other direction. Modelling of ESDD accumulation rate as a function of wind and salt spray mass flux remains open for further investigation.

7.9.4 Insoluble Components of Insulator Pre-contamination

Figure 7.25 shows some typical observations of non-soluble deposit density (NSDD), surface conductivity and flashover voltage of naturally polluted insulators from 40 test stations located along a proposed 1,000-km route in Russia, adapted from Farzaneh et al. (2007).

In this case, the relatively constant level of surface conductivity in Fig. 7.25 correlates with the ESDD. Along the line between 100 and 200 km, the measured NSDD peaked at about 0.14 mg/cm^2 (140 μg/cm^2). This non-soluble deposit was found to depress the flashover voltage per insulator disc from 30 to 24 kV.

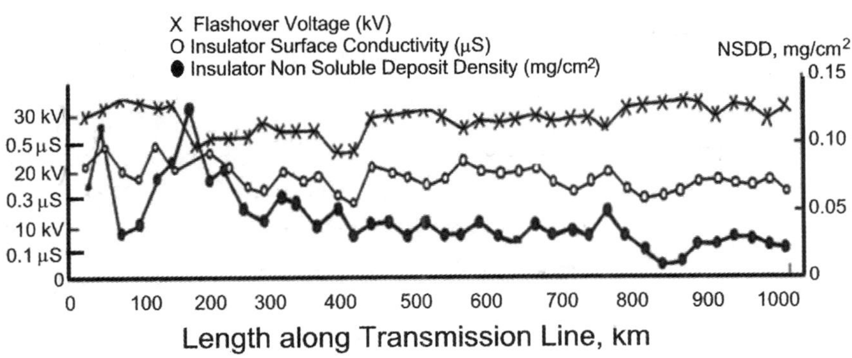

Fig. 7.25 Observed pollution parameters along 1000-km transmission line (Reproduced by permission of IEEE)

It has been suggested that accumulation of natural fog or thin layers of ice can also be treated as NSDD, especially if the insulator surfaces are highly contaminated and the precipitation is relatively clean (Chisholm 2007). In these cases, a thin 50-μm ice layer with a density of about 0.8 g/cm^3 would have a NSDD of 4 mg/cm^2. This is significantly higher than the median levels observed in Fig. 7.25. A high level of non-soluble deposit would exert a strong stabilizing influence on the pollution layer, leading to lower electrical strength.

7.9.5 Natural Ice, Snow or Fog Conductivity

A simplified model can be used to establish the appropriate value of melted-water conductivity for insulation coordination in winter conditions. In the absence of insulator pre-contamination, the probability of exceeding precipitation conductivity, σ, can be approximated by (Farzaneh 2007a):

$$P(\sigma) = \frac{1}{1 + \left(\frac{\sigma}{\sigma_a}\right)^{2.3}} \tag{7.23}$$

where σ_a, the median of precipitation conductivity, is typically 25–30 μS/cm and the log standard deviation of the distribution is 0.8. The median and the exponent values of 2.3 can both be adjusted to match the observed distribution of local measurements. The IEEE Task Force work recommended selecting a precipitation conductivity σ corresponding to $P(\sigma) = 5\%$ (Farzaneh 2007a), obtained from Eq. 7.23.

Statistical distributions of precipitation conductivity can be obtained from a number of sources, such as the National Atmospheric Deposition Program in the USA (NADP 2007). Melted ice and snow in the winter tend to have nearly the same conductivity as precipitation in the summer for most locations. The distribution of snow conductivity for the Ishiuchi test site in Japan (Yasui et al. 1988) is reasonably represented by Eq. 7.23, when the median conductivity, σ_a, is updated to 14 μS/cm.

7.9.6 Number of Affected Insulators in Series and Parallel

The extent of freezing rain storms normally causes accretion on hundreds of insulators all at once. Many parallel insulators share a common line-voltage stress, even if they are supporting conductors at different voltage levels. Failure of any one of them will cause a power system fault affecting thousands of customers. The relevant calculation for the flashover probability P_m of m parallel insulators exposed to the same conditions is:

$$P_m = 1 - (1 - P_1)^m \tag{7.24}$$

where P_1 is the flashover probability of a single insulator.

The problem of having many similar insulators in parallel, exposed to the same conditions, is often compounded by the use of multiple insulators in series. As the most basic example, extra-high-voltage station insulators and arresters are often manufactured as two to four separate sections. High-voltage capacitor banks constructed from a number of lower voltage capacitors in series have also proved vulnerable to icing flashovers in Canada and the USA.

In these cases, a partial flashover of any one section will increase voltage stress on the others. For example, there will be an 11% increase in stress when any one of ten insulators in series has an icing flashover. With the low standard deviation of flashover voltage of less than 5% in icing conditions, the additional stress is very likely to cause additional flashovers. The multi-arc model in Section 7.6.2 (Farzaneh and Zhang 2007) provides a quantitative model to evaluate this effect.

As an additional problem, when one section of a multiple-section surge arrester fails externally from icing, the remaining sections will not have sufficient overvoltage rating to withstand the normal line voltage and they will fail internally.

7.9.7 Effect of Consecutive Flashovers on Reliability Requirements

Normally, power systems are designed to withstand many fault conditions, including single-phase to ground faults anywhere on the network. Protective relaying will operate, causing circuit breakers to open their terminals for periods of 100 to 500 ms, which interrupts the current through nearly all types of arcs, allowing the plasma to cool sufficiently so that re-application of voltage will not cause another fault. In such a context, the air insulation is said to be self-restoring. Air differs from oil or cable insulation in this respect, as any fault in a transformer or cable is permanent. This is an important reason why overhead lines continue to be used in areas of high icing incidence. In fact, the consequences of each icing fault on self-restoring air insulation are far less important than the permanent cable faults from other causes.

Power system faults are usually caused by lightning. Only one in ten to twenty lightning flashes has enough peak current to cause a power system fault, so it is rare that a breaker would have to operate more than two or three times during a storm. In contrast, under icing conditions, every insulator is stressed, and many of them may fail over a short period of time. One utility in Canada reported 57 flashovers of their critical 500-kV system in a two-hour period, whereas this system would not have more than five flashovers in a full season of lightning. With so many breaker operations at once, the stored energy (such as compressed air) that operates circuit breakers can run out, leading to loss of protection or inability to restore power after a fault.

Many backbone EHV power systems are robust enough to withstand the loss of any one station. However, with separation between stations of 20 to 50 km, there are icing storms that can cause problems at two or more stations simultaneously. These are identified and monitored carefully because the consequences of losing two or more EHV stations would generally be a widespread blackout.

7.9.8 Design Examples for Insulation Coordination in Icing Conditions

Engineers in the 1900s and 1910s obtained reliable electrical service with the insulators of that time by limiting the voltage gradient across the dry-arc distance to about 70 kV per meter, corresponding to 10 kV (rms line to ground) per insulator disc on a transmission line. With this ac stress, lightning flashovers became the most frequent cause of failure. Utilities adapted the leakage distance of the insulators to different contamination environments, but, until the development of EHV transmission (at voltages of 345 kV and above), the stress of 10 kV per standard (146-mm spacing by 254-mm diameter) disc was widely used.

Once twenty or more discs are placed in series, the voltage distribution along the insulator string becomes highly nonlinear (Gorur et al. 1999). This means that the ability to resist system-generated switching overvoltage no longer scales linearly with dry-arc distance. When transmission voltages above 345 kV were adopted, other measures were introduced to calculate and control these switching overvoltages. The most common method was a pre-insertion resistor fitted into circuit breakers, although the development of metal oxide surge arresters was also effective.

With these alternate measures in place, the utilities no longer needed to respect the 10 kV per disc guideline. Typical 500-kV lines use 13–14 kV per disc (90 kV/m of dry arc) and some 735-kV lines use a stress as high as 15 kV per disc.

There has also been a wide range of stresses on station post insulators. One utility in North America with good experience at 73 kV/m stress for 230-kV systems built a number of 500-kV stations with 91 kV/m stress in the 1960s, and obtained reliable performance. With improvements in switching surge control, this utility reduced the dry-arc distance in its new stations in the 1970s, giving a stress of 106 kV/m. They had no problems with switching surge flashovers, but instead started to experience winter flashover problems. These were addressed by a retreat to the reduced stress level of 80 kV/m of dry-arc distance in areas of moderate contamination near expressways.

7.10 Mitigation Options to Improve Network Reliability in Winter Flashover Conditions

In addition to the selection of an appropriate dry-arc distance, an appropriate selection of insulators with good performance under icing conditions should consider:

- Preventing ice bridging across sheds with larger shed-to-shed separation using alternating shed or disk diameters.
- Reducing ice accumulation by decreasing insulator diameter.
- Facilitating natural ice shedding using insulators with greater slope, smoother surface finish or semiconducting glaze.
- Implementing more uniform electric field distribution using well-designed metallic end fittings and grading rings, or semiconducting glaze insulators.

It is possible to improve the reliability of individual components such as insulators in a number of ways (Farzaneh and Chisholm 2006). They can be selected with dry-arc and leakage distances suitable for a severe combination of icing and contamination. Surface coatings such as RTV silicone can decrease the surface conductivity. Semiconducting glazes offer improved performance by increasing the surface conductivity enough to mitigate the uneven electric field distribution that occurs under icing conditions. Accessories such as creepage extenders or booster sheds can break up the icing pattern and raise the threshold of ice accretion necessary for insulator bridging. Maintenance options such as power washing are effective when the build-up of insulator contamination is continuous over the course of weeks or months.

There are also methods to ensure that a power system remains reliable under icing flashover conditions. These relate to choices of power system configuration and load dispatch.

Normally, the backbone grid systems are designed with multiple lines between stations. If any one line undergoes a flashover for any reason, then the other lines can pick up their share of load and maintain continuous service. As already mentioned, lightning is the most common cause of line flashover. Transformers and switching equipment at line junctions are thus well protected against both direct and incoming lightning surges, with surge arresters and a large, low-resistance ground grid.

Icing flashovers are more common within transformer stations than on lines. This weakness relates partly to the lower shed-to-shed separation on the station insulators. This distance is 30–50 mm for station posts and more than 120 mm from the bottom of one disc to the top of the disc beneath it for strings of ceramic discs on lines. As shown in Table 7.11, it takes about 10 mm of accretion on a reference cylinder to fully bridge a vertical station post insulator, while 25 mm or more are needed to fully bridge a string of standard ceramic transmission line disc insulators. This means that the risk of having a number of lines unavailable at once is concentrated at stations under freezing rain conditions.

Some utilities exploit the limited extent of freezing rain activity. Since main transformer stations are usually more than 100 km apart, it can be reasonable to plan for storms that will cut the supply to any single station. Network configurations that allow for supply of load from at least two different transformer stations have proved to be robust under a variety of freezing rain conditions.

7.10.1 Improved Insulator Dimensions and Configurations

7.10.1.1 Increase of Dry-Arc Distance

Test results in the range of 0.3 to 3 m show that icing flashover strength increases linearly with dry-arc distance (Farzaneh and Zhang 2007). This means that increased insulation length remains the most reliable way to improve performance. In cases where the dry-arc distance has proved to be inadequate for local icing and contamination conditions, insulators with increased dry-arc distance can be substituted. As mentioned in Section 7.9.8, one utility has changed from 2.9-m dry-arc distance to 3.3 or 3.8 m, depending on local conditions, for its 500 kV network.

7.10.1.2 Change of Insulator Diameter

With their small diameter, compared to porcelain insulators, polymeric line insulators may perform better in some icing conditions. For moderate accretion, however, the icing performance of polymeric insulators with closely spaced sheds will generally be worse than that of cap-and-pin insulators, due to ice bridging. Polymer insulators designed with alternating diameter sheds may also perform somewhat better than uniform-shed designs in spite of the increased diameter, because full bridging occurs at higher accumulation levels. At present, polymer designs specifically for improved icing performance are promising. Manufacturers can provide a wide range of shed diameters and spacing, compared to porcelain posts or discs, and they are studying the optimal choices of these dimensions.

In one example supported by laboratory tests with heavy icing, cap-and-pin insulator strings outperformed two different types of non-ceramic insulator (Gutman et al. 2002). However, tests with moderate levels of ice accretion, still leading to full ice bridging on the non-ceramic insulators (Farzaneh et al. 2005), led to the opposite conclusion. Modelling also suggests that the reduced diameter of the NCI is an advantage (Farzaneh et al. 2005). For design purposes, test results from polymer insulators should be benchmarked against the performance of IEEE-standard disc strings, as shown in Table 7.11 above.

7.10.1.3 Combination of Insulators with Different Diameters

Strings of alternate-diameter insulator discs, such as those shown in Fig. 7.26, take advantage of profiles designed to increase the effective shed-to-shed spacing (Cherney 1986). This is a practical retrofit option for existing lines, as well as a good choice for new designs.

When upgrading or replacing damaged insulators in a station, it may be important to select bolt-in replacements that can be installed without re-configuring the bus work. Station post insulators with the same dry-arc distance but different, alternate-diameter profiles and leakage distances are shown in Fig. 7.27.

Under a certain range of icing conditions, the alternate-diameter profiles will perform better than standard profiles. However, there is a limit. More ice will form on insulators of larger diameter. This will degrade rather than enhance performance under very heavy icing conditions. This effect of insulator diameter is also considered when selecting replacement insulators with greater cantilever strength to resist mechanical effects of ice accretion.

7.10.1.4 Insulator Discs and Profiles with Extended Leakage Distance

Fog-type insulators with a high ratio of leakage distance to dry-arc distance give good performance in marine fog conditions. These are also applied to improve

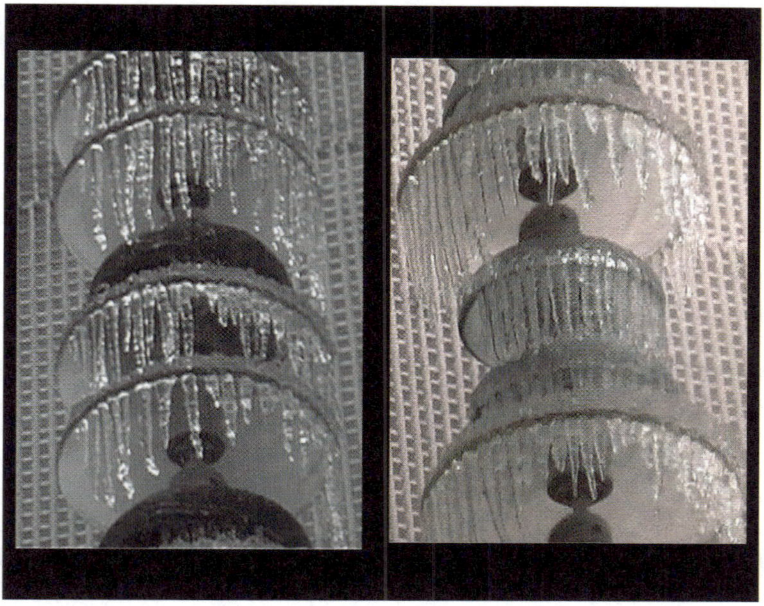

Fig. 7.26 Ice accretion on bell-bell-disc and bell-disc-disc suspension insulator strings (Reproduced by permission of IEEE)

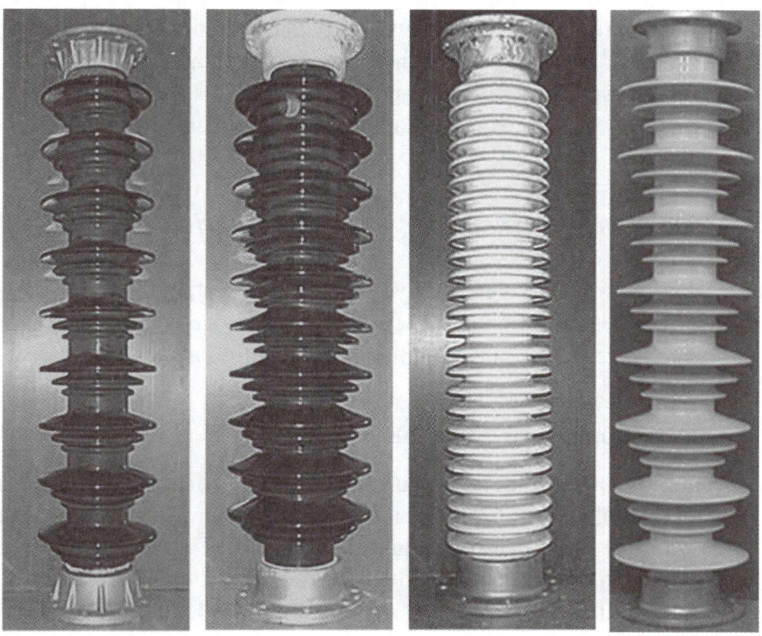

Fig. 7.27 Alternative station post profiles with same 1,540-mm dry-arc distance

performance in any area where contamination outages occur frequently. An extended leakage distance does not have much effect on the ice-bridging performance, but it will improve the cold-fog performance. Retrofit of insulators with increased leakage distance is only useful in areas where ice and snow accretion is not severe.

7.10.1.5 V-Strings

Suspension strings may sometimes be converted to V-strings when upgrading existing transmission lines to restrain conductor motion. Under icing conditions, two references (Cherney 1986; EPRI 1982) rank the electrical performance of pre-contaminated V-strings best among a series of competitive options. The relatively high amount of ice accretion needed to achieve full bridging was cited as an advantage in both cases.

7.10.1.6 Horizontal Strings

One problem with V-strings is that, while one of the two strings will be inclined into the wind and will perform better, the other will be inclined in the other direction and may require less ice or snow to bridge the dry-arc distance. It has been noted that the performance of horizontal strings of insulators shows marked differences from vertical or V-strings. Under icing conditions, a series of air gaps spaced at the shed-to-shed distance (often 146 mm) has high electrical strength. On the other hand, Yasui et al. (1988) noted that the heaviest snow layers build up in the space between pairs of horizontal strings of insulators, were separated by about 0.5 m. It is common to use two to four separate insulator strings in this horizontal configuration at strain (dead-end) towers, especially for EHV and UHV lines. This is an application where the reduced diameter of non-ceramic insulators may yield an advantage.

7.10.2 Alternatives for Insulator Surface Materials

Two mitigation options that are effective in clean-fog testing also give some improvement in cold-fog performance. These options, silicone coatings and semiconducting ceramic glaze, function in completely different ways.

7.10.2.1 Silicone Coatings or Materials

The application of room-temperature vulcanized (RTV) silicone coatings or materials to ceramic insulators gives an increase in surface impedance. This increase is remarkable (about six orders of magnitude) above 0 °C, and much less (about one order of magnitude) below freezing. However, a tenfold increase in surface resistance should still be sufficient to raise the voltage needed for arc re-ignition by a factor of three. In practice, gains of about 25% in flashover strength per meter of leakage distance are gained in cold-fog conditions when silicone coating is applied to porcelain insulators (Hora et al. 1983; Chisholm 2005). The improvement stems from the high surface impedance offered by the coating even after contamination with salt.

However, the RTV advantage reduces or reverses as soon as insulators are fully bridged in heavy ice conditions. Ice bridging on silicone rubber coated insulators occurs just as rapidly as on uncoated ones. The only difference is in the initial formation of the ice. Due to the higher hydrophobicity of the coated insulator surface, water droplets form a connected network of ice globules whereas, on uncoated insulators, water filming produces a layer of ice. Time to full bridging is nearly the same, and the ice is retained for a longer period on RTV-coated porcelain during the dangerous melting phase.

7.10.2.2 Semiconducting Glaze

Several manufacturers have offered ceramic insulators glazed with metallic elements that provide a semiconducting surface. The use of semiconductive glaze insulators leads to a decrease in surface impedance. A normal conduction current of 1 mA at normal service voltage is an accepted compromise between good electrical performance and low power losses. The power dissipation is of the order of 70 to 400 W, depending on the line-to-ground voltage. With this heat input, the insulator warms up slightly in low-wind conditions. However, 400 W power dissipation distributed on a 4-m insulator does not melt much ice. With a latent heat of fusion of 334 Joules per gram, there would be 1.2 grams of ice melted per second on a 4-m insulator, and it would take nearly five hours to melt a typical ice accumulation of 20 kg.

The semiconducting glaze layer on contaminated insulators also limits the voltage across dry bands by providing a parallel, lower-impedance path for current flow in the semiconductive layer. This prevents the initial formation of arcs. It has proved very difficult to flashover semiconductive glaze post insulators in cold-fog tests and conditions. Instead, the glaze can stabilize a heavy pollution layer so well that the insulator overheats, possibly leading to broken sheds. This thermal consideration, rather than cold-fog flashover performance, limits the use of semiconductive glaze insulators in heavy-pollution areas where the winter-maximum ESDD may exceed $0.7\ \mathrm{mg/cm^2}$.

The improved performance of semiconducting glaze insulators in icing conditions is actually a function of the improved voltage gradient along the insulator surface rather than ice melting. When any portion of the surface is bridged by ice, the glaze conducts enough current to limit voltage gradients, preventing the initiation of arcing. The parallel resistance of the glaze and ice layer makes the electric field more uniform. In both laboratory tests and field exposure, as seen in Fig. 7.28, semiconductive glaze insulators develop important ice-free zones as large pieces of ice fall away.

Semiconducting glaze post insulators have proven to be very effective in mitigating icing problems and their application is particularly effective at 500 kV

Fig. 7.28 Operation of 500-kV semiconducting glaze insulators under icing conditions in West Virginia (Reproduced by permission of Dominion Virginia Power)

and 735 kV. The ice-free zones that formed on 500-kV insulators in Fig. 7.28 are more than 1 m long, sufficient to withstand ac system voltages with a 30% margin.

This effect would also occur on semiconductive glaze disc strings for line insulation, but the benefits have not yet been demonstrated in laboratory testing.

7.10.3 Reduced System Voltage

Many power systems can be operated at reduced voltage, and utilities sometimes carry out tests to ensure that the system remains stable in brownout conditions. Typical voltage reductions of 5% are available. Initiating a deliberate reduction in transmission system operating voltage during critical ice or snow conditions can be an effective means of increasing insulator reliability (Hara and Phan 1979; Cherney 1980).

Considering a case study with 50 parallel insulators under full ice bridging conditions, and the model of Eq. 7.7, a 5% reduction of the operation voltage results in a decrease of the flashover probability from $P_{n=50} = 49.5\%$ (maximum operation voltage) to $P_{n=50} = 11.7\%$. A voltage reduction of 10%, if available, would decrease the flashover probability to $P_{n=50} = 1.2\%$.

Utility operators have traditionally resisted the use of system-wide voltage reductions under adverse conditions that may lead to a number of faults, because this measure often reduces the transient stability margin of the power system. However, with newer high-power electronic controls and for specific components such as dc links, it may be feasible to implement this measure more effectively now or in the near future.

7.10.4 Retrofit of Insulator Accessories

7.10.4.1 Grading Rings

The selection of grading rings has a strong influence on ice accretion and ice flashover performance of polymer insulators and EHV posts, where voltage distribution without rings is highly nonlinear. The grading of electric field to produce uniform icing does not necessarily lead to improved flashover performance. In some cases, grading rings can be used instead to control the ice accretion to create an air gap, such as that shown in Fig. 7.28, which is large enough to withstand normal service voltage. Tests on full-scale insulators properly equipped with grading rings at normal service voltage are needed to quantify the improvement in flashover voltage from these effects.

7.10.4.2 Booster Sheds and Creepage Extenders

These accessories for improving the wet or contamination flashover performance of bushings and post insulators have also proved to provide shelter during ice accretion. Properly applied, each booster shed provides an air gap in the ice and also adds to the dry-arc distance. Section 7.6.5 shows that each air gap tends to increase flashover strength by 10–20 kV. One problem found in testing is that some flexible polymer booster sheds fold up or collapse under heavy ice accretion. Since they gather considerably more ice than the insulator (from the large diameter), this restricts the useful range of mitigation.

7.10.5 Preventive Maintenance

7.10.5.1 Insulator Inspection and Replacement

Electrical flashovers on insulators covered with atmospheric ice tend to cause more damage than other kinds of flashovers. This is a combination of several factors. The expensive high-strength post insulators and bushings located at substations tend to be more vulnerable to icing flashovers than line insulators. The fault current at a substation is much higher than the fault current on a line. The transient mechanical forces on the station insulators during this fault are higher. Also, it is common that the fault-current arc becomes trapped near the ice or snow layer, keeping this intense heat source very close to the insulator surface where, in the usual case, it moves off into the air nearby.

Utilities which experience an atmospheric icing event or a season leading to electrical flashovers should implement a rigorous program to inspect and replace all defective substation and line insulators.

7.10.5.2 Insulator Washing

Washing ice-covered insulators under voltage using hot pressurized water was tested with good results in laboratory conditions (Sklenicka and Vokalek 1996). However, using this method in actual conditions can be very complex, because wetting unwashed ice-covered insulators can initiate a flashover. Several utilities in North America and Japan have reported the long-term success of winter insulator washing programs, including the use of de-ionized water as an electrical barrier.

7.10.5.3 Heat Lamps

High-powered infrared heat lamps aimed upward have been used to melt ice on insulators and other station components. This approach has high operating costs for energy, and also for heat-lamp replacement caused by falling pieces of ice, but it is especially effective for substations near waterfalls or other continuous sources of spray.

7.10.5.4 Stripping Ice with Hot Water or Steam

Hydro-Québec explored the use of steam to strip ice from insulators and switchgear (Lanoie et al. 2005). Steam offers the same electrical advantages as de-ionized water in terms of low electrical conductivity, but has much higher energy costs.

7.11 Conclusions and Recommendations

Generally, the ice storms that lead to electrical problems on power networks differ from those that lead to mechanical problems. Normally, porcelain or polymer insulators have a surface leakage distance that is two to four times longer than the connection length. The ice or snow layer *becomes the insulator* during accumulation, sometimes by bridging the dry-arc distance, but always withstanding system voltage at temperatures well below freezing. Electrical flashovers occur only when the ice or snow layer warms up near the melting point, at which point the layer conductivity increases dramatically. Modest levels of atmospheric ice accumulation can lead to flashovers at normal system operating voltage under adverse combinations of surface pre-contamination, accumulation and melting conditions.

Atmospheric ice accumulation tends to stabilize or dissolve any pre-existing pollution on insulator surfaces. The ice holds the pollution until temperatures melt and ionic impurities are rejected to the surface. Flashovers occur because the wet ice surface has a high electrical conductivity. The flashover process can be modelled mathematically using many of the same approaches that have been used for contamination flashovers.

Many of the mitigation options used by electrical utilities in areas where they have severe pollution flashover problems are also somewhat or fully effective in mitigating atmospheric icing problems. Some utilities have simply accepted the risk of multiple flashovers in their largest EHV stations by setting up system operating procedures to bypass any one station under adverse winter weather conditions. After a problem event leading to atmospheric icing flashovers, the electrical power system reliability needs to be restored to its initial state by inspecting and replacing insulators that have been damaged. To reduce the risk of flashover in future events of equal severity, insulators with similar connection lengths but different profiles, or with semiconductive glaze, can be substituted; insulators can be washed more frequently in the winter; surface coatings (RTV silicone) or accessories can be applied; or the line or station can be re-built with insulators of longer dry-arc distance.

References

Cherney EA (1980) Flashover Performance of Artificially Contaminated and Iced Long-Rod Transmission Line Insulators. IEEE Trans. Power Apparatus & Systems, vol PAS-99: 46–52

Cherney EA (1986) Ice Bridging Flashover of Contaminated 500 kV Line Insulation in Freezing Rain. Ontario Hydro Report 86-80-H

Chisholm WA (2005) Ten Years of Application Experience with RTV Silicone Coatings in Canada. In: Proc of 2005 INMR World Conference & Exhibition on Insulators, Arresters & Bushings, Hong Kong, no 17: 1–20

Chisholm WA, Ringler KG, Erven CC, Green MA, Nigol O, Tam Y, Kuffel J, Boyer A, Pavasars IK, Macedo FX, Sabiston JK, Caputo RB (1996) The Cold Fog Test. IEEE Trans. Power Delivery, vol 11(4): 1874–1880

Chisholm WA (2007) Insulator Leakage Distance Dimensioning in Areas of Winter Contamination using Cold-Fog Test Results. IEEE Trans. Dielectrics and Electrical Insulation vol 14(6):1455–1461

Chrzan KL, Vokalek J, Sklenicka V, Petrusch W, Kindersberger J (2003) Pollution flashover of long rod insulators with different profiles. In: Proc 13th International Symposium on High Voltage Engineering (ISH), Delft: 227–231

CIGRE TF 33 04 09 (1999) Influence of Ice and Snow on the Flashover Performance of Outdoor Insulators – Part I: Effects of Ice. Electra no 187: 91–111

CIGRE TF 33 04 09 (2000) Influence of Ice and Snow on the Flashover Performance of Outdoor Insulators – Part II: Effects of Snow. Electra no 188: 55–69

CRC (2007) Handbook of Chemistry and Physics, 87th Edition. CRC Press, Boca Raton FL

EPRI (1982) Transmission Line Reference Book 345 kV and Above. EPRI, Palo Alto

Farzaneh M, Kiernicki J (1997) Flashover Performance of IEEE Standard Insulators under Ice Conditions. IEEE Trans. on Power Delivery, vol 12(4): 1602–1613

Farzaneh M, Laforte JL (1998) A laboratory simulation of wet icing build-up on HV insulators. Int. J. Offshore Polar Eng, vol 8(3): 167–172

Farzaneh M, Zhang J (2000) Modelling of DC Arc Discharge on Ice Surfaces. In: IEE Proceedings on Generation Transmission and Distribution, vol 147(2): 81–86

Farzaneh M, Chisholm WA (2006) Guide to Define Design Criteria for Outdoor Station Insulators Taking into Account Pollution and Icing. Canadian Electricity Association Technologies Inc. Report T043700 3326, Montreal

Farzaneh M, Zhang J (2007) A Multi-Arc Model for Predicting AC Critical Flashover Voltage of Ice-Covered Insulators. IEEE Trans. on Dielectrics and Electrical Insulation, vol 14 (6):1401–1409

Farzaneh M, Zhang J, Chen X (1997) Modeling of the AC Discharge on Ice Surfaces. IEEE Trans. on Power Delivery, vol 12(1): 325–338

Farzaneh M, Zhang J, Chen X (1998) DC characteristics of Local Arc on Ice Surface. Atmospheric Research, vol 46: 49–56

Farzaneh M, Zhang J, Volat C (2005) Effect of Insulator Diameter on AC Flashover Voltage of an Ice-Covered Insulator String. IEEE Trans. Dielectrics and Electrical Insulation, vol 13(2): 264–271

Farzaneh M, et al. (2003) Insulator Icing Test Methods and Procedures. IEEE Trans. Power Delivery, vol 18(4): 1503–1515

Farzaneh M, Li Y, Zhang J, Shu L, Xiang X, Sima W, Sun C (2004) Electrical Performance of Ice-Covered Insulators at High Altitudes. IEEE Trans. Dielectrics and Electrical Insulation, vol 11: 870–880

Farzaneh M, et al. (2005a) Selection of Station Insulators with Respect to Ice or Snow – Part I: Technical Context and Environmental Exposure: A Position paper prepared by IEEE Task Force on Icing Performance of Station Insulators. IEEE Trans. Power Delivery, vol 20(1): 264–270

Farzaneh M, et al. (2005b) Selection of Station Insulators with Respect to Ice or Snow – Part II: Methods of Selection and Options for Mitigation: A Position paper prepared by IEEE Task Force on Icing Performance of Station Insulators. IEEE Trans. Power Delivery, vol 20(1): 271–277

Farzaneh M, et al. (2007a) Selection of Line Insulators with Respect to Ice and Snow Part I: Context and Stresses. IEEE Trans. Power Delivery, vol 22(4): 2289–2296

Farzaneh M, et al. (2007b) Selection of Line Insulators with Respect to Ice and Snow Part II: Selection Methods and Mitigation Options. IEEE Trans. Power Delivery, vol 22(4): 2297–2304

Farzaneh-Dehkordi J, Farzaneh M, Zhang J (2004) Experimental study: Mathematical Modelling of Flashover on Extra-High Voltage Insulators Covered with Ice. Hydrological Processes, vol 18: 3471–3480

Fikke SM, Hanssen JE, Rolfseng L (1993) Long Range Transported Pollution and Conductivity of Atmospheric Ice on Insulators. IEEE Trans. on Power Delivery, vol 8(3): 1311–1321

Forrest JS (1936) The electrical characteristics of 132-kV line insulators under various weather conditions. IEE Journal, vol 79: 401–423

Ghosh P, Chatterjee N (1995) Polluted Insulator Flashover Model for ac Voltage. IEEE Trans. Dielectrics and Electrical Insulation, vol 2(1): 128–136

Ghosh P, Chatterjee N (1996) Arc propagation over electrolytic surfaces under power frequency voltage. IEEE Trans. Dielectrics and Electrical Insulation, vol 3(4): 529–536

Gorur RS, Cherney EA, Burnham JT (1999) Outdoor Insulators, ISBN 0-9677611-0-7 (www.insulators.net)

Gutman I, Berlijn S, Fikke S, Halsan K (2002) Development of the Ice Progressive Stress Method Applicable for the Full-Scale Testing of 420 kV Class Overhead Line Insulators. In: Proc of the 6th International Workshop on Atmospheric Icing of Structures, Brno, paper 3–6: 1–6

Hara M, Phan CL (1979) Leakage Current and Flashover Performance of Iced Insulators. IEEE Trans. Power Appratuns and Systems, vol PAS 98(3): 849–859

Hemmatjou H, Farzaneh M, Fofana I (2007) Modeling of the AC Arc Discharge inside Wet Snow. IEEE Trans. Dielectrics and Electrical Insulation, vol 14(6): 1390–1400

Hora M., Korcova I., Sklenicka V., Vokalek J. (1983) Influence of conductive ice on electric strength of HV insulators. In: Proc. Int. Symp. on Pollution Performance of Insulators and Surge Diverters, Madras, vol 1(1.02): 1–4

IEC (1986) Standard IEC 60815 Guide for the Selection of Insulators in Respect of Polluted Conditions. International Electrotechnical Commission, Lausanne

IEC (1991) Standard IEC 60507: Artificial Pollution Tests on HV Insulators to be used on AC Systems. International Electrotechnical Commission, Lausanne

IEC (1996) Standard 60071-2: Insulation co-ordination – Part 2: Application guide 3rd edition. International Electrotechnical Commission, Lausanne

IEC (2006) Standard 60071-1: Insulation co-ordination – Part 1: Definitions principles and rules 8th edition. International Electrotechnical Commission, Lausanne

IEEE (1995) IEEE Standard 4: High Voltage Testing. IEEE Press, Piscataway NJ

IEEE (2000) IEEE Standard 100: The Authoritative Dictionary of IEEE Standards Terms 7th Edition. IEEE Press, Piscataway

Jiang X, Wang S, Zhang Z, Xie S, Wang Y (2005) Study on AC Flashover Performance and Discharge Process of Polluted and Iced IEC Standard Suspension Insulator String. In: Proc of the 14th International Symposium on High Voltage. Beijing, paper D-70: 1–8

Lanoie R, Bouchard D, Lessard M, Turcotte Y, Roy M (2005) Using Steam to De-Ice Energized Substation Disconnect Switch. In: Proc 11th International Workshop on Atmospheric Icing of Structures, Montreal: 353–356

Li S., Zhang R., Tan K (1990) Measurement of Dynamic Potential Distribution during the Propagation of a local Arc along a Polluted Surface. IEEE Trans. Electrical Insulation, vol EI-25(4): 757–761

Looms JST (1988) Insulators for High Voltages. Peter Peregrinus Ltd, London

Lundmark A, Olofsson B (2007) Chloride Deposition and Distribution in Soils along a De-iced Highway – Assessment Using Different Methods of Measurement. Water Air Soil Pollution, vol 182: 173–185

Meier A, Niggli MM (1968) The influence of snow and ice deposits on supertension transmission line insulator strings with special reference to high-altitude operation. In: IEE Conference Publication 44 London: 386–395

NADP (2007) National Atmospheric Deposition Program [online] http://nadp.sws.uiuc.edu/

Ndiaye I, Farzaneh M, Fofana I (2007) Study of the Development of Positive Streamers along an Ice Surface. IEEE Trans. Dielectrics and Electrical Insulation, vol 14(6): 1436–1445

Obenaus F (1935) Die Ubersclagsspannung verschmutzer Isolatoren. Elektrotechnische Zeitschrift, vol 56: 369–370

Peyregne G, Rahal A, Huraux C (1982) Flashover of a Liquid Conducting Film Part 1: Flashover Voltage and Part 2 Time to Flashover – Mechanisms. IEEE Trans. Electrical Insulation, vol EI-17(1): 10–19

Pigini A, Rizzi G, Garbagnati E, Porrino A, Baldo G, Fesavento G (1989) Performance of Large Air Gaps under Lightning Overvoltages: Experimental Study and Analysis of Accuracy of Predetermination Methods. IEEE Trans. Power Delivery, vol 4(2): 1379–1392

Rizk FAM (1981) Mathematical Models for Pollution Flashover. Electra, vol 78, pp 71–103

Rumeli A (1976) Flashover along a Water Column. IEEE Trans. Electrical Insulation, vol EI-11(4): 115–120

Sklenicka V, Vokalek J (1996) Insulators in icing conditions: Selection and measures for reliability increasing. In: Proc of the 7th International Workshop on Atmospheric Icing Structures, Chicoutimi: 72–76

Vlaar J (1991) Thermal and Electrical Properties of Icicles. University of Waterloo 2B Honours Physics Report, SN 88104434

Vuckovic Z, Zdravkivic Z (1990) Effect of Polluted Snow and Ice Accretions on High-Voltage Transmission Line Insulators. In: Proc of the 5th International Workshop on Atmospheric Icing of Structures. Tokyo, no B4-3: 1–6

Wieck H, Gutman I, Ohnstad T (2007) Investigation of Flashover Performance of Snow-covered Breakers. IEEE Trans. Dielectrics and Electrical Insulation, vol 14(6): 1339–1346

Wilkins R (1969) Flashover voltage of HV insulators with uniform surface pollution films. Proceedings of the IEE, vol 116(3): 457–465

World Meteorological Organization (1996) Document 8: Guide to meteorological instruments and methods of observation 6th Edition. World Meteorological Organization, Geneva

World Meteorological Organization (1997) Document 8: Guide to meteorological instruments and methods of observation, Supplement no 1. World Meteorological Organization, Geneva

Yasui M, Naito K, Hasegawa Y (1988) AC Withstand Voltage Characteristics of Insulator String Covered with Snow. IEEE Trans. Power Delivery, vol 3(2): 828–838

Yoshida S, Naito K (2005) Survey of electrical and mechanical failures of insulators caused by ice and/or snow. CIGRE WG B2.03, Electra, vol 222: 22–26

Zhang, J, Farzaneh, M (2000) Propagation of AC and DC Arcs on Ice Surfaces IEEE Trans. on Dielectrics and Electrical Insulation, vol 7(2): 269–276

Chapter 8
Design of Transmission Lines for Atmospheric Icing

Anand Goel

8.1 Introduction

In many parts of the world ice loading is the most important parameter influencing the performance of overhead transmission lines, as each year there are thousands of transmission line failures worldwide caused by excessive ice loadings and utilities have to spend millions of dollars in restoration efforts. A prime example of the damages caused by atmospheric icing in recent years is the ice storm which hit Eastern Canada (Ontario, Quebec and New Brunswick) and the Northeastern United States, in January 1998. The storm caused widespread damage to the power networks (Turcot 2002; Goel 2002; McClure et al. 2002). Over two million people were without electric power for weeks as 1,300 transmission towers and 35,000 distribution structures were destroyed by excessive ice loads. Restoration costs for the network in Quebec alone were over 5 billion dollars.

For the design of overhead lines subjected to ice loads, of equal importance to ice loads themselves is the determination of winds occurring during ice accretions, as the ice accretions not only increase the vertical load on line conductors, they also increase the surface area exposed to the wind loads. However, of the various meteorological parameters which are considered in the design, construction and operation of transmission systems, atmospheric icing is the most difficult to estimate accurately. For a number of years, many analytical, experimental and field studies have been carried out to improve the understanding of various ice accretion processes, the concurrent winds occurring during the ice accretions, and their impact on power network equipment.

This chapter discusses impact of atmospheric icing on transmission lines and presents the salient aspects of how to design transmission lines to withstand loads and other effects caused by atmospheric icing.

Dr. A. Goel
AG Engineering Innovations, 76 Pathlane Road, Richmond Hill,
Ontario, Canada L4B 4C7
e-mail: anandgoel@rogers.com

8.2 Types of Atmospheric Icing Accretion

Atmospheric icing is a term used for various processes where water in various forms in the atmosphere freezes and adheres to objects exposed to the air. In general, the atmospheric icing accretion can be classified in the following types (Cigré 2000; Cigré 2006):

- Precipitation icing
- In-cloud icing and
- Hoar frost

Precipitation icing occurs in several forms such as freezing rain (glaze or rime), wet snow and dry snow, as described below:

Glaze: Glaze has a density of 900–920 $\mathrm{kg/m^3}$. Glaze grows in a clear, smooth structure with no air bubbles. It is usually formed from freezing precipitation, rain or drizzle, or from clouds with large liquid water content and large drop size. The freezing rate of droplets is less than the impingement rate, which causes part of the drop to splash or flow around the conductor before freezing. While glaze contains no air bubbles as such, in strong wind situations it grows in irregular shapes incorporating pockets of air.

Rime: Rime has a density of 300–900 $\mathrm{kg/m^3}$. It is usually classified as soft rime or hard rime:

Soft rime has a density of less than 600 $\mathrm{kg/m^3}$ and grows in a granular structure that is white and opaque with many air bubbles within the structure. It usually grows in a triangular or pennant shape pointed into the wind. The granular structure results from the rate of freezing of individual drops, each drop freezing completely before another one impinges on the surface.

Hard rime has a density ranging from 600 to 900 $\mathrm{kg/m^3}$ and tends to grow in a layered structure with clear ice mixed with ice containing air bubbles. In this case the freezing rate of the droplets is equal to the impingement rate.

Wet snow: Wet snow has a density of 300–800 $\mathrm{kg/m^3}$. It is usually defined as snow which falls at temperatures equal to or above -5° C. Under these conditions, the snow is sticky enough to adhere to surfaces easily and accumulate rapidly. Wet snow tends to build on tops and windward surfaces of structures and in cylindrical layers around conductors. At temperatures below about -2° C, snow particles are usually too dry to adhere to surfaces in appreciable quantities. If the temperature falls below 0° C after the accretion of wet snow, the accumulation freezes into a dense hard layer with strong adhesion. Transmission line problems have occurred due to wet snow events (Kiessling and Ruhnau 1993; Sakamoto 1998).

Dry snow: Dry snow accretes at subfreezing temperatures. This type of accretion appears only when wind speed is very low i.e. below 2 m/s. The density of dry snow is, in general, very low, not exceeding 100 $\mathrm{kg/m^3}$.

In-cloud Icing: In-cloud icing is a process by which supercooled water droplets in a cloud or fog freeze immediately upon impact on objects in the airflow,

Table 8.1 Physical Properties of Atmospheric Ice

Type of ice	Density kg/m^3	Adhesion	Appearance		Cohesion
			Colour	Shape	
Glaze ice	900–920	Strong	Transparent	Cylindrical icicles	Strong
Hard rime	600 – 900	Strong	Opaque to transparent	Eccentric pennants into wind	Very strong
Soft rime	300 – 600	Medium	White	Eccentric pennants into wind	Low to medium
Wet snow	300 – 800	Medium	White	Cylindrical	Medium to strong
Dry snow	100	Low	White	Cylindrical	Low

i.e. overhead lines in mountains above the cloud base. For the structures situated at mountain summits, exposure to supercooled clouds or fog usually results in soft rime (300–600 kg/m^3), however, precipitations resulting in hard rime (600–900 kg/m^3) also occur and are most frequent in early winter (McComber et al. 1995). Sometimes very large in-cloud icing occurs on overhead lines.

Hoar frost: Hoar frost has a density of less than 300 kg/m^3. It is a deposit of interlocking ice crystals formed by direct sublimation of water vapour in the air onto objects. It forms when air with a dew point below freezing is brought to saturation by cooling. Hoar frost is featherlike in appearance and builds occasionally to large diameters with very little weight.

Normally, hoar frost does not constitute a significant loading problem, however, it is a very good collector of supercooled fog or cloud droplets and at subfreezing temperatures with light winds, fog conditions gradually become soft rime of significant volume and weight.

The typical physical properties of atmospheric ice are presented in Table 8.1.

8.3 Ice Accretion on Overhead Line Conductors and Structures

In actual practice, glaze ice can be observed to form on transmission line conductors in a variety of shapes ranging from the classical, smooth cylindrical sheath, through crescent on the windward side, and icicles hanging on the bottom, to large irregular protuberances spaced along the conductor (Fig. 8.1). In most cases glaze on support structures develops as a fairly smooth layer on the windward surfaces with icicles forming below horizontal members as the excess water flows to the bottom and drips off. The shape of the glaze on conductors is dependent upon a combination of factors present during the accretion stage (Cigré 2006), such as:

- Wind speed and variation in wind speeds
- The direction of the wind with respect to the surface
- The turbulence of the wind flow
- The ability of the conductor to rotate
- Variations in air temperature
- The duration of the icing event

Due to the variability of these parameters from one event to the other, and even during the same event, a cylindrical shape with equivalent weight is assumed for design. In Figs. 8.1 to 8.4 photographs of ice accretion on power line conductors are shown. Figure 8.1 shows glaze ice accretion and Fig. 8.2 shows rime ice accretion creating unequal loading on the conductors. Figure 8.3 shows the largest in-cloud icing observed on transmission line conductors. It was observed on top of Mt. Lonahorgi, Norway, in 1961, with an elliptical shape measuring 1.4 m to 0.94 m and weighing approximately 305 kg/m (Fikke 2005). Figure 8.4 shows hoar frost on a distribution line in Manitoba, Canada (Farias 2005). Figure 8.5 is a photograph showing in-cloud ice accretion on a Hydro-Québec transmission line support structure (Lavoie 2005) and Fig. 8.6 shows wet snow on conductors in Japan being removed by a robot (CEATI 2002).

As mentioned above, the total deposit of the ice along the conductor is dependent on the ability of the conductor to rotate. If a conductor is free to rotate, such as a long-span ground wire or single conductor, the ice will build in a circular deposit. If the conductor is not free to turn, such as bundle conductors, the ice will form in a pennant shape extending into the wind.

In the case of rime ice, it has been observed (Leavengood and Smith 1968) that the density of the rime tends to decrease as the duration of the riming condition increases, in other words, as the radial dimension of the rime gets large. In individual cases, this varies according to the variance of wind speed with time, drop sizes and temperature during icing events. In many cases, hard rime has been observed close to the conductor or tower surface gradually changing to soft rime near the outer surface of the ice, leaving large voids as the outer diameter or surface area increase.

Hoar frost results in significant loading only in the special circumstances mentioned in Section 8.2. In such cases, the frost feathers collect supercooled fog droplets and the resulting load is a case of rime icing.

Fig. 8.1 Ice accretion – Glazed ice with icicles (Reproduced by permission of Pioneer Electric Cooperative Inc. USA)

Fig. 8.2 Rime ice accretion creating unequal loading on conductors (Anderson 2005, reproduced by permission of BC Hydro, Canada)

In mountainous terrain, wind flow on the windward side is lifted, causing an increase in the amount of condensation and more cloud contact with conductors and towers, which results in icing. On the leeward side of the mountain, the air is drier and icing is considerably less or may not occur at all. This results in the possibility

Fig. 8.3 In-cloud icing (soft rime), Mt. Lonahorgi, Norway (Photo O. Wist, Fikke 2007, reproduced by permission of S. Fikke, Norway)

Fig. 8.4 Hoar frost on distribution line (Farias 2005, reproduced by permission of Manitoba Hydro, Canada)

of a series of iced spans followed by a series of bare or lightly loaded spans, thus creating unbalanced loads on supports.

Unbalanced loads on transmission line supports, due to icing, can occur in a number of ways with varying degrees of severity. Some of the causes are:

- Unequal ice accumulation
- Ice loads on unequal spans
- Ice shedding, i.e. sudden ice dropping in one or several spans (Morgan and Swift 1964; Kollar and Farzaneh 2005; EPRI 2006)
- Galloping of conductors (Havard 2002; Havard and Van Dyke 2005)

Fig. 8.5 In-cloud rime icing in Quebec (Lavoie 2005, reproduced by permission of Hydro Quebec, Canada)

Fig. 8.6 Wet snow accretion in Japan being removed by a robot (CEATI 2002, reproduced by permission of Fujikura Ltd, Japan)

Fig. 8.7 Damage to transmission structure due to snow creep (Gwilliam and Anderson 1999, reproduced by permission of BC Hydro, Canada)

Loads due to snow creep experienced by some utilities (Gwilliam and Anderson 1999) has shown that substantial damage can occur to lattice towers installed in areas of deep snowfall. The damage can occur due to applied snow pressure which can build up through the snow creep, uneven melting or due to snow slides (Fig. 8.7)

8.4 Ice Load Measurements

In general, ice loads are not measured at meteorological stations where usually only air temperature, barometric pressure, wind speed and direction, relative humidity and rainfall or freezing precipitation are measured. Therefore, other means are used to obtain the information about ice accretion. The following means are used to obtain icing data:

- Measurement by devices that simulate ice accretion on conductors; some devices consist of tubes or cable assemblies installed near ground level for ease of observation (CEATI 2003, COST 727 2006)
- Estimating icing thickness by monitoring of conductor tension and or vertical component of the weight at insulator attachment point (Krishnasamy and Tabatabai 1992)
- Estimate ice thickness based on direct measurement of conductor tension and sag (Seppa 1996)
- Meteorological models based on field data collected during various icing events and dependent on temperature, humidity, precipitation rate and expected wind direction
- Direct measurements of icing thickness (Fig. 8.8), weight of ice samples (Fig. 8.9) or from observation devices installed on line conductors (Cigré 2000, ALA 2004).

At the present time, although there is adequate historical data available to calculate wind loads on overhead lines, the available information on ice and wind-on-ice is generally not adequate. The lack of measured historical data can be overcome by:

- Establishing a network of stations to gather data on ice and wind-on-ice over an extended period of time sufficient for statistical analysis
- Utilizing the available historical weather data such as wind speed and direction, temperature, precipitation, etc., to predict the ice and wind-on-ice loading by way of modelling

The most common method to obtain ice accretion data is by means of ice accretion models (Cigré 2006). Many of these models are essentially based on laboratory and wind tunnel studies and have not been tested under real conditions (Lozowski and Makkonen 2005). One of the models, called Chainé model, was developed by Environment Canada in 1974 (Chainé and Skeates 1974). The model calculates equivalent ice accretion on a conductor and determines the associated wind speed. Ontario Hydro (now Hydro One), in Canada, used this model to develop a

Fig. 8.8 Ice thickness measurement (Lavoie 2005, reproduced by permission of Hydro Quebec, Canada)

Fig. 8.9 Ice thickness measurement on ice removed from conductor (ALA 2004, reproduced by permission of CRREL USA)

database for ice loads and wind-on-ice loads for Ontario Hydro transmission lines (Krishnasamy and Tabatabai 1988).

In the United States, beginning in 1994, a consortium of individuals and government agencies undertook a project to produce U.S. climatology of ice thickness due to freezing rain, in the form of an extreme value analysis. This effort, under the auspices of the American Society of Civil Engineers (ASCE), utilizes data modelling techniques to develop such climatology. A map of extreme ice loads with concurrent wind speeds has been developed, which is included in latest revision of National Electrical Safety Code (NESC) (NESC C2 2002), ASCE Standard 7, Minimum Design Loads for Buildings and Other Structures (ASCE 7 2002), and the ASCE Manual 74 Guidelines for Electrical Transmission Lines Structural Loading (ASCE 74 2005). The map is based on historical weather data gathered from hundreds of weather stations around the country over last 50 years.

Some utilities base their estimation of climatic loading for new transmission lines on the operating experience of existing lines in the area and recorded meteorological data, region weather data, and site observations, as was done for a Canadian utility's proposed future line at Mission Ridge, B.C. (Anderson, 1996).

Establishing programs to gather field data on ice and wind-on-ice has to be long range. As shown in the Extreme Value Analysis example in the Appendix, a minimum of 10 years of data is required for meaningful estimation of ice accretion.

8.5 Standards for Ice Loads

In many industrialized countries, standards for the design of overhead transmission lines have existed for many years such as NESC (National Electrical Safety Code) in the USA, CSA C22.3 Overhead Systems in Canada, DIN VDE 0210 in Germany, NBR 5422 in Brazil, and JEC-127(1979) in Japan. Some of these standards are discussed below.

The National Electrical Safety Code, C2, in the United States (NESC C2-2002), establishes the minimum requirements necessary to ensure the safety of personnel during the installation, operation and maintenance of power lines. Most states in the U.S. have adopted this code and require that its provisions be met as a minimum by the utilities under their jurisdictions. Sections 25 and 26 of the NESC C2 contain specific loading and strength requirements for overhead line support structures. The NESC divides the US into large loading zones (heavy, medium and light) and specifies radial ice thickness/wind pressure/temperature relationships to define the minimum load levels that must be used within each loading zone.

The IEC (International Electro-technical Commission) 60826 (IEC 60826 2003) specifies that the loading and strength requirements of overhead lines be derived from reliability-based principles. These requirements apply to lines operating at 45 kV and above, but can also be applied to lines with lower nominal voltages. This IEC technical report has been developed to serve as the basis for standardization in countries which do not have a standardization program for the design of overhead lines. The standard describes the procedures for the determination of climatic loads, such as wind and ice, based on measurements.

The IEC specifies the yearly maximum ice loads by means of the Gumbel distribution. The analysis of available information on ice loads is carried out and the mean value of the yearly maximum observations and the corresponding standard deviations are obtained. The reference design ice load is obtained based upon the required reliability of the line and the following parameters:

- Mean value of the yearly maximum ice loads
- Standard deviation of the yearly maximum ice loads
- Number of observations
- Conductor and OHGW diameter
- Average conductor/OHGW height above ground

The ice load is specified as weight per unit length. The reference ice load is defined as the load on 30 mm conductor in a 100 m long span at 10 m above the ground.

To address wind and ice load combinations, IEC 60826 suggests considering 65 to 85% maximum wind with maximum ice in the absence of significant statistical data.

For designing the supports under ice accretion, the following four conditions are specified:

- Uniform ice accretion on all conductors and overhead ground wires
- Non-uniform ice formation on one phase conductor or one OHGW
- Non-uniform ice formation on all conductors in adjacent spans
- Non-uniform ice formation on one circuit of a double or multi-circuit line

IEC 60826 states that ice accretion on structures increases the vertical loads on the structures and may control the design of foundations and some support members. The weight of the ice can be calculated using the geometry of the members and relevant thickness of ice accretion.

In Europe, *Standard EN 50341* (EN 50341-1 2001) was established as a regional overhead electrical line standard. The standard consists of a main body with general requirements and common specifications, as well as National Normative Aspects (NNA) for individual CENELEC (European Committee for Electrotechnical Standardization) countries. The standard provides specifications for the design and construction of new overhead lines with respect to human safety, as well as maintenance, operational and environmental considerations. The standard contains both the probabilistic and deterministic approaches. The deterministic approach is based on calibration using the long and successful history of existing overhead lines, while the probabilistic approach is based on IEC 60826.

The ice load is assumed according to the local conditions as specified by NNA and the load cases are similar to those of the IEC 60826.

The ice load is specified as weight per unit length, and the reference ice load is defined as the load on a 30-mm conductor in a 100-m span at 10 m the above ground.

To address wind and ice load combinations, EN 50341 suggests to consider 70% (with wet snow) to 85% (with in-cloud icing) maximum wind with maximum ice in absence of significant statistical data.

Standard CSA C22.3 No.1 Overhead Systems (CSA C22.3 No.1 2001), which forms part of the Canadian Electrical Code, provides requirements for the construction of overhead systems. It covers both the electrical supply and communication circuits. The requirements contained in this standard do not constitute complete design and construction specifications, but rather prescribe the minimum design requirements that are most important for human safety, continuity of service and protection of property. The standard provides climatic loading maps with return periods of 50 years for various regions of the country. The standard presents a choice between deterministic and reliability-based design methods. Reliability-based design methods are covered by CAN/CSA-C22.3 No. 60826, which is largely based on IEC 60826.

The design ice load is assumed to correspond to the extreme radial ice thickness of glaze ice having a 50-year return period. As the data for wind and ice maps are collected at single points, while the transmission line is a linear system that is exposed to large number of extreme load events than a single point location, the CSA standard uses a spatial correction factor. For extreme ice load cases, a spatial factor of 1.5 is used.

In the absence of statistical data, 60% maximum wind is considered with maximum ice (no spatial factor is considered).

Guidelines for Electrical Transmission Line Structural Loading, ASCE Manual 74 (ASCE 74 2005), provides reliability-based guidelines for structural loadings on transmission lines in the United States. The most recent edition of the guidelines was published in 1991. An updated version has now been completed and is due to be published in 2008. The manual addresses transmission line design issues that must be considered to provide reliable, economically viable structures.

The design ice load is specified as the extreme radial ice thickness of glaze ice corresponding to a 50-year return period. Wind and ice combinations are based on

50-year return period contour maps of concurrent 3-second wind in combination with ice.

Regulations for Electric Power Facilities, Japan and Design Standard of Power Line Structures (JEC-127, 1979) (JIEE 1979): The mechanical design of power lines in Japan is governed by the "Regulations for Electrical Power Facilities" containing legal directives established by the Government of Japan. In addition, tower design is carried out using JEC-127, established by the Japanese Institute of Electrical Engineers. In JEC-127(1979) wind and wet snow load maps are presented for 50-year return periods based on statistical analysis. The component stresses calculated for loads determined with stochastic considerations are compared with the elastic limits of components, as specified by the Japanese standards. This is called the deterministic design approach, which will be discussed later in this chapter. However, in practice, as the strength of structural components is governed by one of many design load cases (maximum wind, maximum ice, combined wind and ice, etc.), Japanese utilities review the strength of critical components of structures and increase the strength by changing component size when it is thought appropriate, thus making the practice reliability based to some extent (Sakamoto 1998).

8.5.1 Utility Standards

In general, utility standards in Canada and the USA specify ice loads meeting the requirements of the CSA or NESC standards, respectively, as a minimum, and based on the meteorological observations in their region. In most cases the ice loads specified by utilities in Canada exceed CSA requirements. Combined ice and wind loads are specified based on experience and local data. Similar steps are also taken by utilities in other countries.

8.6 Transmission Line System

A transmission line is an integrated system made up of two subsystems, namely:

- Structural support system
- Wire system

The structural support system, consisting of the supporting structures (towers, poles and their foundations), supports the wire system under various loading conditions.

The wire system consists of the conductors and overhead ground wires, including the components connecting the wires to the supporting structures.

The basic task of the structural support system is to withstand the loads generated or acting on the wire system. Moreover, most of the loads generated on an overhead line are generated on or by the wire system, except for special accidental loads, as will be discussed later.

8.6.1 Loads in a Transmission Line System

The loads on a transmission system are caused by:

- Weather-related events
- Construction and maintenance activities
- Accidental loads

Weather-related events of interest are extreme ice, extreme wind or ice with accompanying wind, and high intensity winds such as tornadoes and downdrafts. Coincident temperatures are also important as they can significantly impact the loading on transmission systems.

Construction-related loads are those that act upon the structures due to the assembly and erection of the structures, and during the installation of conductors, ground wires, insulators, etc. Similarly, maintenance loads are those that act as a result of scheduled or emergency inspection, and/or the replacement of all or part of the structures, or all or part of the conductors, ground wires insulators, and hardware.

Accidental loads are caused by events difficult to predict. Accidental events include component breakage resulting from natural hazards such as hurricanes, tornadoes, landslides, and washouts, or man-made hazards such as sabotage. They could also occur due to material defects, wear, fatigue, or impact from motorized vehicles and machinery.

The forces acting on a typical steel pole structure are shown in Fig. 8.10. As discussed above, they result from the loads generated by the weather, accidental causes, and/or construction and maintenance tasks. For static analysis, forces are defined by their intensity (or magnitude) and their direction of action (also referred to as line of action). The resultant forces are shown at each conductor and ground wire suspension point in parallel and alignment with the vertical, longitudinal, and transverse direction relative to the transmission line. Therefore, the effects of each

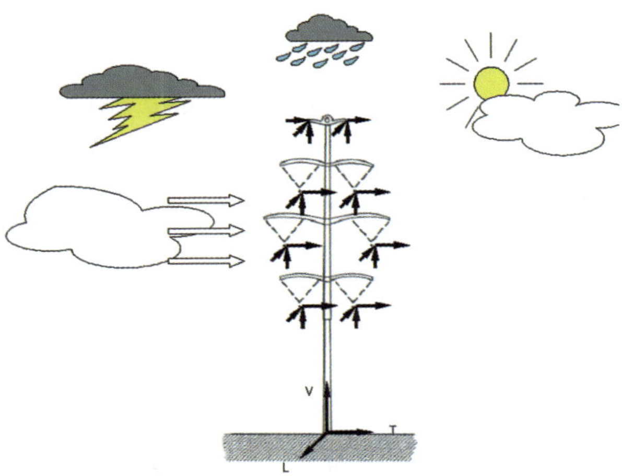

Fig. 8.10 Steel pole – Loading events and structural forces

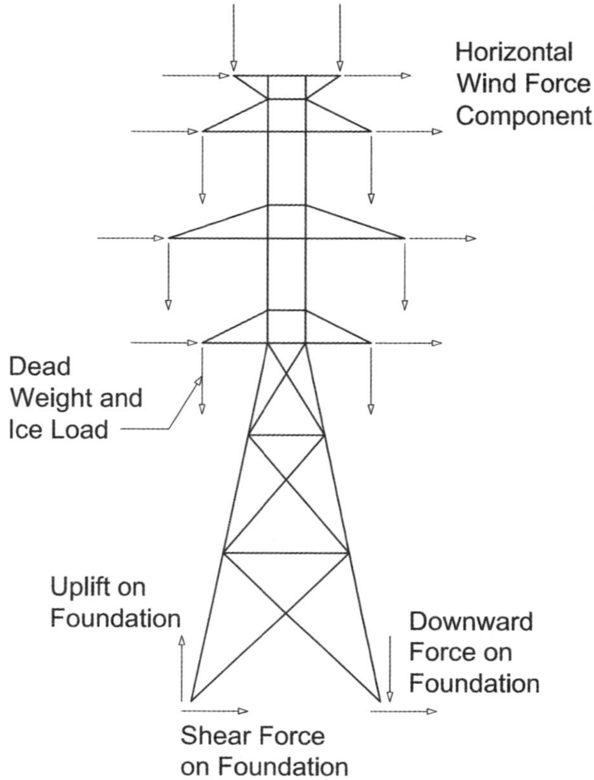

Fig. 8.11 Tangent lattice tower loading events and structural forces

loading event, i.e. the forces, are represented by three orthogonal (perpendicular) components acting at each wire suspension point (the applied loads).

Figure 8.11 is a schematic of loads acting on a double-circuit steel lattice tangent (suspension) tower. In this case it has been assumed that longitudinal loads due to conductor tensions in adjacent spans balance each other out and, hence, no longitudinal loads act on the structure.

Generally, the loading directions considered in the analysis, design, and management of transmission lines are aligned with the geometric axes of the structure, and these will vary at angle-support structures according to directional changes in the transmission line. The structure resists by safely and reliably transferring the loads from the wire support points to the foundation, and ultimately to the surrounding ground. The soil characteristics of the surrounding ground determine whether the foundation is adequate to support the loads safely and reliably.

As mentioned above, in designing transmission lines to withstand the effects of climatic loads, it is necessary to consider the following random and interacting parameters:

- Ice
- Wind
- Combination of wind and ice

The external loads on transmission line conductors can be divided into:

- Vertical loads on conductors due to ice
- Transverse loads on bare conductors due to wind
- Transverse loads on ice-covered conductors due to wind
- Longitudinal loads on conductors due to unbalanced ice accretion on adjacent spans

The external loads on supporting structures can be divided into:

- Load due to ice and wind on conductors
- Loads due to ice and wind on the structure itself (In practice, however, the ice load on structure itself is usually neglected)

Weather-related loads are not only affected by the amount of ice accretion on the conductors and wind speed and direction, but also by the terrain and topography of the area where the line is located.

As mentioned earlier, the actual shape of ice on a conductor may vary depending upon the type of ice, conductor shape, wind conditions, etc. (Fig. 8.12). However, for the convenience of calculating ice loads an equivalent radial ice on the conductor is assumed (Fig. 8.13).

8.6.2 Calculation of Loads on Conductors

8.6.2.1 Ice Loads on Conductors

In general, only glaze and rime-ice accretions are considered in determining ice loads on conductors. In some situations, loads due to wet snow also have to be taken into consideration.

Vertical loads on conductors due to ice are calculated from

$$P_i = w\frac{\pi}{4}[(D + 2t)^2 - D^2]L \tag{8.1}$$

Fig. 8.12 Typical ice accretion on conductor (Reproduced by permission of CRREL USA)

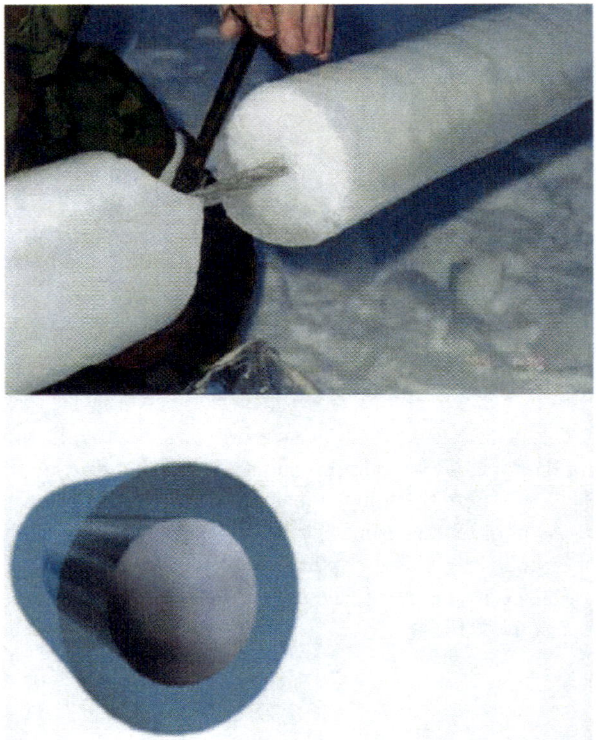

Fig. 8.13 Radial ice thickness (Reproduced by permission of AJ Eliasson, Iceland, and Osmose Utilities Services, USA)

where

P_i = ice load
w = ice density
D = conductor diameter
t = equivalent radial thickness of the cylindrical ice deposit
L = conductor span length

8.6.2.2 Wind Loads on Conductors

The transverse wind load on a bare conductor is calculated from

$$P_w = \frac{1}{2}\rho_a V_m{}^2 C_{DC} L D \tag{8.2}$$

where

P_w = wind load
ρ_a = air density
V_m = mean wind speed at conductor height
C_{DC} = conductor drag coefficient

It should be noted that at most standard weather stations the wind speed is recorded at 10 m above the ground level. When using the data to assess loads on transmission line conductors and structures, which are usually at higher levels, the data are appropriately adjusted to account for the variation of wind speed with height. The relationship commonly used is given by (ASCE 74 2005, IEC 60826 2003)

$$V_h = \left(\frac{H_h}{H_r}\right)^{\alpha} V_r \tag{8.3}$$

where:

V_h = mean wind speed at height H_h
H_r = reference height at which wind speed was recorded
V_r = mean wind speed recorded at height H_r
H_h = height at which load is calculated,
α = power law exponent representative of terrain

8.6.2.3 Wind Loads on Iced Conductors

The transverse wind load on an iced conductor, P_{wi}, is calculated as in eq. (8.2) where the exposed diameter is adjusted to include the effect of the ice deposit

$$P_{wi} = \frac{1}{2}\rho_a C_{DC} V_m^2 (D + 2t) L \tag{8.4}$$

8.7 Design Methodology

The following steps are generally followed for the design of transmission lines:

- Collect preliminary line design data and available climatic data
- Select the reliability level in terms of return period of climatic loads
- Select the security requirements (failure containment)
- List safety requirements
- Calculate climatic variables corresponding to selected return period
- Calculate loads related to security requirements
- Calculate loads related to construction and maintenance requirements
- Determine the suitable strength coordination
- Select design loads and strength factors
- Calculate characteristic strength requirements
- Design components

8.7.1 Design Process

The main objective in designing a transmission line structure is to provide an economic structure that will, with the desired level of reliability, remain functional throughout its design life. This objective is expressed as

$$\text{Load effects} < \text{Resistance} \tag{8.5}$$

Where load effects are the forces, moments, deflections, stresses, and other internal measures resulting from the action of the loads, which are to be compared with the resistance parameters such as strength, stiffness, and any other measure of load-carrying capacity.

The main steps are:

Define loads: Define the loads that act on the structure. This requires conversion of environmental loads, such as ice or wind, into forces acting on the structure.

Define member capacities: Define the capacities of the structural components considering their different possible failure modes, and assuming that these are not governed by connection failure. This requires data on the mechanical properties of materials and on strength, stiffness and stability, including detailed information on dimensions and construction practices along with appropriate theories to develop flexure, shear, buckling and other capacities.

Conduct structural analysis: An analysis of the model of the assumed structure is then carried out to determine the effects of loading, for comparison with serviceability limits or capacity limits of the structural components.

Check adequacy of the design and repeat process if necessary: If any of the component resistance is less than the load effects or any of the serviceability limits are exceeded, then appropriate changes are made in the structural model and the above steps are repeated until an acceptable solution is obtained.

There are the following two design approaches:

- Deterministic design approach
- Reliability-based design approach

8.8 Deterministic Design Approach

Traditionally, a deterministic design approach is used where single values for load effects, Q_n, and resistance, $\mathrm{R_n}$, are assumed. The maximum load effect Q_n is calculated assuming the maximum anticipated value of the loads (ice, wind, combined wind and ice), and a nominal strength of the component material R_n is assumed. There are two traditional calculation approaches to the deterministic design:

- Working Stress design (also called allowable stress design) and
- Load Factor design

In the Working Stress design, adequate safety is obtained by specifying a Factor of Safety (F.S.) such that

$$Q_n \leq \frac{R_n}{FS} \tag{8.6}$$

In the Load Factor design, the maximum load effect Q_n is multiplied by a load factor (LF), selected on the basis of experience, importance of line, variation of the load over the design life of the line, or the utility's standard practice. The design condition is expressed as

$$(LF)Q_n \leq R_n \tag{8.7}$$

In general, the Load Factor Design approach is followed by utility design engineers. The Load Factor (LF) is chosen to cover uncertainties stemming from:

- Likelihood of the occurrence of the specified load
- Dispersion of the structural strength
- Deterioration of strength during service life
- Structure function (suspension, line angle or dead end structure, foundation)

The sequence of failure is implicitly taken into consideration by the choice of the load factor. For example, the load factor for tangent structures is 1.1 (say), for angle and dead-end structure it is increased to 1.2. The foundations are designed for an additional load factor of 1.2.

The design process consists in defining all anticipated loading conditions and using appropriate LFs, and then carrying out the structural analysis for each of the loading conditions.

In general, the deterministic approach has the following drawbacks:

- There is no rational method to select the safety factor or the LF
- The approach does not lead to uniform safety factor values for various design situations
- The performance level or reliability of the designs is unknown
- It is difficult to adjust to local conditions
- There is no control of sequence of failure

The shortcomings of the deterministic design were demonstrated by Pohlman (2001) who analyzed an existing lattice tower structure design of one utility with the 3D modelling techniques used in the forensic investigation of structures in line failures. In this technique, after the designed structure is properly modelled, loading events of increasing magnitude are applied from different directions until the actual critical capacity for each key member is reached. The results reported from the analysis are presented in Table 8.2 (Pohlman 2001).

In this particular case the following observations were made:

- The legs have high probability of survival during the 50-year design life

Table 8.2 Analysis results for an existing lattice structure (Reproduced by permission, © [2001] IEEE)

Key member	Governing load case	Return period (years)	Annual probability of failure	Probability of failure in 50 years
Tension chord in conductor cross-arm	Ice	110	0.0090	37%
Tension chord in OHGW cross-arm	Ice	35	0.0286	77%
Tower leg	Wind	115	0.0087	35%
Foundation	Wind	25	0.0400	87%

Table 8.3 Results from calibration exercise of Pennsylvania Light Company Structures (Reproduced by permission of CIGRÉ)

Member	Load case	Reliability (safety) index		Probability of failure	
		Tangent structure	Running angle structure	Tangent structure	Running angle structure
OHGW arm	Ice load	2.79	2.99	0.002560	0.001350
Conductor cross-arm	Ice load	4.43	3.87	0.000005	0.000059
Leg point 1	Ice and wind	2.12	1.87	0.020182	0.030430
Leg point 2	Ice and wind	2.21	4.10	0.013850	0.000021

- The loss of OHGW is more likely event than the loss of conductors
- The foundation is a weak link in the system

This example, although demonstrating the main drawbacks of the deterministic design approach, is not representative of the general utility practice. A calibration exercise was carried out on two tubular steel H-Frames, one a tangent structure and the other a small running structure of the Pennsylvania Power and Light Company's 500 kV system (Di Gioia et al. 1982). Several critical structure members were selected under different loading conditions. A unique reliability (safety) index as defined in Section 8.9 was calculated from each member –load condition combination. The results are shown in Table 8.3 (Di Gioia et al. 1982).

Similar results were obtained in studies conducted to establish reliability levels of Ontario Hydro's (now Hydro One) transmission structures (Goel 1986).

8.9 Reliability-based Design (RBD) Approach

The reliability-based design approach recognizes and takes advantage of the statistical variability of load and strength, and quantifies the usually very low probability of coincidence of maximum load with minimum strength. Thus, the principal

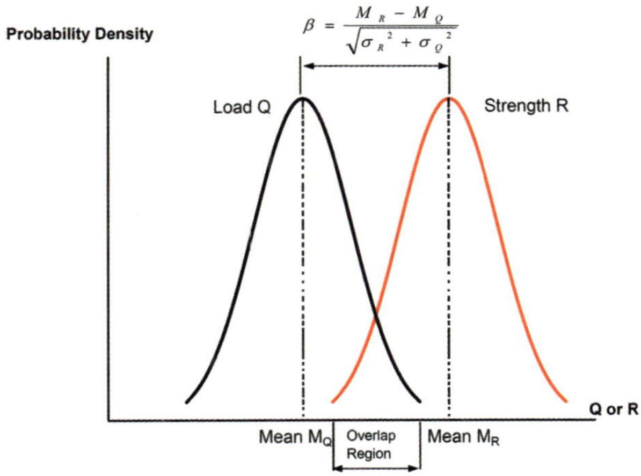

Fig. 8.14 Probability density function for load effect Q and resistance R

difference between the RBD and the deterministic approach lies in the application of the reliability theory, which allows uncertainties to be quantified and manipulated in a consistent manner. Here, uncertain quantities such as load effect Q and strength R are modeled as random variables, while the risk is quantified as probability of failure P_f. The basic reliability problem is to evaluate P_f from statistical data for Q and R, which typically include the mean (M_R and M_Q) and the standard deviation (σ_R and σ_Q), which provides a quantitative measure of the magnitude of uncertainty about the mean value (Fig. 8.14).

The overlap region in Fig. 8.14 represents the area where:

$$\text{Load Effect, } Q > \text{Resistance, } R$$

The failure function g is defined as

$$g = R - Q \tag{8.8}$$

Therefore, if $g < 0$, failure occurs, and if $g \geq 0$, the design is adequate.

If a normal Gaussian distribution is assumed for Q and R, then it can be assumed that the function g will also have normal distribution and its mean value and standard deviation can be expressed as

$$g_m = R_m - Q_m \text{ and} \tag{8.9}$$

$$\sigma_g = \sqrt{\sigma_R{}^2 + \sigma_Q{}^2} \tag{8.10}$$

The probability of failure P_f can be evaluated as follows

$$
\begin{aligned}
P_f &= \text{Prob}\,(g < 0) \\
&= \phi(-g_m/\sigma_g) \\
&= \phi(-\beta)
\end{aligned}
\tag{8.11}
$$

Where the reliability index β (also called Safety Index) (Eq. 8.11) expresses a margin to the specified limit state and provides a quantitative expression of reliability, i.e. a measure of survival. It is given as

$$
\beta = \frac{R_m - Q_m}{\sqrt{\sigma_R{}^2 + \sigma_Q{}^2}}
\tag{8.12}
$$

If a log-normal distribution is assumed for Q and R instead of a Gaussian distribution, then

$$
\beta = \frac{\ln\left[\frac{R_m}{Q_m}\sqrt{\frac{\ln(1+V_Q{}^2)}{\ln(1+V_R{}^2)}}\right]}{\sqrt{\ln(1+V_R{}^2)+\ln(1+V_Q{}^2)}}
\tag{8.13}
$$

where

$$
V_Q = \frac{\sigma_Q}{Q_m} \text{ and } V_R = \frac{\sigma_R}{R_m}
\tag{8.14}
$$

When both V_Q and V_R are less than about 0.3, the following approximation can be used (EPRI 1987)

Table 8.4 Relationship between reliability (safety) index and probability of failure

Reliability (safety) index β	Probability of failure $P_f = \phi(-\beta)$
1.0	0.159
1.2	0.115
1.4	0.0808
1.6	0.0548
1.8	0.0359
2.0	0.0228
2.2	0.0139
2.4	0.00820
2.6	0.00466
2.8	0.00256
3.0	0.00135
3.2	0.000687
3.4	0.000337
3.6	0.000159
3.8	0.0000723
4.0	0.0000316

$$\beta = \frac{\ln\left[\frac{R_m}{Q_m}\right]}{\sqrt{({V_R}^2 + {V_Q}^2)}} \tag{8.15}$$

Table 8.4 lists values of β and the corresponding probability of failure. The critical member exhibiting the lowest reliability index β are those that limit the reliability of the structure.

8.10 Return Period

The return period of a climatic load event is defined as the mean time interval T for the load intensity to exceed a specified value. If the exceedance events are statistically independent, T can be obtained with P being the probability of exceedance event within unit period

$$P = \frac{1}{T} \tag{8.16}$$

Hence, a 50-year return period ice load indicates a probability of exceedance of 1/50, or 0.02 in any one year.

However, the probability of exceedance of a 50-year period ice load during the 50-year design life of a structure is

$$1 - \left(1 - \frac{1}{T}\right)^{50} = (1 - 0.02)^{50} = 0.364 \text{ i.e. } 36.4\%$$

8.11 Variability of Component Resistance

The resistance of a component is a function of:

- Mechanical properties (stiffness and strength) of its constituent materials
- Geometry of the component
- Assumptions involved in the derivation of equations relating the mechanical properties, geometry, and the governing failure state

and for the existing lines;

- Quality of construction
- Age and maintenance

The component resistance R and the nominal resistance R_n (as specified in the material property handbooks) can be related as

$$\mathrm{R} = \mathrm{R_n}\ (\mathrm{MxFxC}) \tag{8.17}$$

where

M = measure of the material variability (including the effects of age and maintenance)
F = measure of fabrication (dimensional) variability (including the effect of quality of construction)
C = measure of calculation (assumption in derivation of equations representing uncertainties in strength theory)

If M, F, and C are uncorrelated random variables, the mean value and the standard deviation for strength can be approximated by

$$R_m = R_n M_m F_m C_m \tag{8.18}$$

and

$$\Omega_R = \sqrt{\Omega^2{}_M + {\Omega_F}^2 + {\sigma_C}^2} \tag{8.19}$$

where parameters M_m, F_m and C_m, and Ω_M, Ω_F and Ω_C are the mean values and COVs (mean/standard deviation) of M, F and C determined from tests or other rational methods. Table 8.5 summarizes some of these statistical parameters estimated from steel member data reported in the literature (Goodwin et al. 1982).

Cigré Study Committee B2 Overhead Line (Menezes and Ferreira de Silva 2000) carried out a study on the variability of the material properties of typical angle members of steel towers. The report indicated that the mean value of the compressive capacity for 350 MPa leg members could be as large as 120% of the nominal capacity with a COV (standard deviation/mean value) of 7%.

It has also been demonstrated (IEC 60826 2003) that if a load Q_T (load having a return period T) is associated with strength of components corresponding to the 10% exclusion limit, the resulting reliability is almost constant and equal to 1/(2T) in the normal range of variation of load and resistance. Thus, the accuracy of the estimates of the load Q_T becomes quite important to design lines according to a target reliability level.

In the IEC 60826 (IEC 60826 2003) and CSA standards (CSA C22.3 No 1 2001), the nominal or characteristic strengths considered for an entire line are determined based on the desired failure sequence of components subjected to the critical load

Table 8.5 Statistical parameters for steel members (Reproduced by permission, © [1982] IEEE)

Member type	Limit state	Material factor M		Fabrication factor F		Calculation factor C	
		Mean	COV	Mean	COV	Mean	COV
Tension	Yield	1.05	0.10	1.00	0.05	1.00	0.00
Compression	Axial buckling	1.00	0.06	1.00	0.05	1.21	0.16
Bending	Yield or flexural buckling	1.00	0.01	1.00	0.05	1.16	0.16

intensity during any single occurrence, and the quality and statistical parameters to achieve the target reliability. The reference strength to be associated with load effect Q corresponds to the 10% exclusion limit.

In the ASCE manual 74 (ASCE 74 2005), the nominal strength is the component strength with an exclusion limit ranging from 5% to 10%. Strength factors are specified for different component reliability levels, exclusion limits and coefficients of variation.

EN 50341(EN 50341-1, 2001) specifies partial factors to be used to calculate the factored strength.

Appendix contains an example demonstrating the process for determining the reliability (safety) index.

8.12 Other Loads

As mentioned earlier, transmission line design also includes other loads and considerations in addition to the more common climatic loads. The following discussion pertains to these other considerations and loadings that transmission structures may encounter and must be designed for.

8.12.1 Longitudinal Loads

Unbalanced longitudinal loads due to unequal ice loading on conductors or conductor breakages can cause cascading of transmission structures. The risk of cascading is reduced by containment. The requirements are usually referred to as 'Security requirements'. The following methods are used for containment of the cascading failures.

Design for longitudinal loads: Rigid square-based latticed structures, typical guyed structures and single-shaft pole structures can be designed to resist the longitudinal loads required. Therefore, it is common practice to specify longitudinal loads that provide enough strength to resist cascading at every structure. Although the loss of localized structures adjacent to the origin of the failure can occur, the other structures resist cascading and the failure is contained to few structures.

Use of intermediate anchor structures: Structure types, such as H-frames and narrow-based latticed structures have little inherent ability to withstand the longitudinal loads that enable them to resist cascading. In such cases, some utilities erect anchor structures at regular intervals along the line, so that cascading failure is contained to a section of the line. The anchor structure could be created by adding longitudinal guys to the suspension structure, if guys can be used, or by using an angle structure.

Use of mechanical fuses: In recent years, the use of mechanical fuses has been practiced by some utilities for failure containment. In this case, conductors are supported by slip- or release-type suspension clamps which act as fuses and limit the longitudinal load that can be applied to the structure by the conductors. To achieve this, the design of the slip or release mechanism must ensure serviceability under all operating conditions, such as temperature extremes and ice loads. The method is however not recommended for areas of heavy ice build up, as a clamp release under unbalanced ice loading could result in reduced clearance to the ground, causing danger to the public and utility personnel.

8.12.2 Galloping

Galloping is an aerodynamic instability phenomenon that occasionally occurs in transmission line ground wires and conductors. It usually occurs when moderate winds blow across ice-coated wires. The wires may move at amplitudes ranging from a fraction of a meter to more than the full sag. Galloping may cause one or a combination of the following adverse effects:

- Excessive dynamic wire tensions causing failure of the supporting structure
- Flashover or direct contact between phases or between phase and ground wires, resulting in wire damage and outage
- Excessive conductor sag
- Low-cycle fatigue failure or damage of the wire or hardware

Studies have been conducted to determine the loads produced during galloping. The load magnitude depends on many factors such as span length, wire type, wire tension, rigidity of the wire support hardware, weather related conditions, etc. In a recent paper (Havard 2002), a review of the measurements available from the literature was presented. Tables 8.6 and 8.7 show the summary of the field

Table 8.6 Summary of field measurements for maximum horizontal tension during galloping (Havard 2002)

Conductor (number × cross-sectional area in mm^2)	Span lengths (m)	Static tension (kN)	Dynamic tension pk-pk (kN)	Ratio dynamic/ static tension	Reference
4×410	312, 319	60.3	72.6	1.2	Anjo et al. 1974
4×950	312, 319	91.2	76.5	0.8	-do-
2×620	308	35.3	39.2	1.1	Escarmelle et al. 1997
3×620	308	35.3	73.6	2.1	-do-
4×410	363, 247	23.5	30.6	1.3	Morishita et al. 1984
8×810	230, 190	29.4	31.2	1.1	-do-
6×410	363, 247	23.5	18.8	0.8	-do-
8×410	353, 230, 350	22.6	14.4	0.6	-do-
10×810	230, 190	29.4	11.8	0.4	-do-
28.1 mm dia	80 m dead end	8.2	18.3	2.2	Eliasson 2002
-do-	-do-	7.8	21.1	2.7	-do-
-do-	-do-	7.6	21.2	2.8	-do-
-do-	-do-	7.8	10.2	1.3	-do-

Table 8.7 Summary of field measurements for maximum vertical loads during galloping (Havard 2002)

Conductor (number × cross-sectional area in mm^2)	Span lengths (m)	Static load(kN)	Dynamic load pk-pk (kN)	Ratio dynamic/ static load	Reference
4×410	312,319	20.6	34.3	1.7	Anjo et al. 1974
4×950	312, 319	39.9	24.5	0.6	-do-
34 mm dia	459	10.3	19.5	1.9	Krishnasamy 1984
28 mm dia	418	6.6	7.9	1.2	-do-
41 mm dia	626	6.1	12.3	2.0	-do-
2×30.4 mm dia	312, 308	13.6	3.7	0.3	Brokenshire 1979
2×30.4 mm dia	291, 242	14.0	4.6	0.3	-do-
2×30.4 mm dia	259, 251	10.4	2.4	0.2	-do-
2×36.2 mm dia	232 256	12.0	13.4	1.1	-do-

measurements of maximum horizontal tension and maximum vertical loads during galloping events.

As a result, the following general conclusions were made:

1. The maximum dynamic tension during galloping is up to 2.1 times the static tension on transmission lines, and is up to 2.8 times on distribution lines.
2. The dynamic vertical load changes during galloping are up to 2 times the static value.
3. Due to the large influence of conductor support rigidity on dynamic tension, galloping is more damaging to dead-end structures and structures with rigid conductor supports than most suspension structures having flexible conductor supports.

Support structure damage usually occurs when the fluctuating load produced by galloping is in resonance with the structure, as the Aeolian vibration of wires has been reported to cause. There have been no reported cases of complete structure failure due to galloping loads.

The following are proven measures to reduce or eliminate galloping-related damage and outage:

1. Spacing the individual wires so that the wire paths during galloping do not cross
2. Detuning pendulums or airflow spoilers to control galloping amplitudes
3. Installing interphase spacers between conductors
4. Developing modified conductor design. For example, T2 (*General Cable trademark*) conductors (Fig. 8.15) are composed of two identical bare conductors twisted together with a twist length of about 3 meters. The conductor cross-section is a rotating figure of number 8. Based on wind tunnel testing and field trials, T2 conductors are known to provide improved performance in controlling conductor galloping.

These methods are discussed in greater detail in EPRI 2006.

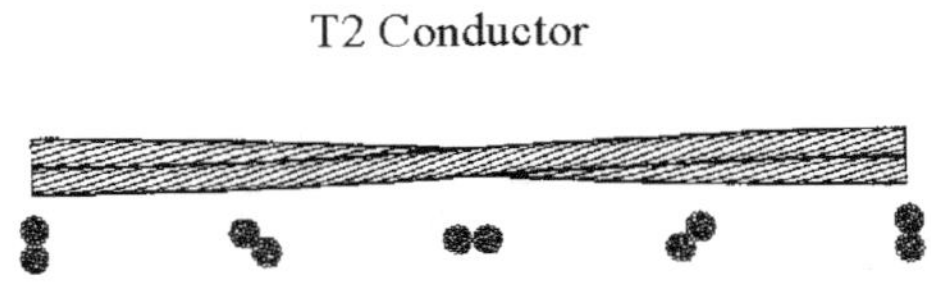

Fig. 8.15 T2 conductor (Reproduced by permission of General Cable: www.generalcable.com)

8.12.3 Construction and Maintenance Loads

Construction loads are those loads that act upon structures due to the assembly and erection of structures themselves, and due to the installation of ground wires, insulators and conductors, and connecting hardware. Maintenance loads are loads that act on structures as a result of scheduled or emergency maintenance activity. Maintenance loads consist of the effects of workers on the structure being maintained, and loads at adjacent structures due to temporary modifications to permit the repair or replacement of the structure being maintained. Workers can be seriously injured if a structure fails; therefore, personnel safety is paramount in establishing construction and maintenance loads. These are also termed 'Safety Requirements'.

8.12.4 Ice Shedding

Ice shedding is the physical phenomenon that occurs when ice accumulated on overhead ground wires and conductors is removed naturally or by other means. The ice shedding mechanism is affected by a number of factors and parameters, such as ice morphology, meteorological conditions, and structural design of lines, as well as wire and conductor characteristics. Sudden ice shedding from the conductors may result in high-amplitude vibrations leading to the reduction of phase spacing, and the application of excessive dynamic forces to support structures, which may cause flashover between conductors or damage to structures. Therefore, it is important to predict conductor swinging (jumping) due to sudden ice shedding along the line length. An approximate formula was suggested by List and Pochop (1963) as follows

$$h_{sw} = \frac{3\sigma^2}{4E\gamma \cos\beta} \tag{8.20}$$

where h_{sw} = the maximum swing (vertical jump) of the midpoint of the conductor after the sudden ice shedding of a complete span:

σ = horizontal tensile stress in conductor loaded with ice
E = elastic modulus
γ = density of the conductor material
β = the angle between the horizontal and the line connecting the supports

Furthermore, since the entire ice load shedding is unlikely to happen at once, the following equation is considered to be sufficient to estimate conductor jump

$$h_{sw} = \frac{\sigma^2}{2E\gamma \cos\beta} \tag{8.21}$$

Although the computed value is approximate, it does serve the purpose of eliminating the risk of flashovers.

After a large number of phase-to-phase flashover faults due to ice shedding on the 132 kV and 275 kV CEGB (now National Grid, UK), tests were conducted (Morgan and Swift 1964) on five span sections of 132 kV to determine the jump height of the conductors after the release of simulated ice loads from centre span. It was concluded that, with heavy ice loads, phase-to-phase flashover can only be prevented by ensuring sufficient horizontal separation between phases. In the 1970s, during the compact line research in the USA (EPRI 1978), experiments and analytical simulations were conducted to study the effects of ice shedding on compact lines. As it is useful to predict both the conductor jump height and the maximum conductor tension during ice shedding, in recent years a number of analytical models have been developed, and considerable work has been done to simulate line behaviour after ice shedding (Roshan Fekr and McClure 1998; McClure and Lapointe 2003; Kollar and Farzaneh 2005; Kalman et al. 2007). However, a general practical method that can be easily used by designers is still not available due to the complexity of the problem and vast number of parameters that influence line behaviour.

Zemljaric and Jakse (2005) have presented an innovative design approach to mitigate inter-span contacts caused by snow shedding in Slovenia. In this case, the I-string insulators in the top cross-arm of a two-circuit 110 kV lattice tower were replaced by V-string insulators.

8.12.5 Bundle Rolling of the Bundle Conductor Lines

Bundle conductor transmission lines passing through mountainous regions are sometimes subjected to the massive accumulation of ice which can lead to the rolling of bundle conductors. The same effect can occur with ice build-up on long spans over valleys or rivers. This form of instability can leave the bundle in the rolled position, causing outage and damage to spacers and conductors. The utilities in Canada and Japan have reported problems due to bundle rolling (Saint-Louis et al. 1993; Manukata et al. 1963; Matsubayashi 1963). The studies carried out in Canada have provided simplified modelling of the phenomenon and a simple tool to predict the critical conditions that can cause the bundle to roll over (Nigol et al. 1977). These studies have indicated that in a region of similar ice accumulation, the longest spans are at the greatest risk, and the resistance to rolling can be increased by using a larger than normal number of spacers. Careful attention must be paid to the spacer clamp design, so that spacers can retain their grip on the conductors during rolling, as the bundle will return to its normal orientation once the ice has melted off. If the clamp slips on the conductor, the restoration of the bundle to its normal orientation will be very difficult. The studies also concluded that the spacers with short articulated arms usually meet the torsional stiffness requirements, and two or three torsion resistant spacers per span are needed in most cases.

8.13 Ice/Snow Accretion Mitigation Techniques

Due to the susceptibility of overhead transmission lines to ice and snow accretion, a variety of research studies and investigations have been carried out to minimize or mitigate the impact of ice accretion (CEATI 2002; Goel 2007). Some of the methods used are:

- Avoid areas susceptible to high ice accretion during the route design
- Design lines and structures to withstand the effects of ice accretion
- Install ice-sensing devices on the transmission lines and remove ice build-up as soon as it exceeds an acceptable threshold
- Install methods to suppress ice accretion. These techniques have mainly been tried in Japan to limit wet snow accretion. They include protrusions at the surface of conductors such as fins, rings and fittings which prevent the conductors from being wrapped in cylinders of snow, and counter-weights attached to conductors to prevent twisting (Oogi et al. 1998).

8.14 Lessons from the 1998 Ice Storm

A paper presented at IWAIS 2002 (International Workshop on Atmospheric Icing of Structures) (McClure et al. 2002) discussed the main lessons learned from the catastrophic ice storm of January 1998 in Quebec, as reported by the Structures Group of the Scientific Commission appointed by the Government of Quebec after the disaster. During this storm, the structural damage to Hydro Quebec's grid consisted of the collapse of more than 600 steel towers and damage to another 100. In addition, about 2,500 wood sub-transmission structures were broken and nearly 700 more had local failures. Some of the observations and recommendations made by the Commission were as follows:

Observations:

1. The overhead ground wires, which accumulate more ice than conductors in comparable climatic conditions, fell to the ground over several hundreds of kilometres of line. It was established that in certain cases the conductors and ground wires had triggered the collapse.
2. Several anchored angle towers, found in sections where lines were damaged by cascade effect, did not have the extra longitudinal or transversal strength required to stop the cascades.

3. The collapse of several transmission and sub-transmission lines was attributed to cascades due to longitudinal loads caused by conductor breakage or problems with fastening or anchoring assemblies for conductors.
4. Collapse in some cases occurred at vertical load levels close to maximum design loads.

Recommendations:

1. Conduct a review of the basic climatic loads and load combinations. The maximum ice loading case must be modified to include the effects of wind during and after ice build-up on conductors. There should be greater design accretion levels for overhead ground wires than for line conductors.
2. The design of anchored angle towers and their hardware should take into account conductor tensile strength so that they can withstand cascading. Cases involving unbalanced residual loads must be more stringent.
3. Improve the mechanical sturdiness of lines. In the absence of a dynamic analysis, anti-cascading towers should be designed to be anchored and to consider dynamic effects in order to prevent cascading. Effective metal anti-cascading towers should be added to all important wood portal lines.
4. Hydro-Quebec should take part in studies on the timeliness and effects of installing mechanical fuses, which would limit axial load in suspension systems for conductors and overhead ground wires.

Similarly, a paper (Rimmer 1999) presented at a CEATI workshop in Montreal, discussed the observations and recommendations after the assessment of the impact of the 1998 storm on the Ontario Hydro transmission network. It was mentioned that the Ontario Hydro transmission lines in Eastern Ontario withstood the onslaught of the ice build-up remarkably well. Most of the major damage occurred on a 1934 steel lattice tower line due for refurbishment during 1998. Other major damages resulted from overhead ground wires sagging into the phase conductors and the wind creating sufficient swing to cause flashover. The paper concluded that asset management is a vital diagnostic process for assessing the health of transmission lines. Design criteria for transmission lines designed since 1960 appear to be adequate, however, lines designed before 1960 may show certain weaknesses, meaning that their design needs to be reviewed and corrected.

8.15 Concluding Remarks

Some basic information regarding the design of transmission lines to withstand the impact of atmospheric icing has been presented in this chapter. In the United States, catastrophic line failures in 1990 and 1991 in Arkansas, Iowa, Minnesota, and North and South Dakota sparked increased interest in transmission line icing issues. Similarly, the January 1998 ice storm in Eastern Canada drove utilities in Canada to review their design standards and practices. Traditionally, deterministic design methods have been used in transmission line design, but with the availability of good climatic data, it is now possible to use reliability-based design. Studies are also being conducted to devise techniques to mitigate ice loading on overhead transmission lines.

Appendix: Extreme Value Analysis

An analysis of meteorological data has shown that the distribution of the maximum radial ice thickness or annual wind velocities could be accurately expressed using an extreme value distribution law such as Fisher Tippet or Gumbel Type 1. The basic formula for the cumulative distribution function (CDF) has the following format (Gumbel 1958; IEC 60826 2003)

$$F(x) = e^{-e^{-a(x-\mu)}} \tag{8.22}$$

where

$$a = \frac{c_1}{\sigma} \tag{8.23}$$

$$\mu = x_m - \frac{c_2}{a} \tag{8.24}$$

The formula expresses the probability $F(x)$ that a random value will be less than a value x in a distribution with a mean value of x_m and standard deviation σ.

The constants c_1 and c_2 depend on the number of observations in a measurement series. For a measurement period of n years, the value z_i is calculated as follows

$$z_i = -\ln\left[-\ln\frac{i}{n+1}\right] \text{ where } 1 \le i \le n \tag{8.25}$$

c_1 is the standard deviation of Z_i values, therefore

$$c_1 = \sigma_z = \sqrt{\frac{1}{n}\sum_{i=1}^{n}\left(z_i^2 - z_m^2\right)} \text{ and} \tag{8.26}$$

c_2 is the average of z_i values, i.e.

$$c_2 = z_m = \frac{1}{n}\sum_{i=1}^{n} z_i \tag{8.27}$$

Table 8.8 Values of constants c_1 and c_2 (Reproduced by permission of the International Electrotechnical Commission (IEC) from its International Standard IEC 60826 ed. 3.0 – 2003)

Number of yearly observations (n)	c_1	c_2
5	0.792778	0.458794
10	0.949625	0.495207
20	1.062822	0.523552
25	1.091450	0.530860
30	1.112374	0.536221
40	1.141315	0.543620
50	1.160661	0.548542
100	1.206489	0.560023
∞	1.282600	0.577200

Calculated values of c_1 and c_2 are given in Table 8.8.
The probability $P(x)$ that the value will be higher than x in any year is given by

$$P(x) = 1 - F(x). \tag{8.28}$$

The return period, T, of the value x is given by

$$T = \frac{1}{P(x)} \text{ and therefore } P(x) = \frac{1}{T}. \tag{8.29}$$

Hereafter, the following equation can be derived

$$x = x_m - \sigma\frac{c_2}{c_1} + \frac{\sigma}{c_1}\left[-\ln(-\ln\left\{1 - \frac{1}{T}\right\})\right] \tag{8.30}$$

Example 1 The following example shows the application of extreme value analysis and the importance of the number of observations n.

The following data (Table 8.9) for annual maximum radial ice thickness is taken from Ontario Hydro research report for Toronto, Canada. (Krishnasamy and Tabatabai 1988; Goel 2005)

From the above data, the mean and standard deviations for the annual maximum ice thickness can be obtained. To study the sensitivity of the number of observations, n , it is assumed that the pool of data (sample size) contains $n = 5$ years only (data numbers 1–5, 6–10 or 26–30), $n = 10$ years (data numbers 1–10 or 11–20), the first 20 years (data numbers 1–20), the last 25 years (data numbers 6–30), as the data for first years (about 50 years) ago may not be based on very sophisticated measurements or all 30 years. The following Table 8.10 presents the results:

Table 8.10 clearly shows the large discrepancy in the mean and standard deviation values if the number of observations is available only for 5 years, however the values are more consistent if a minimum of $n = 10$ observation years are considered.

Table 8.9 Annual maximum radial ice thickness, Toronto, Canada

Data no.	Year	Maximum radial ice thickness (mm)	Data no.	Year	Maximum radial ice thickness (mm)
1	1954	4.1	16	1969	5.1
2	1955	4.1	17	1970	3.3
3	1956	6.6	18	1971	12.2
4	1957	2.0	19	1972	10.2
5	1958	1.8	20	1973	2.5
6	1959	10.7	21	1974	6.9
7	1960	20.6	22	1975	5.4
8	1961	15.7	23	1976	5.3
9	1962	7.9	24	1977	3.3
10	1963	3.8	25	1978	4.8
11	1964	5.3	26	1979	4.1
12	1965	14.0	27	1980	2.3
13	1966	6.6	28	1981	0.0
14	1967	7.6	29	1982	6.9
15	1968	23.1	30	1983	7.1

Using the values of the mean and standard deviations in Table 8.10 and appropriate values of c_1 and c_2 from Table 8.8, the values for the maximum radial ice thickness listed in Table 8.11 are calculated from Eq. 8.30 for various return periods.

Again, a large discrepancies in values are noted if only 5 observation years are considered.

Similar results will be obtained if the extreme value analysis for the hourly wind speed is carried out from the following data (Table 8.12). The wind speed was measured at the reference height of 10 m above ground at the Toronto International Airport.

For the maximum hourly wind speed data for 30 years, the mean value is 67.3 km/h and the standard deviation is 9.27 km/h. With the mean and standard deviations, the maximum wind speed for various return periods can be calculated (Table 8.13).

Table 8.10 Mean value and standard deviation of maximum radial ice thickness

Sample size n	Mean value of the maximum radial ice thickness (mm)	Standard deviation (mm)
5(1–5)*	3.7	1.96
5(6–10)	11.7	6.58
5(26–30)	4.1	3.04
10(1–10)	7.7	6.22
10(11–20)	9.0	6.20
20	8.4	6.10
25(6–30)	7.8	5.63
30	7.1	5.38

* Data numbers

Table 8.11 Maximum radial ice thickness for various return periods

Number of observation years (n)	Maximum radial ice thickness (mm)			
	T = 25	T = 50	T = 100	T = 500
5(1–5)	10.4	12.2	13.9	17.9
5(6–10)	34.5	40.3	461	59.5
5(26–30)	14.6	17.3	20.0	26.2
10(1–10)	25.5	30.1	34.7	45.2
10(11–20)	26.6	31.2	35.8	46.3
20	23.7	27.1	31.7	40.9
25 (6–30)	21.6	25.2	28.8	37.1
30	20.0	23.4	26.8	34.6

Table 8.12 Maximum hourly wind speed, Toronto, Canada

Data no.	Year	Maximum hourly wind speed (km/h)	Data no.	Year	Maximum hourly wind speed (km/h)
1	1954	84.2	16	1969	54.7
2	1955	73.1	17	1970	49.3
3	1956	82.2	18	1971	60.6
4	1957	58.6	19	1972	68.6
5	1958	70.4	20	1973	57.6
6	1959	88.6	21	1974	73.6
7	1960	55.7	22	1975	73.6
8	1961	48.6	23	1976	70.6
9	1962	54.9	24	1977	58.7
10	1963	70.4	25	1978	69.6
11	1964	79.5	26	1979	80.5
12	1965	60.6	27	1980	69.6
13	1966	60.6	28	1981	64.6
14	1967	57.6	29	1982	69.6
15	1968	63.7	30	1983	58.7

Table 8.13 Maximum hourly wind speed for various return periods

Number of observation years (n)	Maximum hourly wind speed (km/h)			
	$T = 25$	$T = 50$	$T = 100$	$T = 500$
30	89.1	95.3	101.1	114.5

Process for Determining the Reliability (Safety) Index

As shown in Fig. 8.16, the tangent tower carries overhead ground wires at two points (label 1) and the conductors are connected at each of the points 2, 3 and 4 of the top, middle and bottom cross-arms, respectively.

The following steps illustrate the procedure to determine the reliability of the main leg member of the tower.

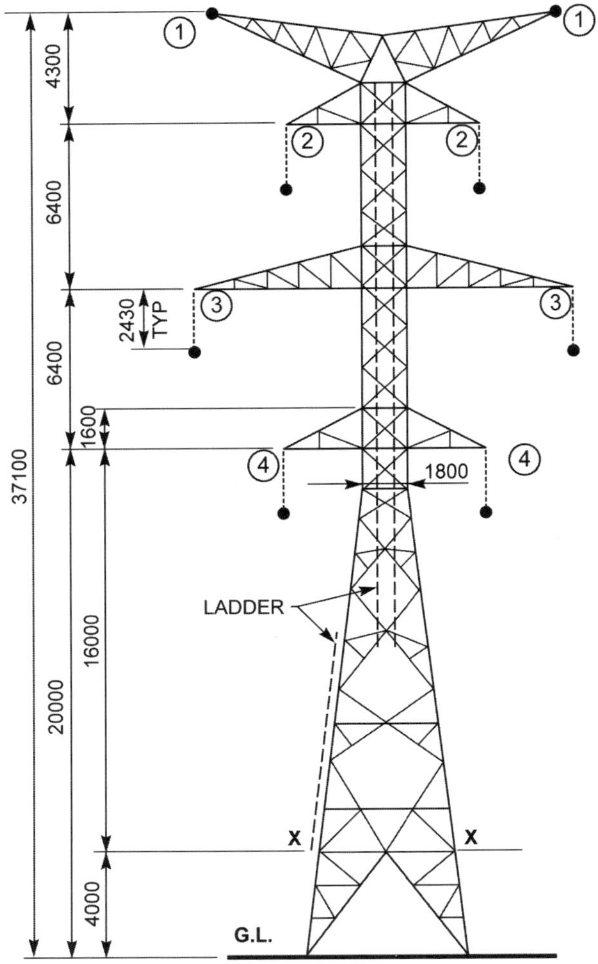

Fig. 8.16 Two-circuit 230-kV tangent tower

Determine the Influence Coefficients:

Overhead Ground Wires:

- Apply a unit load (1 kN) at point 1 in the transverse direction (to the right – see Fig. 8.17 – Case 1) assuming the wind direction will be normal to the overhead ground wire from the left, and obtain the value of the leg load I_{WG}
- Apply a unit load (1 kN) at point 1 in the vertical direction (see Fig. 8.17 – Case 2) and obtain the value of the leg load I_{VG}

Conductors:

- Apply a unit load (1 kN) each at points 2, 3 and 4 in the transverse direction assuming the wind direction will be normal to the conductors from the left

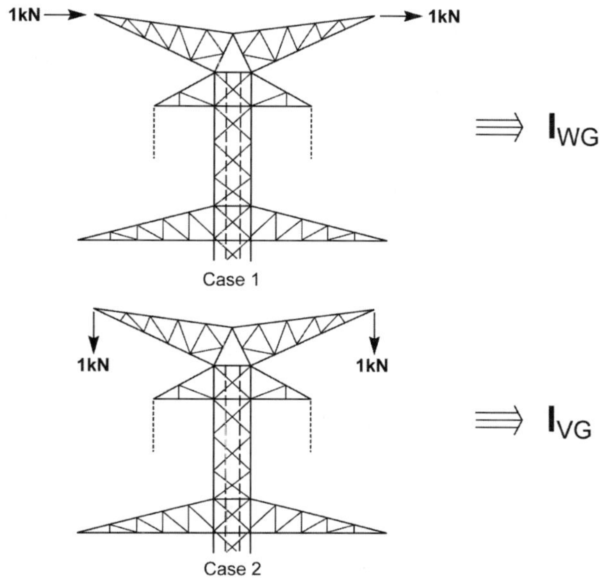

Fig. 8.17 Applied loads to determine influence coefficients at OHGW attachment points 1

and obtain the value of the leg loads I_{WC1}, I_{WC2} and I_{WC3}, respectively (see Fig. 8.18 – Case 3, Fig. 8.19 – Case 5 and Fig. 8.20 – Case 7)

- Apply a unit load (1 kN) each at points 2 , 3, 4 in the vertical direction and obtain the value of the leg loads I_{VC1}, I_{VC2} and I_{VC3} for the top, middle and bottom arms, respectively (see Fig. 8.18 – Case 4, Fig. 8.19 – Case 6 and Fig. 8.20 – Case 8)

Towers:

- Apply a unit pressure load (1 kPa) on the left face of the tower and obtain the value of the leg load I_{WT} (see Fig. 8.21 – Case 9)
- Calculate the leg load in the tower due to self weight I_{VT}

Design variables:

Overhead ground wire:	
Two wires at the two points 1	
Diameter	D_G (mm)
Weight	w_G (N/m)
Wind drag coefficient	G_G
Conductor:	
Single conductor at each of the points 2, 3 and 4	
Diameter	D_C (mm)
Weight	w_C (N/m)

Wind drag coefficient	G_C
Wind speed (with due account of height and gust)	V (m/s)
Radial Ice thickness	t (mm)
Ice density	$900\,\mathrm{kg/m^3}$
Wind span	L (m)
Weight span/wind span	a

The following expressions can be derived for the vertical and transverse loads (all loads in kN) at the OHGW attachment points 1:

$$\mathrm{P_{VG}} = [0.0282(\mathrm{D_G} + \mathrm{t})\mathrm{t} + \mathrm{w_G}]\,\mathrm{aL}/1000$$
$$\mathrm{P_{WG}} = [0.001(\mathrm{D_G} + 2\mathrm{t})\,\mathrm{G_G}(0.613\,\mathrm{V}^2)]\,\mathrm{L}/1000$$

The corresponding expressions for the loads at each of the conductor attachment points 2, 3 and 4 are:

$$\mathrm{P_{VC}} = [0.0282(\mathrm{D_C} + \mathrm{t})\mathrm{t} + \mathrm{w_C}]\mathrm{aL}/1000$$
$$\mathrm{P_{WC}} = [0.001(\mathrm{D_C} + 2\mathrm{t})\mathrm{G_C}(0.613\,\mathrm{V}^2)]\mathrm{L}/1000$$

The resulting wind force on the tower face is:

$$P_{WT} = G_T(0.613V^2)/1000$$

where G_T includes both the gust response factor and drag coefficient for the tower.

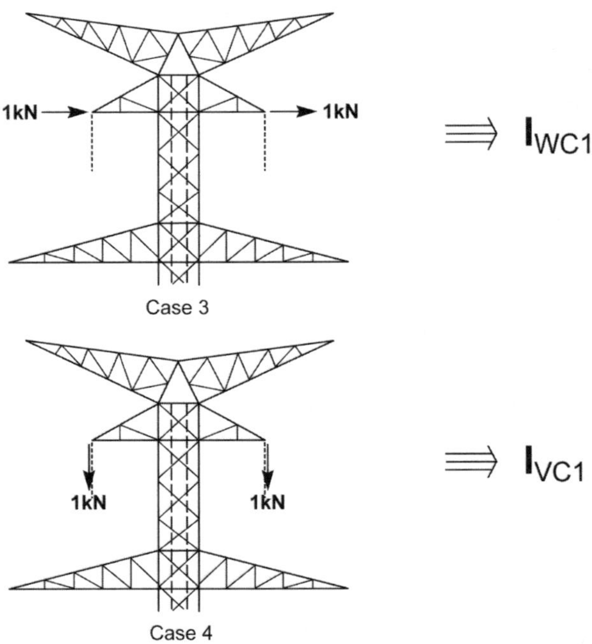

Fig. 8.18 Applied loads to determine influence coefficients at conductor attachment points 2

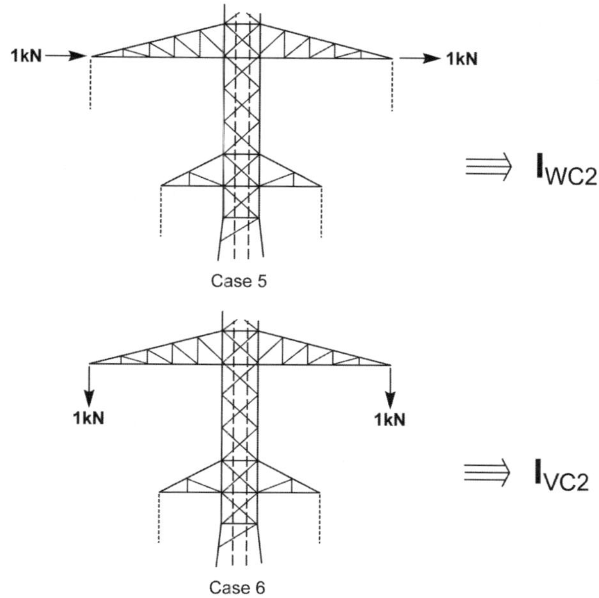

Fig. 8.19 Applied loads to determine influence coefficients at conductor attachment points 3

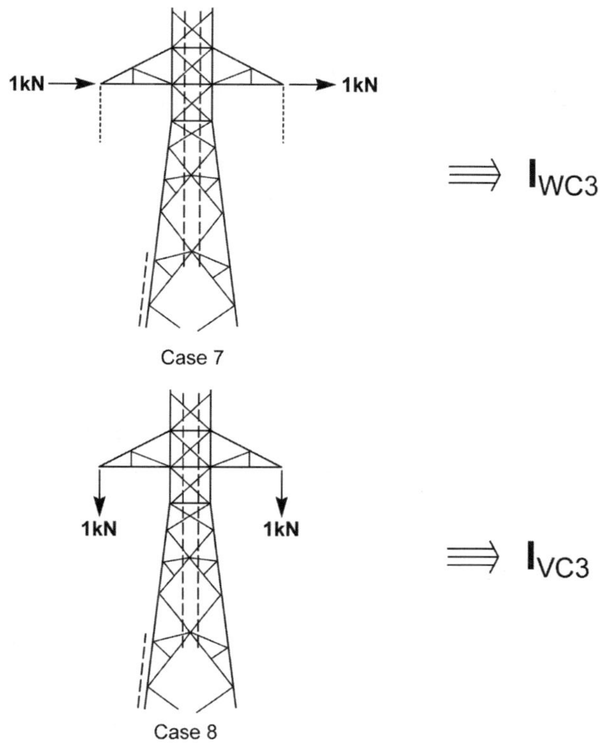

Fig. 8.20 Applied loads to determine influence coefficients at conductor attachment points 4

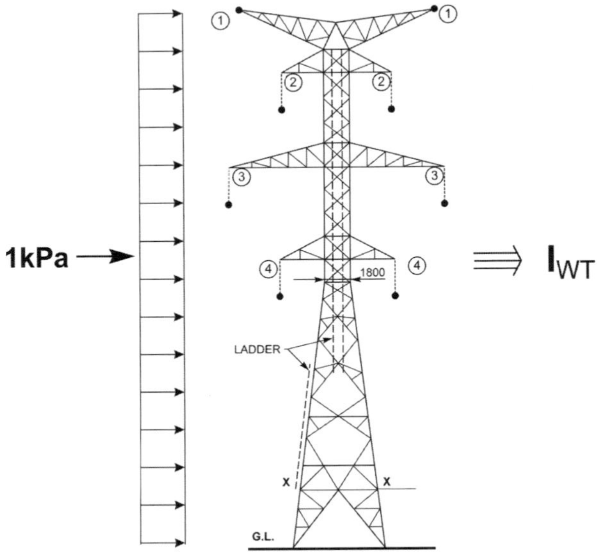

Fig. 8.21 Applied Loads to determine influence coefficient due to wind on tower

The Leg load Q for the tower is then expressed as

$$Q = I_{VG}P_{VG} + I_{WG}P_{WG} + (I_{VC1} + I_{VC2} + I_{VC3})P_{VC} + (I_{WC1} + I_{WC2} + I_{WC3})P_{WC} + I_{WT}P_{WT} + I_{VT}$$

Substituting the appropriate expressions for the various loads P, the leg load Q can be expressed as function of the two main climatic variables, namely the wind velocity V and radial ice thickness t, as follows:

$$Q = f_1 V^2 + f_2 V^2 t + f_3 t + f_4 t^2 + f_5$$

Putting $I_{VC} = I_{VC1} + I_{VC2} + I_{VC3}$ and $I_{WC} = I_{WC1} + I_{WC2} + I_{WC3}$, the expressions for f_1 to f_5 can be derived as follows:

$$f_1 = [(0.001 \times 0.613\, I_{WG}\, D_G\, G_G + 0.001 \times 0.613\, I_{WC}\, D_C\, G_C)\, L + 0.613\, I_{WT}\, G_T]/1000$$
$$f_2 = [(0.001 \times 1.226\, I_{WG}\, G_G + 0.001 \times 1.226\, I_{WC}\, G_C) L]/1000$$
$$f_3 = [(0.0282\, I_{VG}\, D_G + 0.0282\, I_{VC}\, D_C)\, a\, L]/1000$$
$$f_4 = [(0.0282\, I_{VG} + 0.0282\, I_{VC})\, a\, L]/1000$$
$$f_5 = [(I_{VG}\, w_G + I_{VC}\, w_C)\, a\, L]/1000 + I_{VT}$$

It is observed that f_1 to f_5 will have constant values for a particular case under consideration.

The standard deviation of the leg load Q is derived as

$$\sigma^2{}_Q = \left(\frac{\partial Q}{\partial V}\right)^2 \sigma^2{}_V + \left(\frac{\partial Q}{\partial t}\right)^2 \sigma^2{}_t \tag{8.31}$$

The expressions of the partial derivatives $\frac{\partial Q}{\partial V}$ and $\frac{\partial Q}{\partial t}$ are

$$\frac{\partial Q}{\partial V} = 2f_1 V + 2f_2 V t \tag{8.32}$$

and

$$\frac{\partial Q}{\partial t} = f_2 V^2 + f_3 + 2f_4 t \tag{8.33}$$

The coefficient of variation for the leg force Q, V_Q is $\frac{\sigma_Q}{Q}$

The mean value of resistance R , in the absence of specific data, can be assumed to be an appropriate percentage of the nominal value (say 20%) and the coefficient of variation (COV) to be 0.07 for steel members (Menezes and Ferreira de Silva 2000).

Equation (8.15) can then be used to calculate the reliability index β.

Numerical Example:

Let us consider the two-circuit 230 kV tangent tower shown in Fig. 8.16 with the following data:

Towers:

Tower height	37.1 m
Tower weight	92.96 kN
Wind gust effect factor	$G_T = 4.95$

Line design spans:

Horizontal span	290 m
Vertical span	365 m
Vertical span/horizontal span	1.26

Overhead ground wire and conductor:

OHGW:	19#7 Alumoweld	
	Diameter	18.29 mm
	Weight	12.99 N/m
	Drag coefficient G_G	1.32
Conductor:	1924 Kcmil	
	CDiameter	40.64 mm
	Weight	39.41 N/m
	Drag coefficient G_c	1.32

Design loads: Combined ice and wind:

Wind:	Mean speed	19 m/s
	Standard deviation	1.86 m/s
Ice	Mean thickness	23.4 mm
	Standard deviation	5.33 mm

Deterministic Design Practice:

Wind 28 m/s with 19 mm of radial ice thickness with Load Factor of 1.1

Maximum wind:

Wind:	Mean speed	33 m/s
	Standard deviation	3.47 m/s

Deterministic Design Practice:

Maximum Wind 36.5 m/s with Load Factor of 1.1

Leg member capacity:

Nominal value	465 kN
Mean value	556 kN
Standard deviation	38.9 kN

A linear static analysis of the tower yielded the following influence coefficients:

$I_{VG} = 0.474$ kN
$I_{WG} = 5.78$ kN
$I_{VC1} = I_{VC2} = I_{VC3} = 0.474$ kN
$I_{WC1} = 5.03$ kN
$I_{WC2} = 3.91$ kN
$I_{WC3} = 2.79$ kN
$I_{WT} = 44.31$ kN
$I_{VT} = 22.21$ kN

Using the equations derived above, the leg load Q is calculated as 242 kN with a standard deviation of 41 kN.

Using Eq. 8.15, the Reliability Index values (β) are calculated as follows:
Case 1: Combined wind and ice: $\beta = 4.5$
Case 2: Extreme wind: $\beta = 2.1$

The leg load Q according to the deterministic design practice is calculated as:
Case 1: Combined wind and ice: 230 kN (compared with capacity of 465 kN)
Case 2: Extreme wind: 406 kN (compared with capacity of 465 kN)

References

ALA (American Lifelines Alliance) (2004), Extreme ice thickness from freezing rain. Report, American Lifelines Alliance, Washington, DC (www.americanlifelinealliance.org), September

Anderson BC (1996) Estimation of climatic loads for BC Hydro's possible future second 500 kV transmission line at Mission Ridge, B.C. In: Proc 7th International Workshop on Atmospheric Icing of Structures, Chicoutimi

Anderson BC (2005) Private communication

Anjo K, Yamasaki S, Matsubayashi Y, Nakayama Y, Otsuki A, Fujimura T (1974) An experimental study of bundle galloping on the Kasatory-Yama test line for bulk power transmission. CIGRÉ report 22–04, Paris

ASCE 7 (2002), Minimum design loads for buildings and other structures. ASCE Standard 7–02

ASCE 74 (2005) Guidelines for transmission line structural loading. November 2005 draft

Brokenshire RE (1979) Experimental study of the loads imposed on welded steel support structures by galloping 345 kV bundled conductors. IEEE PES Summer Meeting, Vancouver, Paper A 79-551-3

CEATI (2002) De-icing techniques before, during and following ice storms. CEA Technologies Inc. report, CEA Technologies (www.ceatech.ca), Montreal

CEATI (2003) Methods and systems for collecting icing data for real-time monitoring of ice storms. CEA Technologies Inc. report, CEA Technologies (www.ceatech.ca), Montreal

Chainé PH, Skeates P (1974) Ice accretion handbook (freezing and precipitation) Industrial Meteorology – Study VI. Environment Canada, Toronto, Canada

Cigré (2000) Guidelines for field measurements of ice loading on overhead power line conductors. Technical Brochure #179, August

Cigré (2006) Guidelines for meteorological icing models, statistical methods and topographical effects. Technical Brochure #291, April

COST 727 (2006) Atmospheric Icing on Structures: Measurements and data collection on icing, State of the Art, Publication of Meteoswiss, 75,110 p

CSA C22.3 No.1 (2001) Overhead Systems. Canadian Standards Association

DiGioia AM Jr, Pohlman JC, Ralston P (1982) A new method for determining the structural reliability of transmission lines. Cigré Paper 22–08, 1982 Cigre Conference, Paris

Eliasson AJ (2002) Private correspondence

EN 50341-1:2001- Overhead electrical lines exceeding AC 45 kV – Part 1: General requirements – Common specifications. CENELEC, Brussels

EPRI (1978) Transmission line reference book: 115–138 kV compact line design. Electrical Power Research Institute Project 260, Electric Power Research Institute, Palo Alto, California

EPRI (1987) Reliability-based design of transmission line structures. Final Report, EPRI EL-4783, Electric Power Research Institute, Palo Alto, California

EPRI (2006) Transmission line design reference book: Wind-induced conductor motion 2nd edn. Electric Power Research Institute, Palo Alto, California

Escarmelle M, Wolfs M, Lilien JL (1997) Galloping event in Belgium on February 13th 1997. Presentation to CIGRÉ SC22 WG11, Task Force on Galloping, Sendai

Farias A (2005) Private communication

Fikke S (2005), Modern Meteorology and Atmospheric Icing. Presentation at IWAIS 2005 International Workshop on Atmospheric Icing of Structures, Montreal, June

Fikke S (2007) Private communication

Goel A (1986) Reliability of Ontario Hydro's 500 kV narrow base transmission structures. In: Proc of the PMAPS (Probabilistic Methods Applied to Power Systems), Toronto

Goel A (2002) Impact of severe weather events on transmission and distribution facilities: Preparation for extraordinary climatic events. CEATI Workshop, Montreal, October

Goel A (2005) Reliability-based design. CEATI OHLDI WISMIG Meeting, Vancouver

Goel A (2007) Deicing techniques before, during and following ice storms. CEATI Training Seminar, September

Goodwin EJ, Mozer JD, DiGioia AM Jr (1982) Transmission structure design utilizing probability-based load and resistance factor design, IEEE PES Summer Meeting, 1982

Gumbel EJ (1958) Statistics of extremes. New York Columbia Press

Gwilliam M, Anderson B (1999) 500-kV transmission tower damaged by snow creep. In: Proc CEATI Workshop on Transmission Line Asset Maintenance: 149–177

Havard DG (2002) Dynamic loads on transmission line structures during galloping. In: Proc 10th International Workshop on Atmospheric Icing of Structures, Brno, paper 5–1

Havard DG, Van Dyke P (2005) Effects of ice on the dynamics of overhead lines, Part II: Field data on conductor galloping, ice shedding, and bundle rolling. In: Proc 11th International Workshop on Atmospheric Icing of Structures, Montreal: 291–296

IEC 60826 (2003) Design criteria of overhead transmission lines. Edn 3

JIEE (1979) Design standard on structures for transmission lines. JEC-127

Kalman T, Farzaneh M, McClure G (2007) Numerical analysis of the dynamic effects of shock load induced ice shedding on overhead ground wires. Computers and Structures, vol 85, no 7–8: 375–384

Kiessling F, Ruhnau J (1993) Ice loads and their effects on reliability and design of overhead lines. In: Proc 6th International Workshop on Atmospheric Icing of Structures, Budapest

Krishnasamy SG (1984) Weather related loads on transmission lines and their consequences. In: Proc 5th International Workshop on Atmospheric Icing of Structures, Trondheim

Krishnasamy SG, Tabatabai M (1988) Database for weather-related loads on overhead transmission lines in Ontario. Ontario Hydro research report 88-220-K, Toronto

Krishnasamy SG, Tabatabai M (1992) Ice and wind-on-ice load measurement program in Ontario Hydro. International Seminar on Ice load Measurements, Kristiansand, May

Kollar LE, Farzaneh M (2005) Modelling the dynamics of overhead cables with ice. In: Proc 11th International Workshop on Atmospheric Icing of Structures, Montreal: 309–314

Lavoie M (2005) 735 kV Line failures in April 2005, Presentation at CEATI OHLDI WISMIG meeting New Jersey, October

Leavengood DC, Smith TB (1968) Studies of transmission line icing. Meteorology Research report prepared by Meteorology Research Inc for Southern California Edison Co, Los Angeles, Report no MRI 68 FR-801

List V, Pochop K (1963) Mechanical design of overhead transmission lines. SNTL Publisher of Technical Literature, Prague

Lozowski EP, Makkonen L (2005) Fifty years of progress in modelling the accumulation of atmospheric ice on power network equipment. In: Proc 11th International Workshop on Atmospheric Icing of Structures, Montreal: 55–62

Manukata M, Yoshida Y, Ishii H (1963) Determination of spacer intervals in quadruple conductor transmission lines. Sumitomo Electric Technical Review, no 1, Sumitomo Electric Industries Ltd, Japan

Matsubayashi Y (1963) Theoretical considerations of the twisting phenomenon of the bundle conductor transmission lines. Sumitomo Electric Technical Review, no 3, Sumitomo Electric Industries Ltd, Japan

McClure G, Johns KC, Knoll F, Pichette G (2002) Lessons from the ice storm of 1998: Improving the structural features of Hydro-Quebec's power grid. In: Proc 10th International Workshop on Atmospheric Icing of Structures, Brno

McClure G, Lapointe M (2003) Modeling the structural dynamics response of overhead transmission lines. Compters and Structures, vol 81: 825–834

McComber P, Druez J, Laflamme J (1995) Icing rate estimation of atmospheric icing. International Journal of Offshore & Polar Engineering, vol 5, no 5

Menezes R, Ferreira de Silva JB (2000) On the variability of mechanical properties of materials for transmission line steel towers. Electra no 189, April, pp 57–71

Morgan VT, Swift DA (1964) Jump height of overhead conductors after sudden release of ice loads. In: Proc IEE, vol 111, no 10: 1736–1746

Morishita S, Tsujimoto K, Yasui M, Mori N. Shimojima K, Naito K (1984) Galloping phenomena of large bundle conductors: Experimental results of the field test lines. CIGRÉ Paper 22–04

NESC C2 (2002) National electrical safety code C2-2002
Nigol O, Clarke GJ, Havard DG (1977) Torsional stability of bundle conductors. IEEE Paper no F77224-9 January
Oogi I, Ito T, Uino H, Hase N, Matuo H, Takeda K (1998) Conductors for overhead transmission lines in Japan: Overhead transmission lines new technology in Japan. Cigré SC22 Japanese Panel, August: 86–108
Pohlman JC (2001) Transmission line structures. Electrical Power Engineering Handbook, IEEE: 4–13 to 4–22
Rimmer F (1999) Major storm – Impact on maintenance practices. In: Proc CEATI Workshop on Transmission Line Asset Maintenance: 141–147
Roshan Fekr M, McClure G (1998) Numerical Modelling of the dynamic response of ice shedding on electrical transmission lines, Atmospheric Research, vol. 46: 1–11
Saint-Louis M, Hardy C, Bellerive J, Gagne J (1993) Bundle conductor spacers: Hydro-Québec experience. Paper presented at CEA E & O Division meeting, Montreal, March/April
Sakamoto Y (1998) Meteorological loads for transmission lines: Overhead transmission lines new technology in Japan. Cigré SC22 Japanese Panel, August: 10–21
Seppa TO (1996) Transmission line ice measurements with tension monitoring systems. In: Proc 7th International Workshop on Atmospheric Icing of Structures, Chicoutimi: 155–158
Turcot A (2002) Impact of ice storms on TransÉnergie Facilities: Preparation for extraordinary climatic events. Presented at the CEATI Workshop, Montreal, October
Zemlijaric B, Jakse J (2005) Tower-head modification with V-insulator strings to prevent interphase contacts during snow shedding effects. In: Proc 11th International Workshop on Atmospheric Icing of Structures, Montreal: 275–277

Index

D

S